MONOGRAPHS ON STATISTICS AND APPLIED PROBABILITY

General Editors

D.R. Cox, V. Isham, N. Keiding, T. Louis, N. Reid, R. Tibshirani, and H. Tong

1 Stochastic Population Models in Ecology and Epidemiology *M.S. Barlett* (1960)
2 Queues *D.R. Cox and W.L. Smith* (1961)
3 Monte Carlo Methods *J.M. Hammersley and D.C. Handscomb* (1964)
4 The Statistical Analysis of Series of Events *D.R. Cox and P.A.W. Lewis* (1966)
5 Population Genetics *W.J. Ewens* (1969)
6 Probability, Statistics and Time *M.S. Barlett* (1975)
7 Statistical Inference *S.D. Silvey* (1975)
8 The Analysis of Contingency Tables *B.S. Everitt* (1977)
9 Multivariate Analysis in Behavioural Research *A.E. Maxwell* (1977)
10 Stochastic Abundance Models *S. Engen* (1978)
11 Some Basic Theory for Statistical Inference *E.J.G. Pitman* (1979)
12 Point Processes *D.R. Cox and V. Isham* (1980)
13 Identification of Outliers *D.M. Hawkins* (1980)
14 Optimal Design *S.D. Silvey* (1980)
15 Finite Mixture Distributions *B.S. Everitt and D.J. Hand* (1981)
16 Classification *A.D. Gordon* (1981)
17 Distribution-Free Statistical Methods, 2nd edition *J.S. Maritz* (1995)
18 Residuals and Influence in Regression *R.D. Cook and S. Weisberg* (1982)
19 Applications of Queueing Theory, 2nd edition *G.F. Newell* (1982)
20 Risk Theory, 3rd edition *R.E. Beard, T. Pentikäinen and E. Pesonen* (1984)
21 Analysis of Survival Data *D.R. Cox and D. Oakes* (1984)
22 An Introduction to Latent Variable Models *B.S. Everitt* (1984)
23 Bandit Problems *D.A. Berry and B. Fristedt* (1985)
24 Stochastic Modelling and Control *M.H.A. Davis and R. Vinter* (1985)
25 The Statistical Analysis of Composition Data *J. Aitchison* (1986)
26 Density Estimation for Statistics and Data Analysis *B.W. Silverman* (1986)
27 Regression Analysis with Applications *G.B. Wetherill* (1986)
28 Sequential Methods in Statistics, 3rd edition
G.B. Wetherill and K.D. Glazebrook (1986)
29 Tensor Methods in Statistics *P. McCullagh* (1987)
30 Transformation and Weighting in Regression
R.J. Carroll and D. Ruppert (1988)
31 Asymptotic Techniques for Use in Statistics
O.E. Bandorff-Nielsen and D.R. Cox (1989)
32 Analysis of Binary Data, 2nd edition *D.R. Cox and E.J. Snell* (1989)

33 Analysis of Infectious Disease Data *N.G. Becker* (1989)
34 Design and Analysis of Cross-Over Trials *B. Jones and M.G. Kenward* (1989)
35 Empirical Bayes Methods, 2nd edition *J.S. Maritz and T. Lwin* (1989)
36 Symmetric Multivariate and Related Distributions
K.T. Fang, S. Kotz and K.W. Ng (1990)
37 Generalized Linear Models, 2nd edition *P. McCullagh and J.A. Nelder* (1989)
38 Cyclic and Computer Generated Designs, 2nd edition
J.A. John and E.R. Williams (1995)
39 Analog Estimation Methods in Econometrics *C.F. Manski* (1988)
40 Subset Selection in Regression *A.J. Miller* (1990)
41 Analysis of Repeated Measures *M.J. Crowder and D.J. Hand* (1990)
42 Statistical Reasoning with Imprecise Probabilities *P. Walley* (1991)
43 Generalized Additive Models *T.J. Hastie and R.J. Tibshirani* (1990)
44 Inspection Errors for Attributes in Quality Control
N.L. Johnson, S. Kotz and X. Wu (1991)
45 The Analysis of Contingency Tables, 2nd edition *B.S. Everitt* (1992)
46 The Analysis of Quantal Response Data *B.J.T. Morgan* (1992)
47 Longitudinal Data with Serial Correlation—A state-space approach
R.H. Jones (1993)
48 Differential Geometry and Statistics *M.K. Murray and J.W. Rice* (1993)
49 Markov Models and Optimization *M.H.A. Davis* (1993)
50 Networks and Chaos—Statistical and probabilistic aspects
O.E. Barndorff-Nielsen, J.L. Jensen and W.S. Kendall (1993)
51 Number-Theoretic Methods in Statistics *K.-T. Fang and Y. Wang* (1994)
52 Inference and Asymptotics *O.E. Barndorff-Nielsen and D.R. Cox* (1994)
53 Practical Risk Theory for Actuaries
C.D. Daykin, T. Pentikäinen and M. Pesonen (1994)
54 Biplots *J.C. Gower and D.J. Hand* (1996)
55 Predictive Inference—An introduction *S. Geisser* (1993)
56 Model-Free Curve Estimation *M.E. Tarter and M.D. Lock* (1993)
57 An Introduction to the Bootstrap *B. Efron and R.J. Tibshirani* (1993)
58 Nonparametric Regression and Generalized Linear Models
P.J. Green and B.W. Silverman (1994)
59 Multidimensional Scaling *T.F. Cox and M.A.A. Cox* (1994)
60 Kernel Smoothing *M.P. Wand and M.C. Jones* (1995)
61 Statistics for Long Memory Processes *J. Beran* (1995)
62 Nonlinear Models for Repeated Measurement Data
M. Davidian and D.M. Giltinan (1995)
63 Measurement Error in Nonlinear Models
R.J. Carroll, D. Rupert and L.A. Stefanski (1995)
64 Analyzing and Modeling Rank Data *J.J. Marden* (1995)
65 Time Series Models—In econometrics, finance and other fields
D.R. Cox, D.V. Hinkley and O.E. Barndorff-Nielsen (1996)

66 Local Polynomial Modeling and its Applications *J. Fan and I. Gijbels* (1996)
67 Multivariate Dependencies—Models, analysis and interpretation
D.R. Cox and N. Wermuth (1996)
68 Statistical Inference—Based on the likelihood *A. Azzalini* (1996)
69 Bayes and Empirical Bayes Methods for Data Analysis
B.P. Carlin and T.A Louis (1996)
70 Hidden Markov and Other Models for Discrete-Valued Time Series
I.L. Macdonald and W. Zucchini (1997)
71 Statistical Evidence—A likelihood paradigm *R. Royall* (1997)
72 Analysis of Incomplete Multivariate Data *J.L. Schafer* (1997)
73 Multivariate Models and Dependence Concepts *H. Joe* (1997)
74 Theory of Sample Surveys *M.E. Thompson* (1997)
75 Retrial Queues *G. Falin and J.G.C. Templeton* (1997)
76 Theory of Dispersion Models *B. Jørgensen* (1997)
77 Mixed Poisson Processes *J. Grandell* (1997)
78 Variance Components Estimation—Mixed models, methodologies and applications
P.S.R.S. Rao (1997)
79 Bayesian Methods for Finite Population Sampling
G. Meeden and M. Ghosh (1997)
80 Stochastic Geometry—Likelihood and computation
O.E. Barndorff-Nielsen, W.S. Kendall and M.N.M. van Lieshout (1998)
81 Computer-Assisted Analysis of Mixtures and Applications—
Meta-analysis, disease mapping and others *D. Böhning* (1999)
82 Classification, 2nd edition *A.D. Gordon* (1999)
83 Semimartingales and their Statistical Inference *B.L.S. Prakasa Rao* (1999)
84 Statistical Aspects of BSE and vCJD—Models for Epidemics
C.A. Donnelly and N.M. Ferguson (1999)
85 Set-Indexed Martingales *G. Ivanoff and E. Merzbach* (2000)
86 The Theory of the Design of Experiments *D.R. Cox and N. Reid* (2000)
87 Complex Stochastic Systems
O.E. Barndorff-Nielsen, D.R. Cox and C. Klüppelberg (2001)
88 Multidimensional Scaling, 2nd edition *T.F. Cox and M.A.A. Cox* (2001)
89 Algebraic Statistics—Computational Commutative Algebra in Statistics
G. Pistone, E. Riccomagno and H.P. Wynn (2001)
90 Analysis of Time Series Structure—SSA and Related Techniques
N. Golyandina, V. Nekrutkin and A.A. Zhigljavsky (2001)
91 Subjective Probability Models for Lifetimes
Fabio Spizzichino (2001)
92 Empirical Likelihood *Art B. Owen (2001)*

Empirical Likelihood

Art B. Owen

CHAPMAN & HALL/CRC
Boca Raton London New York Washington, D.C.

Library of Congress Cataloging-in-Publication Data

Owen, Art. (Art B.)
    Empirical likelihood / Art Owen.
        p. cm.— (Monographs on statistics and applied probability ; 92)
    Includes bibliographical references and index.
    ISBN 1-58488-071-6 (alk. paper)
    1. Estimation theory. 2. Probabilities. 3. Mathematical statistics. I. Title. II. Series.
    QA276.8 .O94 2001
    519.5′44—dc21   2001028680

This book contains information obtained from authentic and highly regarded sources. Reprinted material is quoted with permission, and sources are indicated. A wide variety of references are listed. Reasonable efforts have been made to publish reliable data and information, but the author and the publisher cannot assume responsibility for the validity of all materials or for the consequences of their use.

Neither this book nor any part may be reproduced or transmitted in any form or by any means, electronic or mechanical, including photocopying, microfilming, and recording, or by any information storage or retrieval system, without prior permission in writing from the publisher.

The consent of CRC Press LLC does not extend to copying for general distribution, for promotion, for creating new works, or for resale. Specific permission must be obtained in writing from CRC Press LLC for such copying.

Direct all inquiries to CRC Press LLC, 2000 N.W. Corporate Blvd., Boca Raton, Florida 33431.

Trademark Notice: Product or corporate names may be trademarks or registered trademarks, and are used only for identification and explanation, without intent to infringe.

Visit the CRC Press Web site at www.crcpress.com

© 2001 by Chapman & Hall/CRC

No claim to original U.S. Government works
International Standard Book Number 1-58488-071-6
Library of Congress Card Number 2001028680
Printed in the United States of America 1 2 3 4 5 6 7 8 9 0
Printed on acid-free paper

TO PATRIZIA, GREGORY, AND ELLIOT

Contents

Preface		**xiii**
1	**Introduction**	**1**
	1.1 Earthworm segments, skewness and kurtosis	2
	1.2 Empirical likelihood, parametric likelihood, and the bootstrap	4
	1.3 Bibliographic notes	6
2	**Empirical likelihood**	**7**
	2.1 Nonparametric maximum likelihood	7
	2.2 Nonparametric likelihood ratios	10
	2.3 Ties in the data	11
	2.4 Multinomial on the sample	12
	2.5 EL for a univariate mean	16
	2.6 Coverage accuracy	17
	2.7 One-sided coverage levels	19
	2.8 Power and efficiency	20
	2.9 Computing EL for a univariate mean	21
	2.10 Empirical discovery of parametric families	24
	2.11 Bibliographic notes	25
	2.12 Exercises	27
3	**EL for random vectors**	**29**
	3.1 NPMLE for IID vectors	29
	3.2 EL for a multivariate mean	30
	3.3 Fisher, Bartlett, and bootstrap calibration	31
	3.4 Smooth functions of means	35
	3.5 Estimating equations	39
	3.6 EL for quantiles	43
	3.7 Ties and quantiles	45
	3.8 Likelihood-based estimating equations	48
	3.9 Transformation invariance of EL	50
	3.10 Side information	51
	3.11 Sandwich estimator	55
	3.12 Robust estimators	56

	3.13	Robust likelihood	59
	3.14	Computation and convex duality	60
	3.15	Euclidean likelihood	63
	3.16	Other nonparametric likelihoods	66
	3.17	Bibliographic notes	70
	3.18	Exercises	74

4 Regression and modeling — 79

	4.1	Random predictors	79
	4.2	Nonrandom predictors	83
	4.3	Triangular array ELT	85
	4.4	Analysis of variance	87
	4.5	Variance modeling	90
	4.6	Nonlinear least squares	91
	4.7	Generalized linear models	95
	4.8	Poisson regression	101
	4.9	Calibration, prediction, and tolerance regions	104
	4.10	Euclidean likelihood for regression and ANOVA	106
	4.11	Bibliographic notes	107
	4.12	Exercises	108

5 Empirical likelihood and smoothing — 111

	5.1	Kernel estimates	111
	5.2	Bias and variance	113
	5.3	EL for kernel smooths	114
	5.4	Blood pressure trajectories	116
	5.5	Conditional quantiles	117
	5.6	Simultaneous inference	118
	5.7	An additive model	121
	5.8	Bibliographic notes	124
	5.9	Exercises	125

6 Biased and incomplete samples — 127

	6.1	Biased sampling	127
	6.2	Multiple biased samples	130
	6.3	Truncation and censoring	135
	6.4	NPMLE's for censored and truncated data	139
	6.5	Product-limit estimators	141
	6.6	EL for right censoring	143
	6.7	Proportional hazards	147
	6.8	Further empirical likelihood ratio results	147
	6.9	Bibliographic notes	149
	6.10	Exercises	153

7 Bands for distributions — 155
- 7.1 The ECDF — 156
- 7.2 Exact calibration of ECDF bands — 157
- 7.3 Asymptotics of bands — 158
- 7.4 Bibliographic notes — 160

8 Dependent data — 163
- 8.1 Time series — 163
- 8.2 Reducing to independence — 165
- 8.3 Blockwise empirical likelihood — 168
- 8.4 Spectral method — 173
- 8.5 Finite populations — 174
- 8.6 MELE's using side information — 176
- 8.7 Sampling designs — 177
- 8.8 Empirical likelihood ratios for finite populations — 179
- 8.9 Other dependent data — 180
- 8.10 Bibliographic notes — 180
- 8.11 Exercises — 183

9 Hybrids and connections — 185
- 9.1 Product of parametric and empirical likelihoods — 185
- 9.2 Parametric conditional likelihood — 186
- 9.3 Parametric models for data ranges — 187
- 9.4 Empirical likelihood and Bayes — 188
- 9.5 Bayesian bootstrap — 188
- 9.6 Least favorable families and nonparametric tilting — 189
- 9.7 Bootstrap likelihood — 191
- 9.8 Bootstrapping from an NPMLE — 191
- 9.9 Jackknives — 192
- 9.10 Sieves — 194
- 9.11 Bibliographic notes — 196
- 9.12 Exercises — 199

10 Challenges for EL — 201
- 10.1 Symmetry — 201
- 10.2 Independence — 207
- 10.3 Comparison to permutation tests — 208
- 10.4 Convex hull condition — 209
- 10.5 Inequality and qualitative constraints — 210
- 10.6 Nonsmooth estimating equations — 211
- 10.7 Adverse estimating equations and black boxes — 213
- 10.8 Bibliographic notes — 213
- 10.9 Exercises — 215

11 Some proofs — 217
- 11.1 Lemmas — 217
- 11.2 Univariate and Vector ELT — 219
- 11.3 Triangular array ELT — 222
- 11.4 Multi-sample ELT — 223
- 11.5 Bibliographic notes — 226

12 Algorithms — 229
- 12.1 Statistical tasks — 230
- 12.2 Smooth optimization — 232
- 12.3 Estimating equation methods — 234
- 12.4 Partial derivatives — 238
- 12.5 Primal problem — 241
- 12.6 Sequential linearization — 244
- 12.7 Bibliographic notes — 247

13 Higher order asymptotics — 249
- 13.1 Bartlett correction — 249
- 13.2 Bartlett correction and smooth functions of means — 250
- 13.3 Pseudo-likelihood theory — 251
- 13.4 Signed root corrections — 253
- 13.5 Large deviations — 255
- 13.6 Bibliographic notes — 257
- 13.7 Exercises — 258

Appendix — 261
- A.1 Order and stochastic order notation — 261
- A.2 Parametric models — 262
- A.3 Likelihood — 264
- A.4 The bootstrap idea — 266
- A.5 Bootstrap confidence intervals — 267
- A.6 Better bootstrap confidence intervals — 269
- A.7 Bibliographic notes — 271

References — 273

Author index — 287

Subject index — 293

Preface

Empirical likelihood is a nonparametric method of inference based on a data-driven likelihood ratio function. Like the bootstrap and jackknife, empirical likelihood inference does not require us to specify a family of distributions for the data. Like parametric likelihood methods, empirical likelihood makes an automatic determination of the shape of confidence regions; it straightforwardly incorporates side information expressed through constraints or prior distributions; it extends to biased sampling and censored data, and it has very favorable asymptotic power properties. Empirical likelihood can be thought of as a bootstrap that does not resample, and as a likelihood without parametric assumptions.

This book describes and illustrates empirical likelihood inference, a subject that is ready for a book although it is still undergoing active development. Most of the published literature has emphasized mathematical study of asymptotics and simulations of coverage properties. This book emphasizes analyzing data in ways that illustrate the power and flexibility of empirical likelihood inference. The presentation is aimed primarily at students and at practitioners looking for new ways to handle their data. It is also aimed at researchers looking for new challenges.

The first four chapters form the core of the book. Chapters 5 through 8 extend the ideas to problems such as smoothing, biased sampling, censored and truncated data, confidence bands, time series, and finite populations. Chapter 9 relates empirical likelihood to other methods, and presents some hybrids. Chapter 10 describes some challenges and results near the research frontier. Chapters 11 through 13 contain proofs, computational details, and more advanced theory, respectively. An appendix collects some background material. A course in empirical likelihood could be designed around Chapters 1 through 4, supplemented with those other topics of most interest to the instructor and students.

Much more could have been written about some aspects of empirical likelihood. Applications and theory relevant to survival analysis and to econometrics come to mind, as do recent developments in kernel smoothing and finite population sampling.

The mathematical level of the text is geared towards students and practitioners, and where possible, the presentation stays close to the data. In particular, measure theoretic subtleties are ignored. Some very short and simple proofs are embedded in the text. Longer or more technical theoretical discussions are confined to their own chapters. Finally, the most difficult results are only outlined, and the reader is referred to the literature for the details. A parallel triage has been applied to computational issues.

The worked examples in the text all use real data, instead of simulated, synthetic, or hypothetical data. I believe that all of the statistical problems illustrated are important ones, although in some examples the method illustrated is not one for which the data were gathered. The reader is asked to indulge some statistical license here, and to imagine his or her own data in the place of the illustrating data.

There is an empirical likelihood home page. At the time of writing the URL for that page is:

http://www.stanford.edu/~owen/empirical

The web site is for software, images, and other information related to empirical likelihood.

As an undergraduate, I studied statistics and computer science at the University of Waterloo. The statistics professors there instilled in me a habit of turning first to the likelihood function, whenever an inference problem appeared. I arrived at Stanford University for graduate study at a time when there was a lot of excitement about nonparametric methods. Empirical likelihood is a way of remaining in both traditions.

The idea for empirical likelihood arose when Rupert Miller assigned a problem in survival analysis from Kalbfleisch & Prentice (1980). The problem was to work out the nonparametric likelihood ratio inferences for the survival function as described in Thomas & Grunkemeier (1975). Around that time, there was a debate among some statistics students as to whether nonparametric confidence intervals for a univariate mean should point in the direction that the data seemed to be skewed, or in the opposite direction. There were intuitive arguments and existing methods to support either choice. I looked into empirical likelihood to see if it might point the way, and was surprised to find a nonparametric analog of Wilks's theorem, with the same distribution as in parametric settings. I now call these ELTs (empirical likelihood theorems) after a referee remarked that they are not Wilks's theorem.

Empirical likelihood has been developed by many researchers, as is evident from the bibliographic notes in this text. It is hard to identify only a few contributions from the many, and leave some others out. But it would be harder still not to list the following: Peter Hall, Tom DiCiccio, and Joe Romano obtained some very difficult and significant results on higher order asymptotics. These include Bartlett correctability, signed root corrections, pseudo-likelihood theory, bootstrap calibration, and connections to least favorable families. Jing Qin is responsible for many very creative problem formulations mixing empirical and parametric likelihood, combining multiple biased samples, and with Jerry Lawless, establishing results on using empirical likelihood with side information. Per Mykland has shown how to handle dependent data in a martingale setting. Empirical likelihood for censored and truncated data has been investigated by Gang Li, Susan Murphy, and Aad van der Vaart. Yuichi Kitamura developed connections between empirical likelihood and modern econometrics, and has studied the large deviations properties of empirical likelihood.

I would like to thank Kirsty Stroud, Naomi Lynch, Hawk Denno, and Evelyn Meany of Chapman & Hall/CRC Press for watching over this book through the production process. It is a pleasure to acknowledge the National Science Foundation for supporting, in part, the writing of this book. I also thank the referees and editors who handled the early empirical likelihood papers. They were constructive and generous with their comments. I thank Dan Bloch, Richard Gill, Ker-Ai Lee, Gang Li, Hal Stern, and Thomas Yee, who sent me some data; Judi Davis for entering some data; Ingram Olkin for some tips on indexing; Tomas Rokicki, Eric Sampson, Rob Tibshirani, and C. L. Tondo for some pointers on LaTeX and related topics; Philip Gill, Michael Saunders, and Walter Murray for discussions over the years on nonlinear optimization; George Judge for conversations about econometrics; and Balasubramanian Narasimhan, who kept the computers humming and always seemed to know what software tool I should learn next. I owe a debt to Jiahua Chen, David Cox, Nancy Glenn, Fred Hickernell, David Hinkley, Kristopher Jennings, Li-Zhi Liao, Terry Therneau, and Thomas Yee for help in proofreading. Of course, I am responsible for any flaws that remain. Finally, and most of all, I thank my wife, Patrizia, and my sons, Gregory and Elliot, for their patience, understanding, and encouragement while this book was being written.

Art Owen
Stanford, CA, U.S.A.
May 2001

CHAPTER 1

Introduction

Empirical likelihood is a nonparametric method of statistical inference. It allows the data analyst to use likelihood methods, without having to assume that the data come from a known family of distributions.

Likelihood methods are very effective. They can be used to find efficient estimators, and to construct tests with good power properties. Those tests can in turn be used to construct short confidence intervals or small confidence regions.

Likelihood is also very flexible, as will be seen in examples in this text. When the data are incompletely observed, or distorted, or sampled with a bias, likelihood methods can be used to offset or even correct for these problems. Likelihood can be used to pool information from different data sources. Knowledge arising from outside of the data can be incorporated, too. This knowledge may take the form of constraints that restrict the domain of the likelihood function, or it may be in the form of a prior distribution to be multiplied by the likelihood function.

In parametric likelihood methods, we suppose that the joint distribution of all available data has a known form, apart from one or more unknown quantities. In a very simple example, there might only be one observed data value X from a Poisson distribution. Then $\Pr(X = x) = \exp(-\theta)\theta^x/x!$ for integers $x \geq 0$, where $\theta \geq 0$ is unknown. The unknown θ, called the parameter, is commonly a vector of values, and of course there is usually more than a single number in the data set.

A problem with parametric likelihood inferences is that we might not know which parametric family to use. Indeed there is no reason to suppose that a newly encountered set of data belongs to any of the well studied parametric families. Such misspecification can cause likelihood-based estimates to be inefficient. What may be worse is that the corresponding confidence intervals and tests can fail completely.

Many statisticians have turned to nonparametric inferences to avoid having to specify a parametric family for the data. In addition to empirical likelihood, these methods include the jackknife, the infinitesimal jackknife, and especially, several versions of the bootstrap. These nonparametric methods give confidence intervals and tests with validity not depending on strong distributional assumptions.

Each method has its advantages, as outlined in Chapter 1.2. For now, we note that the advantages of empirical likelihood arise because it combines the reliability of the nonparametric methods with the flexibility and effectiveness of the likelihood approach.

This first chapter begins by looking at some data, and then describes the advantages offered by empirical likelihood. Subsequent chapters develop the method, and explore numerous aspects of empirical likelihood. Some subchapters focus on theory, others on computation, and others on data analysis by empirical likelihood.

The name "empirical likelihood" was adopted because the empirical distribution of the data plays a central role. It was not called nonparametric likelihood, so as not to assume that it would be the only way to extend nonparametric maximum likelihoods to likelihood ratio functions. Alternative nonparametric likelihood ratio functions have indeed been developed. Some of these are discussed at various points throughout the text. In hindsight, the other methods also give a central role to the empirical distribution function, but they are not true likelihoods. Thus the empirical likelihood presented here is distinguished more by being a likelihood than by being empirical.

1.1 Earthworm segments, skewness and kurtosis

The common garden earthworm has a segmented body. The segments are known as somites. As a worm grows, both the number and length of its somites increase. Figure 1.1 shows a histogram of the number of somites on each of 487 worms

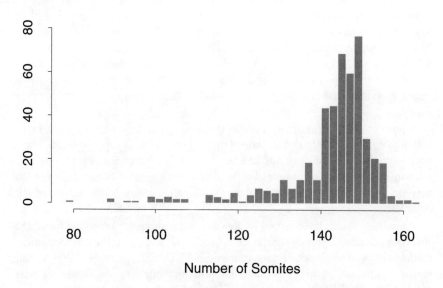

Figure 1.1 *Shown are the number of somites on each of 487 earthworms, as reported by Pearl & Fuller (1905).*

gathered near Ann Arbor, Michigan, in the autumn of 1902. It is visually apparent from Figure 1.1 that worms with many fewer segments than the average are more common than worms with many more segments. We say that the distribution has a heavier tail to the left than it has to the right. Such a mismatch between the tails of a distribution is known as skewness.

Having observed a feature in some data, a valuable next step is to measure it numerically. The alternative is to use terms like "mildly skewed", "very skewed", or "extremely skewed", which may mean different things to different people. The usual way to quantify the skewness of a random variable X is through the coefficient of skewness

$$\gamma = \frac{E((X - E(X))^3)}{E((X - E(X))^2)^{3/2}}. \tag{1.1}$$

As a reference, normally distributed data has $\gamma = 0$, as has any symmetric distribution for which $E(|X|^3)$ exists.

We can estimate γ for the somite distribution by replacing each expectation in (1.1) by the corresponding average over the 487 observed values. The result is $\hat{\gamma} = -2.18$. The estimate $\hat{\gamma}$ is the skewness of the empirical distribution, which places probability $1/487$ on each of the 487 observed values.

The skewness is often accompanied by the kurtosis

$$\kappa = \frac{E((X - E(X))^4)}{E((X - E(X))^2)^2} - 3. \tag{1.2}$$

The 3 in (1.2) is there to make $\kappa = 0$ for normally distributed data. A positive value of the kurtosis describes a distribution with heavier (or fatter) tails than normal distributions have. A negative value describes lighter tails than normal. Later in the text, we will see that skewness and kurtosis play prominent roles in theoretical analysis of confidence intervals

The sample version of the kurtosis for the worm data is $\hat{\kappa} = 5.86$. Because $\hat{\gamma} < 0$ and $\hat{\kappa} > 0$, these values describe a distribution with tails heavier than the normal distribution on average, and a left tail heavier than the right one.

The actual skewness and kurtosis of the somite distribution are unlikely to match their sample values. Figure 1.2 shows joint empirical likelihood confidence regions for γ and κ from these data. The computation and theoretical justification of such confidence regions are deferred to later chapters, and are in fact the main topics of this text. In brief, the empirical likelihood is a multinomial with support on all 487 observed values, and the confidence regions are determined by contours of that likelihood function.

The confidence regions in Figure 1.2 make it particularly clear that the true skewness and kurtosis could not plausibly be 0, though this was perhaps obvious from Figure 1.1. They also show that there is considerable uncertainty in these values, especially the kurtosis. Further, they show that the plausible values of γ extend farther below $\hat{\gamma}$ than they do above it. The plausible values of κ extend farther above $\hat{\kappa}$ than below it.

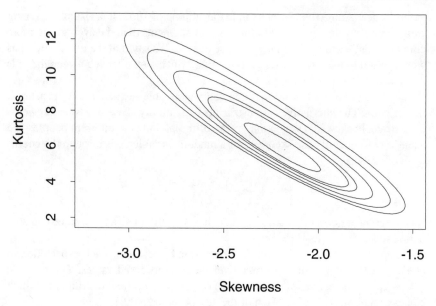

Figure 1.2 *Shown are empirical likelihood confidence regions for the skewness and kurtosis of the somite data. The contours correspond to confidence levels of 50%, 90%, 95%, 99%, 99.9%, and 99.99%.*

Notice also that Figure 1.2 displays a dependency between the skewness and kurtosis. The true (γ, κ) could reasonably be different distances from $(\hat{\gamma}, \hat{\kappa})$ in different directions. The true point can differ more from the sample point on an approximately northwest to southeast axis than it can along an approximately northeast to southwest axis. In hindsight, this is very reasonable. The relative frequency of worms with small somite counts could be lower (or higher) than in the sample, making both γ and κ closer to (or, respectively, farther from) zero than the sample values.

1.2 Empirical likelihood, parametric likelihood, and the bootstrap

The data analysis in Chapter 1.1 would be hard to match with the parametric likelihood methods described in the Appendix. The most popular parametric likelihood methods begin by assuming a normal distribution $N(\mu, \sigma^2)$ for the data. A normal distribution has $\gamma = \kappa = 0$ and so cannot be used to draw inferences on the skewness and kurtosis, as was done for the worm data.

When data are not normally distributed, methods based on the normal distribution do give asymptotically reliable inferences for the mean μ, because of the central limit theorem, if the data distribution has a finite variance. The true variance of the sample mean is σ^2/n, the estimated value of σ^2 approaches the true one, and then confidence intervals with width based on the estimated variance are asymptotically reliable.

The behavior of normal theory methods for the variance $\sigma^2 = E((X-E(X))^2)$ for $X \in \mathbb{R}$ is more subtle. Unlike skewness and kurtosis, the normal distribution has a parameter σ^2 for the variance. The variance is usually estimated by an unbiased quantity $s^2 = S/(n-1)$ or by the maximum likelihood estimate (MLE) $\hat{\sigma}^2 = S/n$ where $S = \sum_{i=1}^{n}(X_i - \bar{X})^2$ and $\bar{X} = (1/n)\sum_{i=1}^{n} X_i$. Both $\hat{\sigma}^2$ and s^2 converge to σ^2 as $n \to \infty$. It can be shown that $\text{Var}(s^2) = \sigma^4(2/(n-1) + \kappa/n)$. The normal distribution forces $\kappa = 0$. If the sampling distribution has $\kappa \neq 0$, then the normal theory model does not use an asymptotically correct estimate of $n\text{Var}(s^2)$, or of $n\text{Var}(\hat{\sigma}^2)$, and the resulting confidence intervals for σ^2 do not approach their nominal coverage levels as n increases.

If the normal distribution is inadequate, perhaps some other distribution is better. It would be hard enough to come up with a parametric family that fit the data and allowed both skewness and kurtosis to vary freely. It would be harder still to find a parametric family in which inferences for the skewness and kurtosis would be reliable, even if the data did not come from a member of that parametric family.

The use of nonparametric methods is in line with John Tukey's quote "It is better to be approximately right, than exactly wrong". But when we contemplate replacing a parametric method by a nonparametric one, we need to consider that sometimes the improved generality comes at a cost of reduced power. Simulations and theory presented in this text suggest that empirical likelihood tests have especially good power properties.

Bootstrap analysis of the data in Chapter 1.1 could be more reliable than parametric likelihood analysis. The bootstrap is described in the Appendix. We could resample the worm data, computing the skewness and kurtosis each time, and plot the results. We would obtain a point cloud of approximately the shape of the regions in Figure 1.2. But there would still be a difficult task in picking out a confidence region from that cloud. Given 1000 points in the plane, we might try to identify the central 950 of them. We could select a rectangle or an ellipse containing 950 of the points. Each of these choices requires us to impose a shape for the region, and still requires us to choose a center, aspect ratio, and possibly an orientation for the region. This problem has also been approached by constructing polygons with vertices at resampled points, and through density estimation of the points (see Chapter 1.3) but so far the methods are not satisfactory.

The main advantages of empirical likelihood, relative to the bootstrap, stem from its use of a likelihood function. Not only does empirical likelihood provide data-determined shapes for confidence regions, it can also easily incorporate known constraints on parameters, and adjust for biased sampling schemes. Unlike the bootstrap, empirical likelihood can be Bartlett corrected, improving the accuracy of inferences. Likelihoods also make it easier to combine data from multiple sources, with possibly different sampling schemes.

The main disadvantage of empirical likelihood relative to the bootstrap is also due to its use of a likelihood function. It can be computationally challenging to optimize a likelihood, be it empirical or parametric, over some nuisance parameters, with other parameters held fixed at test values. In parametric settings this issue is often avoided by making a quadratic approximation to the log likelihood function around the MLE. That option is also available for empirical likelihood.

With parametric likelihoods it is usually not difficult to compute the likelihood itself, even when it is hard to maximize over some nuisance parameters. Most of the statistics considered in this text are defined through estimating equations. In this setting the empirical likelihood is also easy to compute, reducing to a convex optimization for which the unique global optimum can be easily found by iterated least squares.

The bootstrap and empirical likelihood can be combined effectively. One way is to use empirical likelihood to determine a nested family of confidence regions, and the bootstrap to pick out which one to use for a given level of confidence. Another combination is to resample from a distribution that maximizes the empirical likelihood subject to some constraints. Parametric and empirical likelihoods can also be combined to good effect on some problems.

1.3 Bibliographic notes

Owen (1988b) proposed empirical likelihood for the univariate mean and some other statistics, extending earlier work of Thomas & Grunkemeier (1975) who employ a nonparametric likelihood ratio idea to construct confidence intervals for the survival function. That work in turn builds on nonparametric maximum likelihood estimation which has a long history in survival analysis (see Chapter 6).

Empirical likelihood has much in common with some other statistical methods, as described in more detail throughout this text. In particular, the Bayesian bootstrap of Rubin (1981) (see Chapter 9.5), the nonparametric tilting bootstrap of Efron (1981) (Chapter 9.6), a survey sampling estimator of Hartley & Rao (1968) (Chapter 8.10), and the method of sieves of Grenander (1981) (Chapter 9.10) are all closely related to empirical likelihood.

The earthworm data were taken from Pearl & Fuller (1905). In addition to counting somites, they also measured the lengths of the worms, and found that there was very little correlation between the length of a worm and the number of its somites.

Hall (1987) proposes bootstrap confidence regions formed through a kernel density estimate applied to the resampled points. He finds that some oversmoothing is required to get convex contours. Owen (1990b) constructs some polygonal regions based on ideas from Stahel (1981) and Donoho (1982). For every bootstrap point, find the largest rank it ever attains in a linear projection of all bootstrap points. Then sort the B bootstrap points by this rank, and find the $(1-\alpha)B$ points having the smallest rank. The smallest polygon containing those points (their convex hull) is the $100(1-\alpha)\%$ confidence region. It appears that a very large B is required to get smooth region boundaries.

CHAPTER 2

Empirical likelihood

This chapter develops empirical likelihood inference through a nonparametric likelihood ratio function. The result is an approach using a parametric family that is a multinomial distribution on all n observed data values. The focus is on setting confidence intervals for the mean of a scalar random variable. Later chapters extend the approach to other tasks.

2.1 Nonparametric maximum likelihood

We begin by defining the empirical cumulative distribution function, and showing that it is a nonparametric maximum likelihood estimate (NPMLE).

For a random variable $X \in \mathbb{R}$, the cumulative distribution function (CDF) is the function $F(x) = \Pr(X \le x)$, for $-\infty < x < \infty$. We use $F(x-)$ to denote $\Pr(X < x)$ and so $\Pr(X = x) = F(x) - F(x-)$. The notation $1_{A(x)}$ represents the value 1 if the assertion $A(x)$ is true, and 0 otherwise.

Definition 2.1 *Let $X_1, \ldots, X_n \in \mathbb{R}$. The empirical cumulative distribution function (ECDF) of $X_1, \ldots, X_n$ is*

$$F_n(x) = \frac{1}{n} \sum_{i=1}^{n} 1_{X_i \le x}$$

for $-\infty < x < \infty$.

Definition 2.2 *Given $X_1, \ldots, X_n \in \mathbb{R}$, assumed independent with common CDF F_0, the nonparametric likelihood of the CDF F is*

$$L(F) = \prod_{i=1}^{n} \left(F(X_i) - F(X_i-) \right).$$

Definition 2.2 reflects a very literal interpretation of the notion of likelihood. The value $L(F)$ is the probability of getting exactly the observed sample values $X_1, \ldots, X_n$ from the CDF F. One consequence is that $L(F) = 0$ if F is a continuous distribution. To have a positive nonparametric likelihood, a distribution F must place positive probability on every one of the observed data values.

Theorem 2.1 proves that the nonparametric likelihood is maximized by the ECDF. Thus the ECDF is the NPMLE of F.

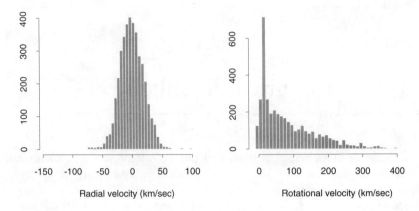

Figure 2.1 *These histograms display the velocities of 3932 stars, from Hoffleit & Warren (1991), as described in the text.*

Theorem 2.1 *Let $X_1, \ldots, X_n \in \mathbb{R}$ be independent random variables with a common CDF F_0. Let F_n be their ECDF and let F be any CDF. If $F \neq F_n$, then $L(F) < L(F_n)$.*

Proof. Let $z_1 < z_2 < \cdots < z_m$ be the distinct values in $\{X_1, \ldots, X_n\}$, and let $n_j \geq 1$ be the number of X_i that are equal to z_j. Let $p_j = F(z_j) - F(z_j-)$ and put $\hat{p}_j = n_j/n$. If $p_j = 0$ for any $j = 1, \ldots, m$, then $L(F) = 0 < L(F_n)$, so we suppose that all $p_j > 0$, and that for at least one j, $p_j \neq \hat{p}_j$. Now $\log(x) \leq x - 1$ for all $x > 0$ with equality only when $x = 1$. Therefore

$$\log\left(\frac{L(F)}{L(F_n)}\right) = \sum_{j=1}^{m} n_j \log\left(\frac{p_j}{\hat{p}_j}\right)$$

$$= n \sum_{j=1}^{m} \hat{p}_j \log\left(\frac{p_j}{\hat{p}_j}\right)$$

$$< n \sum_{j=1}^{m} \hat{p}_j \left(\frac{p_j}{\hat{p}_j} - 1\right)$$

$$\leq 0,$$

and so $L(F) < L(F_n)$. □

Figure 2.1 shows histograms of the radial and rotational velocities of some stars from the bright star catalogue. Stars rotate around the center of our galaxy, with a velocity that depends in part, upon their distance from the center. The radial velocity of a star is the speed with which it appears to be moving away from us, with negative values for stars getting closer. The rotational velocity of a star is its velocity, perpendicular to the line connecting it to the sun.

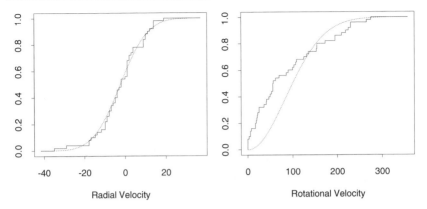

Figure 2.2 *Each plot compares an empirical (solid) and parametric (dashed) CDF for the velocities of 50 stars. Radial velocities are compared to a normal distribution on the left. Rotational velocities are compared to a scaled square root of a $\chi^2_{(2)}$ distribution on the right.*

The step functions in Figure 2.2 shows the NPMLE $F_n(x)$, for velocities of the first 50 stars in the data set. This sample size reduction was made so that the bumpy nature of the NPMLE would be visually apparent. The left plot of Figure 2.2 shows a smooth curve based on the parametric model $X_i \sim N(\mu, \sigma^2)$, for radial velocities X_i. Under this model

$$F(x) = F(x; \mu, \sigma) = \int_{-\infty}^{x} \frac{1}{\sqrt{2\pi}\sigma} \exp\left(\frac{-(z-\mu)^2}{2\sigma^2}\right) dz,$$

and the curves shown are $F(x; \hat{\mu}, \hat{\sigma})$ for parametric MLE's $\hat{\mu}$ and $\hat{\sigma}$.

It may be surprising to find that radial velocities are nearly normally distributed. This would happen if the velocities of the stars relative to the sun had a spherical Gaussian distribution. In that case the rotational velocities would be the square roots of scaled $\chi^2_{(2)}$ distributions. The right plot in Figure 2.2 shows that such a model fits the data poorly. The parametric CDF in that plot is that of the square root of a $\chi^2_{(2)}$ distribution scaled to have mean equal to the sample mean squared rotational velocity.

In parametric models, when $\hat{\eta}$ is the MLE of η, and we are interested in η through some function $\theta(\eta)$, the MLE of θ is $\hat{\theta} = \theta(\hat{\eta})$. In the nonparametric setting, we suppose that we are interested in F through $\theta = T(F)$, where T is a real-valued function of distributions. The true unknown parameter is $\theta_0 = T(F_0)$. Proceeding by analogy, we take the NPMLE of θ to be $\hat{\theta} = T(F_n)$. Thus if we are interested in the mean $\theta_0 = \int x dF_0(x)$ of X when X has the distribution F_0, then by analogy, the NPMLE of θ_0 is the mean of F_n. This mean is of course

$\bar{X} = (1/n) \sum_{i=1}^{n} X_i$. For a subset $A \subset \mathbb{R}$, the NPMLE of $\Pr(X \in A)$ is the sample fraction of X_i in A.

For the radial velocities, the parametric MLE leads to an estimate of 0.589 for $\Pr(X < 0)$, while the nonparametric one gives an estimate of 0.560, in close agreement. Both parametric and nonparametric MLE's estimate $E(X)$ as -2.42. It is visually apparent that the parametric and nonparametric MLE's of tail probabilities for the rotational velocities differ sharply.

Either the parametric or nonparametric MLE can be best, depending on our goals and some assumptions on the data. If we are interested primarily in the probability that $X < x_0$ then the parametric MLE is likely to be best when the true distribution is close to the parametric distribution. If the underlying distribution does not follow the parametric one, then the NPMLE will ordinarily be better, at least for large enough n.

For this data set, the parametric model gives a reasonable fit for the radial velocities, but not for the rotational ones. Empirical CDFs are not very good at showing differences in the tails of a distribution. A QQ plot of all 3932 radial velocities shows a nearly normal distribution, but with heavier than normal tails.

2.2 Nonparametric likelihood ratios

In parametric inference we may base hypothesis tests and confidence regions on the likelihood ratio. If $L(\eta)$ is much smaller than $L(\hat{\eta})$, then we reject the hypothesis that $\eta_0 = \eta$, and exclude η from our confidence region for η_0. Wilks's theorem provides that $-2\log(L(\eta_0)/L(\hat{\eta}))$ tends to a chisquared distribution as $n \to \infty$, under mild regularity conditions, allowing us to decide just how small $L(\eta)$ must be in order for η to get rejected. The degrees of freedom in the chisquared distribution are usually equal to the dimension of the set of η values. When we want a confidence region for θ we take the image of a confidence region for η. That is

$$\{\theta(\eta) \mid L(\eta) \geq cL(\hat{\eta})\},$$

where the threshold c is chosen using Wilks's theorem, with degrees of freedom equal to the dimension of the set of θ values.

We may also use ratios of the nonparametric likelihood as a basis for hypothesis tests and confidence intervals. For a distribution F, define

$$R(F) = \frac{L(F)}{L(F_n)},$$

through the nonparametric likelihood $L(F)$ of Definition 2.2. We proceed by analogy with parametric likelihood. Suppose that we are interested in a parameter $\theta = T(F)$ for some function T of distributions. This F is a member of a set $\mathcal{F}$ of distributions. In some cases we may take $\mathcal{F}$ to be the set of all distributions on $\mathbb{R}$. More often, we use a smaller set of distributions. Define the profile likelihood

TIES IN THE DATA

ratio function:

$$\mathcal{R}(\theta) = \sup \{R(F) \mid T(F) = \theta, F \in \mathcal{F}\}. \tag{2.1}$$

Empirical likelihood hypothesis tests reject $H_0 : T(F_0) = \theta_0$, when $\mathcal{R}(\theta_0) < r_0$ for some threshold value r_0. Empirical likelihood confidence regions are of the form

$$\{\theta \mid \mathcal{R}(\theta) \geq r_0\}. \tag{2.2}$$

In many settings, the threshold r_0 may be chosen using an empirical likelihood theorem (ELT), a nonparametric analogue of Wilks's theorem.

2.3 Ties in the data

If $X_i = X_j$ for $i \neq j$, we say that X_i and X_j are tied. Let us first consider data having no ties. If the distribution F places probability $p_i \geq 0$ on the value $X_i \in \mathbb{R}$, then $\sum_{i=1}^n p_i \leq 1$, and $L(F) = \prod_{i=1}^n p_i$ and so

$$R(F) = \frac{L(F)}{L(F_n)} = \prod_{i=1}^n np_i. \tag{2.3}$$

For data possibly containing some ties, suppose that the distinct value z_j arises $n_j \geq 1$ times in the sample, and has probability p_j under F. Let k be the number of distinct values in the data. Then instead of (2.3) we find

$$R(F) = \prod_{j=1}^k \left(\frac{p_j}{\hat{p}_j}\right)^{n_j} = \prod_{j=1}^k \left(\frac{np_j}{n_j}\right)^{n_j}. \tag{2.4}$$

The theory of empirical likelihood is much simpler through equation (2.3) than equation (2.4). The computation can also be simpler with equation (2.3), though when the number of ties is enormous, so that $k \ll n$, equation (2.4) might lead to faster algorithms. Fortunately, we have the choice. If we use (2.3) instead of the true likelihood ratio (2.4) we get the same profile likelihood ratio function $\mathcal{R}(\theta)$. This holds for any family $\mathcal{F}$ of distributions and for whatever function $T(F)$ is used to define θ.

To see this, we may apportion the probabilities p_j for a distribution F into observation specific weights $w_i \geq 0$ for $i = 1, \ldots, n$. Choose the weights so that p_j is the sum of w_i over all i with $X_i = z_j$. Then a distribution putting weight w_i on observation X_i reproduces F, and hence any $T(F)$.

We define the likelihood of F in terms of these weights as $\prod_{i=1}^n w_i$. When there are ties, this likelihood value is not unique. The value θ enters a confidence region if and only if for some F having $T(F) = \theta$, the largest value of $\prod_{i=1}^n w_i$ exceeds a threshold. So we only need to consider the maximum of $\prod_{i=1}^n w_i$ over weights generating the p_j. This maximum arises when $w_i = p_{j(i)}/n_{j(i)}$, with $j(i)$ determined by $X_i = z_{j(i)}$.

The maximum of $\prod_{i=1}^{n} w_i$ for a given F is

$$\prod_{j=1}^{k}\left(\frac{p_j}{n_j}\right)^{n_j} = L(F) \times \prod_{j=1}^{k} n_j^{-n_j}.$$

When we use nonparametric likelihoods through ratios such as $L(F)/L(F_n)$ the factor $\prod_{j=1}^{k} n_j^{-n_j}$ cancels. Thus we may proceed computationally and theoretically as if there were no ties, writing

$$R(F) = \prod_{i=1}^{n} nw_i, \tag{2.5}$$

where $w_i \geq 0$, $\sum_{i=1}^{n} w_i \leq 1$, and F puts probability $\sum_{j:X_i=X_j} w_j$ on X_i. Equations (2.5) and (2.3) are, of course, equivalent.

Equation (2.5) describes a function on the n-dimensional set of weights

$$\mathbb{S}_{n-1} = \left\{(w_1, \ldots, w_n) \mid w_i \geq 0, \sum_{i=1}^{n} w_i = 1\right\}. \tag{2.6}$$

The set $\mathbb{S}_{n-1}$ is called the probability simplex, or simply the simplex. Because $w_1 + \cdots + w_n = 1$, the weight set is actually $n - 1$ dimensional and so the subscript is $n - 1$ not n. For $n = 3$ the allowable points (w_1, w_2, w_3) are interior to the equilateral triangle with corners at $(1, 0, 0)$, $(0, 1, 0)$ and $(0, 0, 1)$. Figure 2.3 shows contours of $R(F)$ within this triangle. Empirical likelihood confidence regions are usually constructed as the image under a statistical function $T(F)$, of the interior of an $n - 1$ dimensional contour of $R(F)$.

There is nothing special about one-dimensional data in the arguments above. Ties can be ignored for $X_i \in \mathbb{R}^d$, for any $d \geq 1$. In settings more complicated than n IID observations, where we wish to prove that ties can be ignored, we return to this approach of putting probabilities p_j on the distinct observed values and weights w_i on the data points.

It is intuitively reasonable that we should ignore ties. Suppose that we generate tie-breaker random variables $U_i \sim U(0, 1)$ independently of each other and of the X_i. Now form the observations $(X_i, U_i) \in \mathbb{R}^{d+1}$. Because the U_i have a continuous distribution, there will be no ties among the U_i, and hence none among the (X_i, U_i). Now consider a function T on the distribution of (X_i, U_i) pairs, where T completely ignores the U values. Because empirical likelihood ignores ties, we get the same confidence regions for T on the (X_i, U_i) pairs as we do on the X_i data alone. Any other answer would be unreasonable. We know that the U_i contain no information and so their presence should not change our answer.

2.4 Multinomial on the sample

A natural starting point for nonparametric inference is the mean of a scalar random variable, which we take up here. Developing empirical likelihood confi-

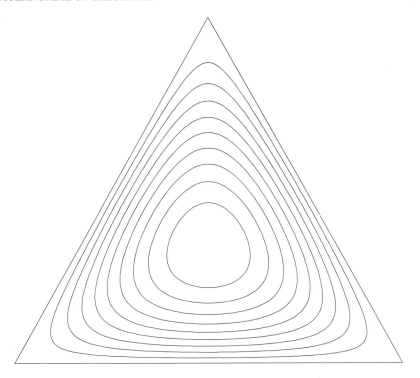

Figure 2.3 *Shown are contours of the empirical likelihood ratio function $R(F)$ for the case of $n = 3$ observations. The likelihood ratio values plotted are 0.1 to 0.9 by steps of 0.1. The bounding triangle has $R(F) = 0$, and the maximum value of $R(F)$ is 1.0 at the center of the triangle.*

dence intervals by analogy with parametric likelihood methods gives a degenerate confidence interval for the mean. To see this, consider the distribution $F = (1-\varepsilon)F_n + \varepsilon\delta_x$ where δ_x is a distribution taking the value x with probability one, and x is not one of the observed X_i values. The likelihood ratio for this F is

$$R(F) = \frac{\prod_{i=1}^{n}(1-\varepsilon)/n}{\prod_{i=1}^{n}1/n} = (1-\varepsilon)^n,$$

and the mean is $(1-\varepsilon)\bar{X} + \varepsilon x$. For any threshold $r_0 < 1$, taking a small enough ε makes $R(F) > r_0$. Then sending x to $\pm\infty$ causes the empirical likelihood ratio confidence interval for the mean to have infinite length.

We can eliminate this problem by changing the set $\mathcal{F}$ of candidate distributions. If X is known to be a bounded random variable, with $-\infty < A \leq X \leq B < \infty$ then by taking $\mathcal{F}$ to be the set of all distributions for which X satisfies these bounds, a nondegenerate confidence interval results. The practical difficulty is that even if we know that X is bounded, we might not know a good bound to use in practice. For example, we might be convinced that the height of a human being is a bounded random variable, yet it might not be easy to specify an upper bound to use in practice.

If we are sampling a bounded random variable, then the sample minimum $A_n = \min_{1 \le i \le n} X_i$ and maximum $B_n = \max_{1 \le i \le n} X_i$ will approach the tightest possible bounds A and B, as n increases. We may obtain finite length confidence intervals by taking $\mathcal{F} = \mathcal{F}_n$ to be the set of distributions of random variables X for which $A_n \le X \le B_n$.

Now suppose that $F \in \mathcal{F}_n$ and $\int x dF(x) = \mu$. Let w_i be the weight that F places on observation X_i. When constructing the profile empirical likelihood function for the mean, we may suppose that $\sum_{i=1}^{n} w_i = 1$. If instead, we have $\sum_{i=1}^{n} w_i < 1$, then F puts probability $1 - \sum_{i=1}^{n} w_i > 0$ on the interval (A_n, B_n) exclusive of sample points there. This probability can be "reassigned" to data points in such a way that the new distribution $\widetilde{F}$ has the same mean as F but has $L(\widetilde{F}) > L(F)$. This reassignment can, for example, be done by increasing the weights w_i for the largest and smallest sample values. The result is that we get the same profile empirical likelihood ratio function for the mean by taking $\mathcal{F}_n$ to be the distributions with $\sum_{i=1}^{n} w_i = 1$ as we do taking $\mathcal{F}_n$ to be all distributions over the interval $[A_n, B_n]$.

Using the distributions with $\sum_{i=1}^{n} w_i = 1$, we may write the profile empirical likelihood ratio function for the mean as

$$\mathcal{R}(\mu) = \max \left\{ \prod_{i=1}^{n} nw_i \;\middle|\; \sum_{i=1}^{n} w_i X_i = \mu, w_i \ge 0, \sum_{i=1}^{n} w_i = 1 \right\}$$

and the resulting empirical likelihood confidence region for the mean as

$$\{\mu \mid \mathcal{R}(\mu) \ge r_0\} = \left\{ \sum_{i=1}^{n} w_i X_i \;\middle|\; \prod_{i=1}^{n} nw_i \ge r_0, w_i \ge 0, \sum_{i=1}^{n} w_i = 1 \right\}.$$

This region is an interval, as shown in Chapter 2.5.

Empirical likelihood inferences for the mean may be recognized as parametric likelihood inferences for the mean, using a data-determined parametric family. The parametric family involved is the multinomial distribution on the observed values of X_i.

For continuous F, this parametric family will have n parameters $w_1, \ldots, w_n$. Since they sum to 1, we can reduce this to $n - 1$ parameters. Most asymptotic results for parametric likelihood are framed in a setting where there are a finite number of parameters and a sample size n tending to infinity. When the number of parameters grows with n, the parametric MLE might not even approach the true parameter value. By having the number of parameters grow as quickly as the sample size, empirical likelihood appears to be very different from parametric likelihoods.

When F is discrete, the empirical likelihood ratio function is that of a multinomial, but on the observed values only. If there are a finite number of possible values of X, then as n increases, eventually all the distinct values have been seen at least once, and empirical likelihood reduces to a parametric likelihood. If F is a discrete distribution for which infinitely many values have positive probability, then the empirical likelihood will be based on a random data-determined multinomial with an ever-increasing number of parameters.

Simply requiring a modestly large likelihood ratio forces all but a vanishingly small amount of the probability to be placed on the sample. Lemma 2.1 quantifies this effect.

Lemma 2.1 *Suppose that distribution F places probability $p_0 = 1 - \sum_{i=1}^{n} w_i$ on the set $\mathbb{R} - \{x_1, \ldots, x_n\}$, and that $\mathcal{R}(F) \geq r_0 > 0$. Then $p_0 \leq (1/n) \log(1/r_0)$.*

Proof. The largest possible value for $\mathcal{R}(F)$, under the problem constraints, arises with all weights equal to $(1-p_0)/n$. Thus $r_0 \leq \mathcal{R}(F) \leq (1-p_0)^n$, from which

$$p_0 \leq 1 - r_0^{1/n}$$
$$= 1 - \exp(n^{-1} \log r_0)$$
$$\leq 1 - (1 + n^{-1} \log r_0)$$
$$= \frac{1}{n} \log\left(\frac{1}{r_0}\right).$$

$\square$

Anticipating that a 95% confidence interval corresponds to $-2\log(r_0)$ close to $\chi_{(1)}^{2,.95} = 3.84$, we consider $\log(1/r_0)/n = 1.92/n$. To contribute a point to the 95% confidence interval, a distribution has to put more than $1 - 1.92/n$ probability on the sample. Our restriction to distributions that reweight the data might push about $2/n$ more probability onto the sample than would otherwise have been there. Thus the empirical profile likelihood ratio function itself does most of the work in reducing the class of functions to those supported on the sample.

We have seen that the restriction to distributions that reweight the sample only changes $O(1/n)$ of the probability. This probability is small because confidence regions typically have diameter of order $n^{-1/2}$. But changes to $O(1/n)$ of the probability of F can have arbitrarily large effects on nonrobust statistics like the mean, so some clipping of the range of F is necessary. Clipping to the sample is perhaps the simplest choice.

In most settings, empirical likelihood is a multinomial likelihood on the sample. There are some exceptions, such as those where boundedness arises naturally in the structure of the problem and need not be imposed. For example, when we are interested in the mean of a bounded function of X such as $1_{X \geq 0}$ or $\sin(X)$, then $\mathcal{F}$ can be the set of all distributions on $\mathbb{R}$. Similarly, inferences for the median of X do not require us to restrict the family of distributions.

2.5 EL for a univariate mean

Nondegenerate intervals still need to be calibrated, so that we can approximate a desired level of confidence such as 95%. The following univariate empirical likelihood theorem (ELT) is the basis for such calibration.

Theorem 2.2 (Univariate ELT) *Let $X_1, \ldots, X_n$ be independent random variables with common distribution F_0. Let $\mu_0 = E(X_i)$, and suppose that $0 < \mathrm{Var}(X_i) < \infty$. Then $-2\log(\mathcal{R}(\mu_0))$ converges in distribution to $\chi^2_{(1)}$ as $n \to \infty$.*

Proof. See Exercises 2.4 and 2.5 for a sketch, Chapter 11.2 for a proof. □

Two features of Theorem 2.2 are noteworthy. First, the chisquared limit is the same as we typically find for parametric likelihood models with one parameter. Second, there is no assumption that X_i are bounded random variables. They need only have a bounded variance, which constrains how fast the sample maximum and minimum can grow as n increases.

Theorem 2.2 provides an asymptotic justification for tests that reject the value μ_0 at the α level, when $-2\log \mathcal{R}(\mu_0) > \chi^{2,1-\alpha}_{(1)}$. The unrejected values of μ_0 form a $100(1-\alpha)\%$ confidence region, with the same asymptotic justification. Details of the proof and some simulations both suggest that the $\chi^{2,1-\alpha}_{(1)}$ threshold should perhaps be replaced by $F^{1-\alpha}_{1,n-1}$. The $F_{1,n-1}$ distribution is the square of a $t_{(n-1)}$ distribution while the $\chi^2_{(1)}$ distribution is the square of a $N(0,1)$ distribution. As $n \to \infty$, we have $t_{(n-1)} \to N(0,1)$ in distribution, and so $F^{1-\alpha}_{1,n-1} - \chi^{2,1-\alpha}_{(1)} \to 0$. Thus, as n increases, the difference between the two calibrations disappears. The F calibration usually gives better results in simulations.

It is easy to see that the empirical likelihood confidence region for a mean is always an interval. If μ_1 and μ_2 are in the confidence region, then there are weights $w_{ij} \ge 0$, $i = 1, \ldots, n$, $j = 1, 2$, with $\sum_{i=1}^n w_{ij} X_i = \mu_j$ and $\sum_{i=1}^n w_{ij} = 1$, and $-2 \sum_{i=1}^n \log(n w_{ij}) \le \chi^{2,1-\alpha}_{(1)}$. Now for $0 < \tau < 1$, let $\mu_\tau = \mu_1 \tau + \mu_2(1-\tau)$. The nonnegative weights $w_i = w_{i1}\tau + w_{i2}(1-\tau)$ sum to 1, satisfy $\sum_{i=1}^n w_i X_i = \mu_\tau$, and

$$-2 \sum_{i=1}^n \log(n w_i) = -2 \sum_{i=1}^n \log\Big(\tau n w_{i1} + (1-\tau) n w_{i2}\Big)$$
$$\le -2\Big[\tau \sum_{i=1}^n \log(n w_{i1}) + (1-\tau) \sum_{i=1}^n \log(n w_{i2})\Big]$$
$$\le \chi^{2,1-\alpha}_{(1)}.$$

It follows that the empirical likelihood confidence region for the mean contains the line segment connecting any two of its points, and so it is an interval.

Figure 2.4 shows 29 determinations of the mean density of the earth, relative to water. These were made by Cavendish in 1798 and appear in Stigler (1977). The mean of Cavendish's values is 5.420, somewhat below the presently accepted

COVERAGE ACCURACY

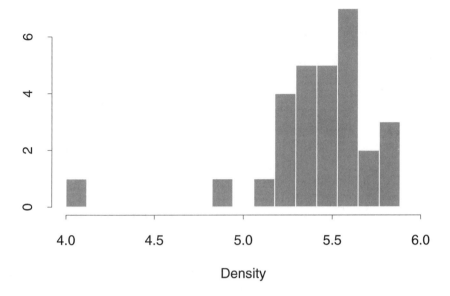

Figure 2.4 *Shown are 29 of Cavendish's measurements of the mean density of Earth, relative to water. Source: Stigler (1977).*

value of 5.517. Figure 2.5 shows the empirical likelihood ratio function for the mean of these data, with the modern value for the density of the earth marked. The modern value lies just inside the 95% empirical likelihood confidence interval, which extends from 5.256 to 5.521.

2.6 Coverage accuracy

A $100(1-\alpha)\%$ empirical likelihood confidence interval is formed by taking those values μ for which $-2\log \mathcal{R}(\mu) \leq \chi^{2,1-\alpha}_{(1)}$, that is $\mathcal{R}(\mu) \geq \exp(-\chi^{2,1-\alpha}_{(1)}/2)$. The probability that μ_0 is in this interval approaches the nominal value $1-\alpha$ as $n \to \infty$. That is, the coverage error

$$\Pr\bigl(-2\log \mathcal{R}(\mu_0) \leq \chi^{2,1-\alpha}_{(1)}\bigr) - \bigl(1-\alpha\bigr) \to 0$$

as $n \to \infty$. The mathematical analysis of coverage error is presented in some works described in the bibliographic notes at the end of this chapter. This section outlines the findings of those works.

Ideally, a confidence interval should have exactly the coverage $1-\alpha$ for any n and any sampling distribution F_0. As discussed in Chapter 2.11, no exact nonparametric confidence intervals exist for the sample mean. As a result nonparametric confidence intervals are asymptotic confidence intervals, as indeed are most para-

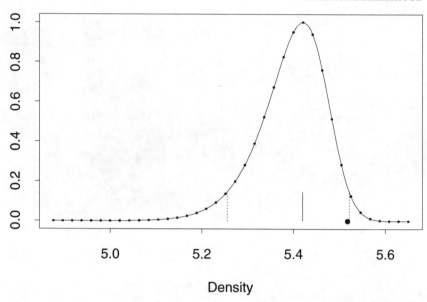

Figure 2.5 *The solid curve shows the empirical likelihood ratio function for the mean of Cavendish's measurements of the density of the earth, relative to water. The modern value of 5.517 is shown as a solid reference point. Two short dotted reference bars delimit the 95% interval and a solid bar shows the sample mean. The points where empirical likelihood was computed are shown as small solid circles connected by interpolation as described in Chapter 2.9.*

metric confidence intervals. The convention used here is that any confidence statement is an asymptotic one, unless explicitly stated otherwise.

Under mild moment conditions, the coverage error for empirical likelihood confidence intervals decreases to zero at the rate $1/n$ as $n \to \infty$. This is the same rate that typically holds for confidence intervals based on parametric likelihoods, the jackknife, and the simpler bootstrap methods. Even the standard confidence intervals, meaning $\bar{X} \pm Z^{1-\alpha/2}s$ where $s^2 = (n-1)^{-1} \sum_{i=1}^{n}(X_i - \bar{X})^2$ and $\Pr(N(0,1) \leq Z^{1-\alpha}) = 1 - \alpha$, have this rate of coverage error. The coverage error in a nonparametric confidence interval for a univariate mean typically takes the form

$$\frac{1}{n}\varphi\left(Z^{1-\alpha/2}\right) \exp\left[A + B\gamma^2 + C\kappa\right] + O\left(\frac{1}{n^2}\right)$$

where φ is the standard normal density function. The quantities γ and κ are the skewness and kurtosis of X, introduced in Chapter 1.1. The constants A, B, C differ for the various confidence interval methods. By $n = 20$, this formula usually predicts the coverage error of a 95% confidence interval for the mean, to within 1 percentage point of its actual coverage error. The exceptions arise for very heavy tailed distributions, where the theory tends to greatly underestimate the coverage level.

For parametric likelihood intervals, the coverage error is typically $O(1/n)$ if the model is true, though it need not converge to 0 as n increases, if the model is not true. A Bartlett correction can be used to reduce parametric coverage errors to $O(n^{-2})$. A Bartlett correction also applies for empirical likelihood. Jackknife, bootstrap, and standard intervals are not Bartlett correctable.

2.7 One-sided coverage levels

For some applications a one-sided confidence interval, corresponding to a one-tailed test, is desired. We may know that $\mu \geq \mu'$ and so be interested in testing $\mu = \mu'$ versus $\mu > \mu'$. Or, when the consequences of small μ are benign while those of large μ are serious we might want a one-sided confidence interval of the form $(-\infty, U)$ for μ.

If (L, U) is a two-sided $100(1 - \alpha)\%$ empirical likelihood confidence interval for μ, then we might consider using $(-\infty, U)$ and (L, ∞) as one-sided $100(1 - \alpha/2)\%$ confidence intervals for the mean. The coverage error in these one-sided intervals decreases to zero, but only at the relatively slow rate $n^{-1/2}$ as $n \to \infty$. Chapter 13 presents methods to modify empirical likelihood, to achieve $O(n^{-1})$ coverage errors for one-sided confidence intervals. The modifications result in shifts of size Δ_L and Δ_R at the left and right endpoints of the interval, respectively, where Δ_L and Δ_R are data determined and equal or very nearly equal to each other.

Confidence intervals based on thresholding a parametric likelihood also typically have $O(n^{-1})$ two-sided coverage errors and $O(n^{-1/2})$ one-sided errors. They also can be modified to have $O(n^{-1})$ coverage errors for one-sided inference.

Every point inside a likelihood interval, parametric or empirical, is deemed to be more likely than every point outside it. When such intervals are modified to equalize coverage errors in the two tails, the result is that some points inside the new interval have lower likelihood than some other points outside of it.

Thus there is a trade-off between separating more likely from less likely parameter values, and equalizing tail errors. We can achieve one goal with high accuracy as $n \to \infty$, but then the other is attained at a slower rate. The examples in this book use confidence regions based on thresholding the empirical likelihood without employing any of the tail area equalization methods described in Chapter 13.

In multi-parameter problems, the goal of equalizing coverage errors between the tails becomes more challenging. Then the shape of the confidence region is more complicated than just a left and right distance from $\hat{\mu}$. Noncoverage events can happen in a continuum of directions, not just two. There is a sense in which empirical likelihood confidence regions have the right shape in higher dimensional problems. There again a shift is required. We must shift the whole set of contours by some vector Δ. See the discussion of pseudo-likelihood in Chapter 13.3.

2.8 Power and efficiency

It is very important to have approximately the right coverage in a confidence interval, or hypothesis test, for otherwise the resulting inferences are not reliable. But there is also a need for efficiency. If a test does not have good power, or a confidence interval is too long, then the data have not been fully utilized. Accuracy and efficiency trade off in confidence interval problems, just as bias and variance often trade off in parameter estimation problems. An interval with nearly the right coverage but highly variable length is not useful. In the extreme case, consider an interval that is equal to all of $\mathbb{R}$ with probability 0.95 and has length 0 with probability 0.05. It has exactly the right coverage but the corresponding test has only 5% power against any alternative.

One way to assess the power of empirical likelihood is through the curvature of $\mathcal{R}$ at the NPMLE $\bar{X}$. For large sample sizes, $\log(\mathcal{R}(\mu)) \doteq -n\sigma_0^{-2}(\mu - \bar{X})^2/2$ for μ near $\bar{X}$, where $\sigma_0^2 = \text{Var}(X_i)$. The greater the (absolute) curvature in this quadratic, the shorter the confidence intervals for a given level of coverage, and hence the greater the power. It can be shown that

$$-2\log \mathcal{R}(\mu_0 + \tau\sigma_0 n^{-1/2}) \to \chi^2_{(1)}(\tau^2), \qquad (2.7)$$

in distribution, where τ^2 is a noncentrality parameter. This means that empirical likelihood inferences will have roughly the same power as parametric inferences, in a family with Fisher information equal to $1/\sigma_0^2$.

Asymptotic comparisons described in Chapter 3.17 show that, to first order, empirical likelihood has the same power as the bootstrap t against alternatives that are $O(n^{-1/2})$ distance from the null hypothesis. Surprisingly, empirical likelihood usually matches the power of parametric likelihood ratio tests to second order, as described in Chapter 3.17.

We might have expected good power properties for empirical likelihood, because likelihood ratio tests are known to be the most powerful tests in multinomial settings, with considerable generality regarding the hypothesis being tested, the alternative of interest, and the competing method under consideration. As Hoeffding (1965) writes:

> If a given test of size α_n is "sufficiently different" from a likelihood ratio test then there is a likelihood ratio test of size $\leq \alpha_n$ that is considerably more powerful than the given test at "most" points in the set of alternatives when n is large enough, provided that $\alpha_n \to 0$ at a suitable rate.

The empirical likelihood setting is not as simple as a multinomial because the support set is random and may increase in cardinality with n. But a version of the universal power optimality of likelihood ratio tests has been established for empirical likelihood. These power results are of the large deviations kind, though they do not necessarily require large sample sizes to be evident. They are described in Chapter 13.5.

Some simulation evidence exists to support these asymptotic results. Simulations can compare power of methods directly, or indirectly by measuring the length of confidence intervals. It has been observed empirically and theoretically that nonparametric confidence intervals tend to undercover, approaching their nominal coverage levels from below as $n \to \infty$. Simulations that ignore this phenomenon can assess coverage but are inconclusive regarding power and its trade-off with coverage. Comparisons of coverage alone favor methods with longer intervals (less power) while comparisons of interval length alone favor methods with more severe undercoverage.

Some simulations cited in Chapter 13.6 compare power after first doing a simulation to calibrate coverage levels. These found that empirical likelihood has better power than the other methods considered at most of the alternative hypotheses simulated. Another simulation, described in Chapter 2.11, compared methods by forcing them to use the same confidence interval length in each Monte Carlo sample. Empirical likelihood obtained competitive coverage whether it or the other method chose the interval length.

2.9 Computing EL for a univariate mean

Empirical likelihood inferences for the univariate mean require the following computational chores: To test whether $\mu = \mu_0$, we need to compute $\mathcal{R}(\mu_0)$. To set confidence limits for μ, we need to find the two values of μ that solve the equation $\mathcal{R}(\mu) = r_0$, given a threshold value r_0. To plot the curve $\mathcal{R}(\mu)$, we need to compute $\mathcal{R}(\mu)$ at numerous points over the range of interest and then interpolate them. The computations are described at a high level, but with some practical details to ease the job of implementing them.

We begin by describing how to compute $\mathcal{R}(\mu)$. Let the ordered sample values be $X_{(1)} \le \ldots \le X_{(n)}$. First we eliminate the trivial cases. If $\mu < X_{(1)}$ or $\mu > X_{(n)}$ then there are no weights $w_i \ge 0$ summing to 1 for which $\sum_i w_i X_i = \mu$. In such cases we take $\log \mathcal{R}(\mu) = -\infty$, and $\mathcal{R}(\mu) = 0$ by convention. Similarly if $\mu = X_{(1)} < X_{(n)}$ or $\mu = X_{(n)} > X_{(1)}$ we take $\mathcal{R}(\mu) = 0$, but if $X_{(1)} = X_{(n)} = \mu$, we take $\mathcal{R}(\mu) = 1$.

Now consider the nontrivial case, with $X_{(1)} < \mu < X_{(n)}$. We seek to maximize $\prod_i nw_i$, or equivalently $\sum_{i=1}^n \log(nw_i)$ over $w_i \ge 0$ subject to the constraints that $\sum_{i=1}^n w_i = 1$ and $\sum_{i=1}^n w_i X_i = \mu$. We write the latter constraint as $\sum_{i=1}^n w_i (X_i - \mu) = 0$. The objective function $\sum_{i=1}^n \log(nw_i)$ is a strictly concave function on a convex set of weight vectors. Accordingly a unique global maximum exists. We also know that the maximum does not have any $w_i = 0$, so it is an interior point of the domain.

We may proceed by the method of Lagrange multipliers. Write

$$G = \sum_{i=1}^{n} \log(nw_i) - n\lambda \sum_{i=1}^{n} w_i (X_i - \mu) + \gamma \left(\sum_{i=1}^{n} w_i - 1 \right),$$

where λ and γ are Lagrange multipliers. Setting to zero the partial derivative of G with respect to w_i gives

$$\frac{\partial G}{\partial w_i} = \frac{1}{w_i} - n\lambda (X_i - \mu) + \gamma = 0.$$

So

$$0 = \sum_{i=1}^{n} w_i \frac{\partial G}{\partial w_i} = n + \gamma,$$

from which $\gamma = -n$. We may therefore write

$$w_i = \frac{1}{n} \frac{1}{1 + \lambda(X_i - \mu)}.$$

The value of λ may be found by numerical search. We know that $\lambda = \lambda(\mu)$ solves

$$\frac{1}{n} \sum_{i=1}^{n} \frac{X_i - \mu}{1 + \lambda(X_i - \mu)} = 0. \tag{2.8}$$

The left side of (2.8) equals $\bar{X} - \mu$ at $\lambda = 0$. It is strictly decreasing in λ, as may be found by differentiation. Monotonicity of (2.8) makes a bisection approach workable, but bisection is slow. Safeguarded search methods, like Brent's method or some versions of Newton's method, are preferable. They combine the reliability of bisection, with a superlinear rate of convergence to the solution.

To begin the search for $\lambda(\mu)$ we need an interval known to contain $\lambda(\mu)$. We know that every $w_i > 0$, and so every $w_i < 1$. A bracketing interval may be found by alternately setting to 1 the weight on the minimum and maximum observations. Thus we may start the search knowing that

$$\frac{1 - n^{-1}}{\mu - X_{(n)}} < \lambda(\mu) < \frac{1 - n^{-1}}{\mu - X_{(1)}}. \tag{2.9}$$

The algorithm then successively refines the interval for λ given by (2.9) until the endpoints agree to a user-specified tolerance. For example, the user might be satisfied if the two values of $\log(\mathcal{R}(\mu))$ from the endpoints of the interval for $\lambda(\mu)$ agree to within 10^{-6}.

To set a confidence interval for μ, we need to locate upper and lower limits μ_+ and μ_- for which $\mathcal{R}(\mu_\pm) = r_0 \in (0, 1)$. We know that

$$X_{(1)} \leq \mu_- \leq \bar{X} \leq \mu_+ \leq X_{(n)}, \tag{2.10}$$

and these bounds can be used in two separate safeguarded searches. We could search for the μ solving $\mathcal{R}(\mu) = r_0$ using a search to find $\mathcal{R}(\mu)$ at each candidate μ.

Such a nested search for μ is slower than necessary, though not necessarily slow in an absolute sense. A faster approach is to reformulate the problem as optimizing $\sum_{i=1}^n w_i X_i$ subject to the constraints $\sum_{i=1}^n w_i = 1$ and $\sum_{i=1}^n \log(nw_i) = \log(r_0)$. In this formulation we take the Lagrangian to be

$$G = \sum_{i=1}^n w_i X_i - \eta\left(\sum_{i=1}^n \log(nw_i) - \log(r_0)\right) - \gamma\left(\sum_{i=1}^n w_i - 1\right).$$

Some calculus, like that above, shows that $w_i = \eta/(X_i - \gamma)$ and so

$$w_i = w_i(\gamma) = \frac{(X_i - \gamma)^{-1}}{\sum_{j=1}^n (X_j - \gamma)^{-1}}.$$

To find μ_- we solve $\sum_{i=1}^n \log(nw_i(\gamma)) = \log(r_0)$, searching for γ between $-\infty$ and $X_{(1)}$. To find μ_+ we search for γ between $X_{(n)}$ and ∞.

The endpoints in the search for γ are more delicate than in the search for λ. One endpoint is infinite and the other gives an infinite value for $\mathcal{R}$. In practice we have to search first for endpoints near the ones given above before beginning the safeguarded search.

To display $\mathcal{R}(\mu)$ we need to compute it at several values. Let $\mu_{(i)} = \bar{X} + i\delta$ for some $\delta > 0$ and integer $i \geq 0$. A good strategy to compute the right side of the empirical likelihood ratio curve is to compute $\mathcal{R}(\mu_{(i)})$ for i increasing from 0, where $\mathcal{R}(\mu_{(0)}) = 1$, until $\log(\mathcal{R}(\mu_{(i)}))$ is too small to be of interest, but in any case stopping before $\mu_{(i)} > X_{(n)}$. For example a limit of $\log(\mathcal{R}) = -25.0$ corresponds to a nominal $\chi^2_{(1)}$ value of $-2 \times 25 = 50$. Such a χ^2 value in turn corresponds to a p-value of about 1.5×10^{-12}, and we seldom need to consider p-values smaller than this. When searching for $\lambda(\mu_{(i)})$, a good starting value is $\lambda(\mu_{(i-1)})$, and we may begin with $\lambda(\mu_0) = 0$. To compute the left side of the empirical likelihood ratio curve, we repeat the process above for i decreasing from 0 until $\mathcal{R}(\mu_{(i)})$ is very small.

If δ is too small then too many steps are required to compute the curve. If δ is too large, then not enough points appear in the curve. The value of δ can be found by trial and error. It is usually satisfactory to have about 20 of the profile points between the 95% confidence interval endpoints. Those endpoints are roughly $4sn^{-1/2}$ apart where s is the sample standard deviation. So $\delta = 0.2 \times sn^{-1/2}$ is usually reasonable. When n is small or the sample is very skewed it may be necessary to use a smaller value of δ.

Most plotting systems will connect the points $(\mu_{(i)}, \mathcal{R}(\mu_{(i)}))$ by straight lines. This can give an unsatisfactory appearance to the curve, often with a prominent triangular peak at the MLE. The function $\log(\mathcal{R}(\mu))$ is approximately quadratic around $\hat{\mu} = \bar{X}$. A better looking curve is obtained by fitting an interpolating spline through $(\mu_{(i)}, \log(\mathcal{R}(\mu)))$. If the spline curve has values (x_j, y_j) on a fine grid of x_j values, then a plot linearly interpolating the $(x_j, \exp(y_j))$ points usually gives a reasonable version of $\mathcal{R}(\mu)$ versus μ. On rare occasions with badly spaced μ_i or unusual behavior in $\mathcal{R}(\mu_i)$ the spline can show a Gibbs effect in which the exponentiated spline produces likelihood ratios over 1.0. The remedy is to insert more likelihood evaluations, or to resort to linear interpolation.

2.10 Empirical discovery of parametric families

Suppose that F_0 is in fact inside a known parametric family. It is natural to wonder whether the empirical likelihood function can discover this fact and match the parametric inferences. It cannot. There is no unique parametric family through F_0 to discover.

Suppose, for example, that X_i are normally distributed with mean $\mu_0 = 1$ and variance $\sigma_0^2 = 1$. Then F_0 belongs to the following families among others:

$$
\begin{aligned}
N(\mu, 1), &\quad \mu \in \mathbb{R}, \\
N(\mu, \mu^2), &\quad \mu \in (0, \infty), \\
N(\mu, e^{1-\mu}), &\quad \mu \in \mathbb{R}, \\
N(1, \sigma^2), &\quad \sigma \in (0, 10).
\end{aligned}
$$

No sample from $N(1, 1)$, however large, can identify one of these models as the true parametric family. They are all equally true, and they have different consequences for how hard it is to learn μ from data. These parametric families are illustrated in Figure 2.6.

A choice of a parametric family requires knowledge from outside of the sample. Such prior information may specify a set of distributions known to include F_0, perhaps based on experience with previous data thought to be similar to the present data. Different investigators could reasonably have different prior information, select different parametric families, and so obtain different answers.

If side information is available, then it can often be used in empirical likelihood. If, for example, we know that $\text{Var}(X) = E(X)^2$ or that $\kappa = 0$, then these facts can be imposed directly as side constraints. It is not necessary to find a parametric family with those constraints built in. See Chapter 3.10.

With empirical likelihood, we ordinarily assume no knowledge outside of the data. Therefore we expect empirical likelihood confidence intervals to be asymptotically at least as long as those for any reasonable parametric family containing F_0. We would like empirical likelihood inferences to behave like parametric inferences in a least favorable family: one in which inference is not artificially easy. Chapters 9.6 and 9.11 discuss connections between empirical likelihood and Stein's concept of least favorable families of distributions.

While empirical likelihood is expected to provide confidence intervals no narrower than a parametric family containing the true distribution F_0, it is possible to find that empirical likelihood confidence intervals are sometimes narrower than parametric ones. This can easily happen if the parametric family is incorrect. For example, if $\kappa < 0$, then the normal theory confidence intervals for the variance will be longer than the empirical ones, for large enough n.

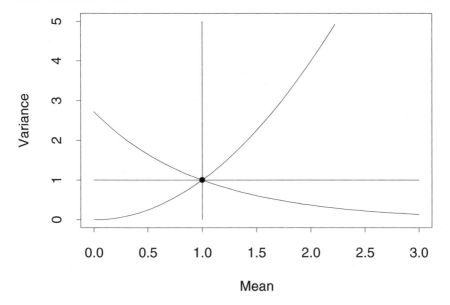

Figure 2.6 *The curves depict four single parameter families through the distribution $N(1, 1)$, shown as a point. For each family, the variance is plotted versus the mean. The families are given in the text. Only the portions for which $0 \leq \mu \leq 3$ and $0 \leq \sigma^2 \leq 5$ are shown.*

2.11 Bibliographic notes

Nonparametric maximum likelihood

The fact that the ECDF is an NPMLE was first noticed by Kiefer & Wolfowitz (1956). The NPMLE idea was used by Kaplan & Meier (1958) to derive the product-limit estimator for the CDF from censored data. Johansen (1978) shows that product-limit estimators are NPMLE's for transition probabilities of continuous time Markov chains. Grenander (1956) constructs an NPMLE for a distribution known to have a monotone decreasing density over $[0, \infty)$. Hartley & Rao (1968) show how to construct the NPMLE from a simple random sample of a finite population.

The star velocities are from the bright star catalogue of Hoffleit & Warren (1991). These data are also available on the Internet. Only those stars for which both velocities are recorded were used. These tend to be the stars nearest the sun. Binney & Merrifield (1998) provide information on stars and their movements.

Likelihood ratios

Theorem 2.2 on empirical likelihood was proved by Owen (1988b), who shows how ties may be ignored, and also presents an ELT for (Fréchet) differentiable statistical functions. An ELT for the mean of bounded scalar random random variable appears in Owen (1985). The discussion of computing is based on Owen (1988b), as is the noncentral χ^2 result at (2.7). Brent's method and other safeguarded searches are described in Press, Flannery, Teukolsky & Vetterling (1993).

Neyman & Scott (1948) provide a famous example of the adverse consequences to likelihood methods when the number of parameters goes to infinity with the sample size. They have $X_{ij} \sim N(\mu_j, \sigma^2)$ for $j = 1, 2$ and $i = 1, \ldots, n$. Then as $n \to \infty$ the MLE $\hat{\sigma}^2 \to \sigma^2/2$ and so is not consistent.

The earliest known use of an empirical likelihood ratio function is Thomas & Grunkemeier (1975). They consider the problem of forming a confidence interval for the survival function based on censored data. The Kaplan-Meier estimate gives the NPMLE, and nonparametric likelihood arguments can be used to form a confidence interval.

Cavendish's data appear in Stigler (1977). That article compares methods of estimating a parameter from data in cases where we now know the true value. Stigler's motivation was to assess whether robust estimators would have helped. One particularly delicate issue is that each scientist's instruments had built-in biases.

Coverage levels

Bahadur & Savage (1956, Corollary 2) show that no nontrivial confidence interval can be computed for the mean if the family $\mathcal{F}$ of possible distributions is sufficiently rich. Their results rule out the existence of exact or even of conservative nonparametric confidence intervals for the mean. They let $\mathcal{F}$ be any set of distributions for $X \in \mathbb{R}$, satisfying three conditions:

1. $\int |x| dF(x) < \infty$ for all $F \in \mathcal{F}$,
2. for each $m \in \mathbb{R}$ there is an $F \in \mathcal{F}$ with $\int x dF(x) = m$, and
3. if $F, G \in \mathcal{F}$, then $\lambda F + (1-\lambda)G \in \mathcal{F}$ for $0 \leq \lambda \leq 1$.

Now suppose that a confidence interval method has probability at least $1 - \alpha$ of covering the mean of F, with this holding for all $F \in \mathcal{F}$. Then for every $F \in \mathcal{F}$ and every $m \in \mathbb{R}$, that method has probability at least $1 - \alpha$ of covering m when sampling from F. This failure cannot be escaped by taking random endpoints for the interval, or by taking a random (but always finite) number of observations. The problem is that F can place a very small probability p on a very large value X_0. The probability p can be small enough that X_0 is unlikely to be seen in a sample of size n. But X_0 can be large enough that pX_0 is not small compared to $E(X)$.

Nonparametric confidence intervals typically approach their asymptotic coverage levels from below. For finite n, the true coverage level is usually, though not always, below the nominal level. This has been observed for empirical likelihood by Owen (1988b), Owen (1990a), and others, for generalized method of moments by Imbens, Spady & Johnson (1998) and others cited therein, and for the bootstrap by Schenker (1985) and others. Kauermann & Carroll (2000) give explicit undercoverage formulas for some confidence intervals based on sandwich estimators of variance. Undercoverage can also arise for asymptotically justified confidence intervals in parametric problems.

As Efron (1988) shows, practically significant errors in coverage levels can correspond to very minor-looking errors in the endpoints of confidence intervals; the noncoverage events are typically near misses. Asymptotic confidence regions give a realistic separation between more and less plausible parameter values, but if we see a special value θ^* just barely outside an asymptotic confidence region, we cannot be sure of the p-value for rejecting θ^*. It is a good practice to plot the confidence interval or region, and then annotate the plot with any values that are special in the context of the data. Still, undercoverage is to be avoided where possible, and it can be greatly alleviated by using bootstrap calibration as described in Chapter 3.3.

Hall (1986) establishes formulas for the coverage error of nonparametric confidence interval methods for the mean. Owen (1990a) provides extensive simulation of various sample sizes and distributions, and finds that by $n = 20$, Hall's formulas are within roughly 1% of the true coverage errors, except for very heavy tailed distributions.

The bibliographic notes for Bartlett correction, signed root corrections and bias shifting of empirical likelihood appear on page 257 at the end of Chapter 13.

Very general power optimality was shown for likelihood ratio tests by Hoeffding (1965). A version for empirical likelihood, discussed in Chapter 13.5. is due to Kitamura (2001), who also presents some simulations. The simulations in Owen (1990a) showed that the bootstrap-t produced the best confidence intervals for the univariate mean. The hardest problem turned out to be covering the mean of the lognormal distribution. In each simulated data set, every competing method constructed a confidence interval. The lengths of these intervals were recorded. Then each method constructed a set of intervals, using the interval lengths chosen by every other method. Empirical likelihood intervals often achieved better coverage when using an interval of a given length than did the method whose nominal 95% interval was of that length.

2.12 Exercises

Exercise 2.1 Another approach to breaking ties is to perturb $X_i \in \mathbb{R}^d$ into $X_i^\varepsilon = X_i + \varepsilon Z_i$, where $Z_i \sim N(0, I_d)$ are independent of each other and the X_i. Let T be a function of the distribution of X_i. Let $\mathcal{R}(\theta, \epsilon) = \max\{\prod_{i=1}^n nw_i \mid T(\sum_{i=1}^n w_i \delta_{X_i}) = \theta, w_i \geq 0, \sum_{i=1}^n w_i = 1\}$. Does $\lim_{\epsilon \to 0} \mathcal{R}(\theta, \epsilon)$ always equal the unperturbed empirical likelihood ratio $\mathcal{R}(\theta, 0)$?

Exercise 2.2 The empirical likelihood ratio is $\prod_{i=1}^{n} nw_i = n^n \times \prod_{i=1}^{n} w_i$. The disadvantage of the second expression is that it is a product of one very large factor and one very small factor. A computer might end up trying to multiply an overflowing number by an underflowing one. For one specific computer, find out how large n has to be for n^n to overflow its floating point representation. Find out how large n must be for $(1/n)^n$ to underflow. In practice accuracy can be lost at values of n smaller than those causing underflow or overflow, and it is better to work with the log likelihood ratio.

Exercise 2.3 Exact confidence intervals are possible for the mean μ in the family $N(\mu, 1)$, and for μ in the family $N(\mu, \sigma^2)$, assuming $n \geq 1$ in the first case and $n \geq 2$ in the second. Explain why this does not contradict the result of Bahadur & Savage (1956) quoted in Chapter 2.11.

Exercise 2.4 This exercise and the next provide a nonrigorous sketch of the proof of Theorem 2.2. Expand equation (2.8) in a Taylor series about $\lambda = 0$. Using the leading terms show that $\lambda \doteq (\bar{X} - \mu)/S(\mu)$ where $S(\mu) = (1/n) \sum_{i=1}^{n} (X_i - \mu)^2$.

Exercise 2.5 Substitute $\lambda = (\bar{X} - \mu)/S(\mu)$ from Exercise 2.4 into an expression for $-2\log \mathcal{R}(\mu_0)$ and show, after a Taylor approximation, that the result is nearly equal to $n(\bar{X} - \mu_0)^2/S(\mu_0)$. A $\chi^2_{(1)}$ limit is then reasonable when $\sqrt{n}(\bar{X} - \mu_0)$ is asymptotically normal with a variance estimated by $S(\mu_0)$.

CHAPTER 3

EL for random vectors

3.1 NPMLE for IID vectors

In this chapter we consider independent random vectors $X_i \in \mathbb{R}^d$ for some $d \ge 1$, assuming that they have a common distribution F_0. It is no longer convenient to describe F_0 by a cumulative distribution function, there being d dimensions along which to cumulate. Instead we describe distributions by the probabilities that they attach to sets. Thus $F(A)$ means $\Pr(X \in A)$ for $X \sim F$ and $A \subseteq \mathbb{R}^d$. We let δ_x denote the distribution under which $X = x$ with probability 1. Thus $\delta_x(A) = 1_{x \in A}$.

We still find that the distribution placing probability $1/n$ on each observation is an NPMLE, under mild changes of notation.

Definition 3.1 *Let $X_1, \ldots, X_n \in \mathbb{R}^d$. The empirical distribution function (EDF) of $X_1, \ldots, X_n$ is*

$$F_n = \frac{1}{n} \sum_{i=1}^n \delta_{X_i}.$$

Definition 3.2 *Given $X_1, \ldots, X_n \in \mathbb{R}^d$, assumed independent with common DF F_0, the nonparametric likelihood of the DF F is*

$$L(F) = \prod_{i=1}^n F(\{X_i\}).$$

Here $F(\{X_i\})$ is the probability of getting the value X_i in a sample from F. Like the univariate Definition 2.2 this is a very literal interpretation of likelihood.

Theorem 3.1 *Let $X_1, \ldots, X_n \in \mathbb{R}^d$ be independent random variables with a common DF F_0. Let F_n be their EDF and let F be any DF. If $F \ne F_n$, then $L(F) < L(F_n)$.*

Proof. The proof is essentially the same as that of Theorem 2.1, except that the distinct values z_j cannot be ordered. That proof made no use of this ordering and so the same argument applies here, too. □

3.2 EL for a multivariate mean

The ideas of Chapter 2 extend naturally to the multivariate mean. As before, the profile empirical likelihood ratio function is

$$\mathcal{R}(\mu) = \max\left\{\prod_{i=1}^{n} nw_i \mid \sum_{i=1}^{n} w_i X_i = \mu, w_i \geq 0, \sum_{i=1}^{n} w_i = 1\right\}, \quad (3.1)$$

except that it is now defined on $\mathbb{R}^d$. The confidence region may still be written

$$C_{r,n} = \left\{\sum_{i=1}^{n} w_i X_i \mid \prod_{i=1}^{n} nw_i \geq r, w_i \geq 0, \sum_{i=1}^{n} w_i = 1\right\} \quad (3.2)$$

only now it is a subset of $\mathbb{R}^d$. The univariate ELT (Theorem 2.2) generalizes to:

Theorem 3.2 (Vector ELT) *Let $X_1, \ldots, X_n$ be independent random vectors in $\mathbb{R}^d$ with common distribution F_0 having mean μ_0 and finite variance covariance matrix V_0 of rank $q > 0$. Then $C_{r,n}$ is a convex set and $-2\log\mathcal{R}(\mu_0)$ converges in distribution to a $\chi^2_{(q)}$ random variable as $n \to \infty$.*

The proof of Theorem 3.2 is given in Chapter 11. Usually V_0 has full rank $q = d$, but if the X_i are confined to a q-dimensional subspace then we need only adjust the degrees of freedom to account for it. In practice, we can use the rank of the sample variance matrix, assuming that $n \gg d$.

Theorem 3.2 suggests that we take $r = \exp(-\chi^{2,1-\alpha}_{(q)}/2)$ in order to get a $100(1-\alpha)\%$ confidence region for the mean. Details of the proof suggest that we might do better by replacing the $\chi^2_{(q)}$ distribution by $(n-1)q/(n-q)$ times an $F_{q,n-q}$ distribution. As $n \to \infty$ these are equivalent. The superiority of F-based calibration has also been observed for parametric likelihoods. In the case $q = 1$, it is like the difference between using the normal distribution and the $t_{(n-1)}$ distribution.

Dippers (Cinclus cinclus) are birds that prey on small aquatic creatures. Figure 3.1 shows the numbers of various types of prey (Caddis fly larvae, Stonefly larvae, Mayfly larvae, and other invertebrates) found at 22 different sites along the river Wye and its tributaries in Wales. The average counts for these prey categories are 92.6, 222.8, 387.0, and 127.6, respectively.

We cannot easily display a four-dimensional confidence region for this mean, but Figure 3.1 does show confidence regions for two pairs of the variables. The normal theory likelihood regions are simply ellipses, while the empirical likelihood regions appear to take on some of the shape of the data points. The normal theory regions shown were computed using the Euclidean likelihood of Chapter 3.15.

For small confidence levels, the empirical likelihood regions are nearly ellipses centered on the sample mean $\bar{X}$. At the highest confidence levels, the empirical likelihood regions approach the convex hull of the data.

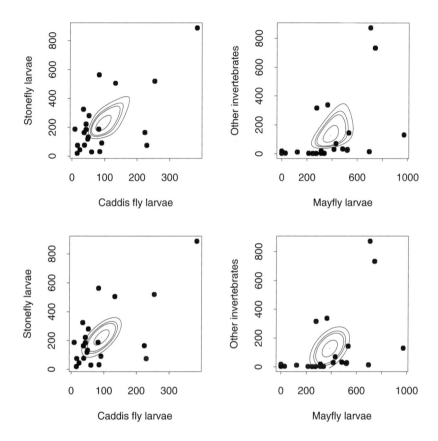

Figure 3.1 *Shown are four sets of confidence regions for a bivariate mean, at nominal levels 50, 90, 95, and 99 percent, based on a $\chi^2_{(2)}$ calibration. The top two panels show empirical likelihood contours, the bottom two panels show contours from a normal distribution likelihood. The left two panels are for the mean counts of Stonefly and Caddis fly larvae. The right two panels are for the mean counts of Mayfly larvae and other invertebrates. The original data points are shown in each plot. Data source: Iles (1993).*

3.3 Fisher, Bartlett, and bootstrap calibration

Theorem 3.2 suggests that the critical value for $-2\log \mathcal{R}(\mu)$ should be $\chi^{2,1-\alpha}_{(d)}$, when the data have a variance of full rank, at least for large enough n. Alternative calibrations can be based on Fisher's F distribution, on a Bartlett correction, or on a bootstrap sample.

Closer inspection of the proof of that theorem gives a lead term of

$$-2\log \mathcal{R}(\mu_0) \doteq n(\bar{X} - \mu_0)'\hat{V}^{-1}(\bar{X} - \mu_0)$$

where

$$\hat{V} = \frac{1}{n}\sum_{i=1}^{n}(X_i - \mu_0)(X_i - \mu_0)'.$$

This is similar to Hotelling's T^2, defined as

$$T^2 = n(\bar{X} - \mu_0)'S^{-1}(\bar{X} - \mu_0),$$

where

$$S = \frac{1}{n-1}\sum_{i=1}^{n}(X_i - \bar{X})(X_i - \bar{X})',$$

is the usual unbiased estimate of $V_0 = \mathrm{Var}(X)$. The relationship is, after some algebra

$$n(\bar{X} - \mu_0)'\hat{V}^{-1}(\bar{X} - \mu_0) = T^2\left(1 + \frac{T^2 - 1}{n}\right)^{-1} = T^2 + O_p\left(\frac{1}{n}\right).$$

For normally distributed X_i, we could use $n(\bar{X} - \mu_0)'V_0^{-1}(\bar{X} - \mu_0) \sim \chi^2_{(d)}$, if we knew V_0. Not knowing V_0 we might use the distribution of Hotelling's T^2, that is $(n-d)T^2/((n-1)d) \sim F_{d,n-d}$. This is also the asymptotic distribution of T^2 when a central limit theorem applies to $\bar{X}$. Based on this asymptotic equivalence to T^2, it would seem appropriate to calibrate empirical likelihood with a critical value of

$$\frac{d(n-1)}{n-d}F^{1-\alpha}_{d,n-d}. \tag{3.3}$$

As $n \to \infty$, the two calibrations become equivalent. The F-based calibration is larger than the one based on the χ^2 and in simulations usually has better coverage. An F calibration can be seen as an adjustment for sampling uncertainty in $\hat{V}$.

Another calibration method for empirical likelihood is to employ a Bartlett correction, as described in Chapter 13.1. A Bartlett correction replaces $\chi^{2,1-\alpha}_{(d)}$ by either

$$\left(1 - \frac{a}{n}\right)^{-1}\chi^{2,1-\alpha}_{(d)}, \quad \text{or,} \quad \left(1 + \frac{a}{n}\right)\chi^{2,1-\alpha}_{(d)} \tag{3.4}$$

for a judiciously chosen constant a. The coverage error in empirical likelihood using the χ^2 or F calibrations is typically of order $1/n$. Bartlett correction reduces the coverage error to $O(1/n^2)$. The appropriate value for a depends on higher moments of X_i, which are generally unknown. If we plug sample versions of those moments into the formula for a obtaining $\hat{a}$, then the asymptotic rate $O(1/n^2)$ also applies to the coverage error using this sample Bartlett correction.

A particularly effective means of calibration is to use the bootstrap. For $b =$

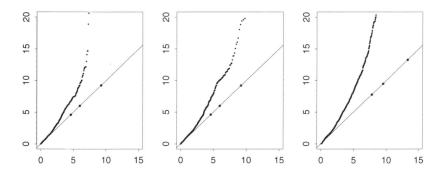

Figure 3.2 *Each figure shows a QQ plot of bootstrap resampled values of* $-2\log\mathcal{R}(\bar{X})$ *plotted against* χ^2 *quantiles. From left to right, the plots are for the bivariate mean of Caddis fly and Stonefly larvae counts, the bivariate mean of Mayfly and other invertebrate counts, and the quadrivariate mean of all four counts. The* χ^2 *degrees of freedom are 4 in the rightmost plot and 2 in the others. A 45° line is included in each plot. Points are marked on it corresponding to nominal 90, 95, and 99 percent coverage levels. Figure 3.13 shows corresponding results for two other nonparametric likelihoods.*

$1, \ldots, B$, and $i = 1, \ldots, n$, let X_i^{*b} be independent random vectors sampled from the EDF F_n of the X_i. This resampling can be implemented by drawing nB random integers $J(i, b)$ independently from the uniform distribution on $\{1, \ldots, n\}$, and setting $X_i^{*b} = X_{J(i,b)}$. Now let $C^{*b} = -2\log(\mathcal{R}^{*b}(\bar{X}))$ where

$$\mathcal{R}^{*b}(\bar{X}) = \max\left\{\prod_{i=1}^n nw_i \mid \sum_{i=1}^n w_i(X_i^{*b} - \bar{X}) = 0, w_i \geq 0, \sum_{i=1}^n w_i = 1\right\},$$

and define the order statistics $C^{(1)} \leq C^{(2)} \leq \cdots \leq C^{(B)}$. For example, to get a 95% bootstrap calibration we might take $B = 1000$ and then use $\{\theta \mid -2\log\mathcal{R}(\theta) \leq C^{(950)}\}$ as the bootstrap-calibrated 95% confidence region.

Figure 3.2 has QQ plots that compare the bootstrap order statistics $C^{(b)}$ with χ^2 quantiles, for the dipper data. The bootstrap empirical likelihood values strongly indicate that a χ^2 calibration is not very reasonable for this small data set. The largest values of $-2\log\mathcal{R}$ including some infinite ones do not fit on the plot. Chapter 3.16 shows corresponding QQ plots for some other nonparametric likelihoods.

It is clear that for data such as these a Bartlett correction will not make much of an improvement. The Bartlett correction multiplies the χ^2 threshold by a constant, corresponding to a reference line through the origin with slope $1 + a/n$ instead of 1. The bootstrap-resampled empirical log likelihoods for the dipper survey data have a QQ plot that is nearly linear close to the origin, but it bends sharply upward between there and the χ^2 quantiles of interest.

The theoretical interest in Bartlett correction is that it shows that a higher order and subtle property of parametric log likelihood ratio functions carries over to the empirical log likelihood. In practice bootstrap calibration is likely to be much more reliable than a Bartlett-corrected χ^2 calibration for moderate n, though the two methods can be combined as described further below.

Figure 3.3 shows bootstrap-calibrated empirical likelihood and normal likelihood regions for the dipper data. Bootstrap calibration makes all the regions larger. The bootstrap-calibrated normal theory regions are always ellipsoidal, and they include some negative values for the mean number of other invertebrates.

There were 5 (of 1000) bootstrap samples for which the Caddis Fly–Stonefly sample mean was outside the convex hull of the resampled Caddis Fly–Stonefly data. There were 4 such cases for the Mayfly–other sample mean, and 23 cases in which the original four-dimensional mean was outside the convex hull of the resampled data. The true value of $-2\log\mathcal{R}$ for such cases is $-\infty$. The computed value was about -1000 in these cases.

For continuously distributed X_i, there is always a hyperplane separating $\bar{X}$ from at least half of the data. The probability of $\bar{X}$ lying outside the convex hull $\mathcal{H}^b$ of $X_1^{*b}, \ldots, X_n^{*b}$ is always at least 2^{-n}, and may be higher for skewed data. If the fraction of bootstrap samples with $\bar{X}$ outside of $\mathcal{H}^b$ is larger than $\alpha/2$, say, then this is a diagnostic that the restriction of empirical likelihood confidence regions to the convex hull may be critical.

Bootstrap and Bartlett alternatives to χ^2 or F calibration apply to other problems than the multivariate mean. Some resampled values of $-2\log\mathcal{R}(\hat{\gamma}, \hat{\kappa})$ for the skewness and kurtosis of the worm somites from Chapter 1 are shown in Figure 3.4. The $\chi^2_{(2)}$ distribution fits the data very well. The sample size is 487, so we might have expected a good fit, but on the other hand, a lot of data might have been thought necessary to gauge uncertainty in a sample kurtosis.

It is also possible to use the bootstrap to estimate the parameter in the Bartlett correction. A regression through the origin of $C^{(b)}$ on $\chi^{2,(b-1/2)/B}_{(2)}$ for the resampled somite empirical likelihoods is shown as a dashed line in Figure 3.4. The regression has a slope of 1.107, corresponding to a Bartlett constant a of 33.1 or 35.6 depending on which formula in (3.4) is used.

The practical value of the Bartlett correction is that, where n is large enough that the Bartlett correction is accurate, bootstrap Bartlett correction might reasonably require fewer resampled likelihood ratios than ordinary bootstrap calibration. The reason is that bootstrap Bartlett correction requires estimation of a mean or ratio of means, instead of an extreme quantile. In the resampled worm data, there were no infinite resampled log likelihoods, but in general such infinities can arise. A safer approach to estimating a is to equate a trimmed mean of the C^b (leaving out the largest values) to the corresponding trimmed mean of the $(1+a/n)\chi^2_{(d)}$ distribution.

SMOOTH FUNCTIONS OF MEANS

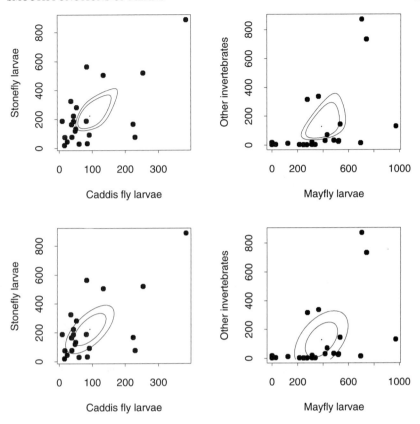

Figure 3.3 *Shown are bivariate 95% confidence regions for the mean, calibrated by χ^2 and by the bootstrap, for the 22 data points in the dipper survey. The top row shows empirical likelihood, the bottom row shows ellipsoidal regions calibrated by normal theory and by the bootstrap. In all four plots, bootstrap calibration gives the larger region.*

3.4 Smooth functions of means

Empirical likelihood methods for the mean require some modifications to work for other parameters like the variance. The variance of X is the mean of $(X-\mu)^2$, in which μ, the mean of X, is usually unknown. If we knew μ then we could form sample values $(X_i - \mu)^2$ and construct an empirical likelihood ratio function for their common mean. Here we consider extending empirical likelihood inferences to statistics that can be written as smooth functions of a finite-dimensional vector of means.

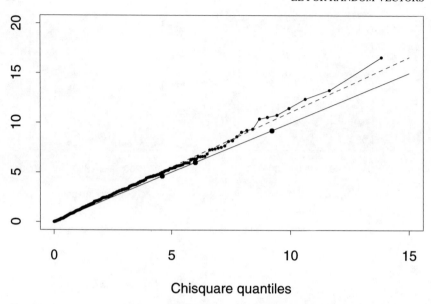

Chisquare quantiles

Figure 3.4 *Shown is a QQ plot of 500 bootstrap resampled values of $-2\log\mathcal{R}(\hat{\gamma}, \hat{\kappa})$ for the earthworm somite counts from Chapter 1, versus $\chi^2_{(2)}$ quantiles. The solid reference line is the 45° line. The reference points on that line correspond to nominal confidence levels of 90, 95, and 99 percent. The dashed reference line is for an estimated Bartlett correction as described in the text.*

For the distribution F with weights w_i the variance is written

$$\sigma^2(F) = \sum_{i=1}^n w_i (X_i - \mu(F))^2 = \sum_{i=1}^n w_i X_i^2 - \left(\sum_{i=1}^n w_i X_i\right)^2, \quad (3.5)$$

where $\mu(F) = \sum_{i=1}^n w_i X_i$. Note that under this definition the NPMLE of the variance is $(1/n)\sum_{i=1}^n (X_i - \bar{X})^2$, or $(n-1)/n$ times the usual unbiased sample variance. Unless n is small the distinction is not practically significant. The profile likelihood ratio function for the variance is

$$\mathcal{R}(\sigma^2) = \max\left\{\prod_{i=1}^n n w_i \mid \sum_{i=1}^n w_i (X_i - \mu(F))^2 = \sigma^2, w_i \geq 0, \sum_{i=1}^n w_i = 1\right\}.$$

Equation (3.5) above shows that σ^2 can be written as a smooth function of the mean of the vector (X, X^2). That is, $\sigma^2 = E(X^2) - E(X)^2$ and $\hat{\sigma}^2 =$

$\hat{E}(X^2) - \hat{E}(X)^2$. Similarly, the correlation between X and Y,

$$\rho = \frac{E(XY) - E(X)E(Y)}{\sqrt{\text{Var}(X)\text{Var}(Y)}},$$

is a smooth function of the mean of (X, Y, X^2, Y^2, XY), and a regression coefficient is a smooth function of means of cross products among the predictors and between the predictors and the response.

To formulate the problem in general, suppose that $X_i \in \mathbb{R}^d$ are independent random variables with common distribution F_0 having mean μ_0. Let h be a smooth function from $\mathbb{R}^d$ to $\mathbb{R}^q$ for $1 \leq q \leq d$. The parameter is $\theta_0 = h(\mu_0)$ and the NPMLE is $\hat{\theta} = h(\bar{X})$. The profile empirical likelihood ratio function is

$$\mathcal{R}(\theta) = \max\left\{\prod_{i=1}^n n w_i \mid h\left(\sum_{i=1}^n w_i X_i\right) = \theta, w_i \geq 0, \sum_{i=1}^n w_i = 1\right\}.$$

Theorem 3.3 *Suppose that $X_i \in \mathbb{R}^d$ are independent with common distribution F_0, having mean μ_0 and variance matrix V_0 of full rank d. Let h be a smooth function from $\mathbb{R}^d$ to $\mathbb{R}^p$ for $1 \leq p \leq d$, and suppose that the $d \times p$ matrix G of partial derivatives of $h(\mu)$ with respect to components of μ has rank $q > 0$ at $\mu = \mu_0$. Define $\theta_0 = h(\mu_0)$,*

$$C_{r,n}^{(1)} = \left\{\sum_{i=1}^n w_i X_i \mid \prod_{i=1}^n n w_i \geq r, w_i \geq 0, \sum_{i=1}^n w_i = 1\right\},$$

$$C_{r,n}^{(2)} = \left\{h(\mu) \mid \mu \in C_{r,n}^{(1)}\right\},$$

and

$$C_{r,n}^{(3)} = \{\theta_0 + G'(\mu - \mu_0) \mid \mu \in C_{r,n}^{(1)}\}.$$

Then as $n \to \infty$,

$$\Pr(\theta \in C_{r,n}^{(3)}) \to \Pr(\chi^2_{(q)} \leq -2\log(r)),$$

and

$$\sup_{\mu \in C_{r,n}^{(1)}} \|h(\mu) - \theta_0 - G'(\mu - \mu_0)\| = o_p(n^{-1/2}).$$

Proof. The linearization $\theta_0 + G'(\mu - \mu_0)$ has mean θ_0 and variance $G'V_0G$ of rank $q > 0$, so the first claim follows by Theorem 3.2. The second claim follows from the definition of the derivative. □

Theorem 3.3 shows that confidence regions $C_{r,n}^{(3)}$ formed by linearizing $h(\mu)$ around $h(\mu_0)$ have an asymptotic $\chi^2_{(q)}$ calibration, where q is the rank of the $\partial h(\mu_0)/\partial \mu$. The confidence regions that one actually uses are of the form $C_{r,n}^{(2)}$

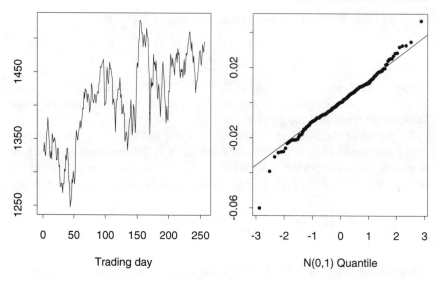

Figure 3.5 *The left plot shows the value of the S&P 500 index for 256 trading days. These values were converted into 255 logarithmic returns. The right plot shows a QQ plot of these returns to the S&P 500 index. Source:* http://finance.yahoo.com

which differs from $C_{r,n}^{(3)}$ by $o_p(n^{-1/2})$. This difference is an asymptotically negligible fraction of the diameter $O(n^{-1/2})$ of the confidence regions.

Theorem 3.3 does not directly say that $\Pr(\theta_0 \in C_{r,n}^{(2)}) \to \Pr(\chi_{(q)}^2 \le -2\log(r))$. That this limit does hold follows from the smooth function of means results described in Chapter 13.1. Those results require stronger conditions than Theorem 3.3, and they also establish Bartlett correctability. The weakest conditions under which $\Pr(\theta_0 \in C_{r,n}^{(2)}) \to \Pr(\chi_{(q)}^2 \le -2\log(r))$ holds have not yet been determined.

Figure 3.5 shows the S&P 500 stock index for 256 trading days, covering the period from August 17, 1999 to August 17, 2000. Letting Y_i denote the value of the index on day i, then $X_i = \log(Y_i/Y_{i-1})$ is the daily (logarithmic) return for the index. Figure 3.5 also shows a QQ plot of the 255 returns for the year starting August 18, 1999 and ending August 17, 2000. We will analyze these data as if the X_i were independent. Chapter 3.17 describes the distribution of financial returns. Chapter 8 describes time series methods of empirical likelihood that could handle the sort of dependencies found in financial data.

Considerable interest attaches to the variance of financial returns. One reason is that an option to buy or sell the underlying quantity at a later date is more valuable if that quantity is highly variable. To convert a daily variance into an annual one, we multiply by 255. The square root of the annualized variance is known as the

Method	Lower	Upper
Empirical Likelihood	18.95%	24.22%
Normal Theory	19.50%	23.22%

Table 3.1 *95% confidence intervals for the volatility of the S&P 500 index.*

volatility. For this data set the sample value of the volatility is

$$\hat{\sigma} = \left(\frac{255}{n} \sum_{i=1}^{n} (X_i - \bar{X})^2 \right)^{1/2} = 0.2116,$$

described as 21.16% volatility.

Logarithmic returns usually have nearly a normal distribution, except for tails that are heavier than those of the normal distribution. Extreme market movements, both up and down, tend to be more extreme than the normal distribution would predict. The sample kurtosis for these data is $\hat{\kappa} = 1.79$. Normal theory confidence intervals can either be based on the normal likelihood, or on the result that $\hat{\sigma}^2/\sigma^2 \sim \chi^2_{(254)}/254$ under a normal distribution. Either form of normal theory confidence interval for the volatility of returns would ordinarily be too narrow.

Table 3.1 shows 95% confidence intervals for the volatility of the S&P 500 index. The intervals obtained using empirical likelihood are in this case almost 42% wider than the normal theory ones. Because returns have heavy tails, we have reason to expect that the normal theory interval is too narrow, whereas asymptotic theory justifies the empirical likelihood interval.

Figure 3.6 shows parametric and empirical profile likelihood ratio functions for the volatility of the S&P 500 index data. The parametric curve is based on a normal distribution for the data. It has asymptotic justification if the normal distribution holds, or more generally, if $\kappa = 0$. The empirical likelihood inferences have an asymptotic justification under the much weaker condition that $\kappa < \infty$. For financial data it is common that the true kurtosis $\kappa > 0$, in which case normal theory confidence intervals are too short. Returns for commodities or some individual stocks can have a kurtosis much larger than that of the S&P 500 index, yet even here taking account of the kurtosis makes a noticeable difference to the confidence interval.

3.5 Estimating equations

Estimating equations provide an extremely flexible way to describe parameters and the corresponding statistics. For a random variable $X \in \mathbb{R}^d$, a parameter

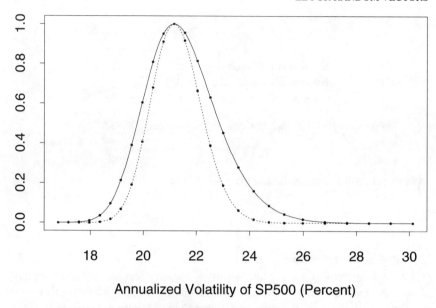

Figure 3.6 *The solid outer curve shows the empirical likelihood ratio function for the annualized volatility of the S&P 500 stock index. The inner curve shows the normal theory likelihood ratio function. Empirical likelihood inferences are asymptotically valid for $\kappa < \infty$ whereas normal theory ones require $\kappa = 0$.*

$\theta \in \mathbb{R}^p$, and a vector-valued function $m(X, \theta) \in \mathbb{R}^s$ suppose that

$$E(m(X, \theta)) = 0. \tag{3.6}$$

The usual setting has $p = s$ and then under conditions on $m(X, \theta)$ and possibly on F, there is a unique solution θ. In this just determined case, the true value θ_0 may be estimated by solving

$$\frac{1}{n} \sum_{i=1}^{n} m(X_i, \hat{\theta}) = 0 \tag{3.7}$$

for $\hat{\theta}$. To write a vector mean by equation (3.6), we take $m(X, \theta) = X - \theta$, and then equation (3.7) gives $\hat{\theta} = \bar{X}$. For $\Pr(X \in A)$ take $m(X, \theta) = 1_{X \in A} - \theta$. For a continuously distributed scalar X and $\theta \in \mathbb{R}$, the function $m(X, \theta) = 1_{X \leq \theta} - \alpha$ defines θ as the α quantile of X. Chapter 3.6 describes tail probabilities and quantiles in more detail.

Equation (3.7) is known as the estimating equation, and $m(X, \theta)$ is called the estimating function. Most maximum likelihood estimators are defined through estimating equations as outlined in Chapter 3.8, as are certain robust statistics known as M-estimators described in Chapter 3.12.

ESTIMATING EQUATIONS

In econometrics, considerable interest attaches to the overdetermined case with $s > p$. In problems with $s > p$ the fact that (3.6) holds is a special feature of F and constitutes important side information. Even when (3.6) holds for the true F_0, it will not ordinarily hold for the NPMLE $\hat{F}$, in which case (3.7) has no solution. The generalized method of moments looks for a value $\hat{\theta}$ that comes close to solving (3.7). An empirical likelihood approach to this problem is described in Chapter 3.10.

The underdetermined case $s < p$ can also be useful. Then (3.6) and (3.7) might each have an $s - p$ dimensional solution set of θ values. Some functions of θ may be precisely determined from the data, while others will not. An example from multivariate calibration is discussed in Chapter 4.9.

Empirical likelihood and estimating equations are well suited to each other. The empirical likelihood ratio function for θ is defined by

$$\mathcal{R}(\theta) = \max\left\{\prod_{i=1}^n nw_i \mid \sum_{i=1}^n w_i m(X_i, \theta) = 0, w_i \geq 0, \sum_{i=1}^n w_i = 1\right\}.$$

Theorem 3.4 *Let $X_1, \ldots, X_n \in \mathbb{R}^d$ be independent random vectors with common distribution F_0. For $\theta \in \Theta \subseteq \mathbb{R}^p$, and $X \in \mathbb{R}^d$, let $m(X, \theta) \in \mathbb{R}^s$. Let $\theta_0 \in \Theta$ be such that $\mathrm{Var}(m(X_i, \theta_0))$ is finite and has rank $q > 0$. If θ_0 satisfies $E(m(X, \theta_0)) = 0$, then $-2\log \mathcal{R}(\theta_0) \to \chi^2_{(q)}$ in distribution as $n \to \infty$.*

Proof. This follows immediately from the Vector ELT, Theorem 3.2. □

The interesting thing about Theorem 3.4 is what is not there. It includes no conditions to make $\hat{\theta}$ a good estimate of θ_0, nor even conditions to ensure a unique value for θ_0, nor even that any solution θ_0 exists. Theorem 3.4 applies in the just determined, overdetermined, and underdetermined cases. When we can prove that our estimating equations uniquely define θ_0, and provide a consistent estimator $\hat{\theta}$ of it, then confidence regions and tests follow almost automatically through Theorem 3.4.

In the underdetermined cases we say that θ is not estimable. Our θ may not be estimable because of a poor choice of $m(x, \theta)$ or an unfortunate distribution F. Assuming that the rank q in Theorem 3.4 is common to all solutions, each solution θ has an asymptotic probability $1 - \alpha$ of being in the confidence region. That region might not shrink down to a single point as $n \to \infty$. Nor should we expect it to.

In the overdetermined case there may be no θ with $E(m(X, \theta)) = 0$. In that case there is the possibility that a confidence region constructed using Theorem 3.4 will be empty. See Exercise 3.13.

The range of problems expressible as estimating equations widens greatly upon the introduction of nuisance parameters. To define the correlation ρ of X and Y, we formulate five estimating equations:

$$0 = E(X - \mu_x) \tag{3.8}$$
$$0 = E(Y - \mu_y) \tag{3.9}$$
$$0 = E((X - \mu_x)^2 - \sigma_x^2) \tag{3.10}$$
$$0 = E((Y - \mu_y)^2 - \sigma_y^2) \tag{3.11}$$
$$0 = E((X - \mu_x)(Y - \mu_y) - \rho\sigma_x\sigma_y), \tag{3.12}$$

for the parameter $\theta = (\mu_x, \mu_y, \sigma_x, \sigma_y, \rho)$. When we are interested only in inferences on ρ and not on the whole θ vector, then the two means and standard deviations in θ are nuisance parameters.

To handle nuisance parameters, write the estimating function as $m(X, \theta, \nu)$, where now $\theta \in \mathbb{R}^p$ is the vector of parameters of interest and $\nu \in \mathbb{R}^q$ is a vector of nuisance parameters, and $m \in \mathbb{R}^s$. The parameters (θ, ν) satisfy the equations $E(m(X, \theta, \nu)) = 0$. Ordinarily $s = p + q$ and the equations jointly define θ and ν. Now we define

$$\mathcal{R}(\theta, \nu) = \max \left\{ \prod_{i=1}^n nw_i \mid \sum_{i=1}^n w_i m(X_i, \theta, \nu) = 0, w_i \geq 0, \sum_{i=1}^n w_i = 1 \right\}$$

and

$$\mathcal{R}(\theta) = \max_\nu \mathcal{R}(\theta, \nu).$$

That is, we maximize over the nuisance parameters. If for a candidate θ there is any value of ν that gives a large likelihood $\mathcal{R}(\theta, \nu)$, then θ is in the confidence region. The function $\mathcal{R}(\theta)$ is called the profile empirical likelihood ratio function. Profile likelihoods are also widely used in parametric models with nuisance parameters. Under mild conditions (Chapter 3.10), $-2\log \mathcal{R}(\theta) \to \chi^2_{(p)}$.

The estimating equation approach includes smooth functions of means and some methods not expressible as smooth functions of means. For example, Huber's M-estimator for the location of a scalar random variable X is defined in Chapter 3.12. It cannot be expressed as a smooth function of a finite number of means. It is convenient if the estimating equations can be arranged so that one of the parameters is the smooth function $h(\mu)$ we are interested in, as is done for the correlation above. If that should prove intractable for $\theta = h(E(X))$, then we may resort to writing equations

$$E(X - \mu) = 0$$
$$E(\theta - h(\mu)) = 0.$$

Because the second estimating equation does not involve X, it would be better handled as simply a constraint on the enlarged parameter vector (μ, θ).

3.6 EL for quantiles

Any value m for which $\Pr(X \le m) \ge 1/2$ and $\Pr(X \ge m) \ge 1/2$ is a median of the distribution of X. Unless $\Pr(X = m) > 0$, a median m satisfies $E(Z) = 0$, where $Z = 1_{X \le m} - 1/2$. We will use this definition because it is convenient. A simple fix for the problem of $\Pr(X = m) > 0$ is described in Chapter 3.7. More generally, for $0 < \alpha < 1$, we will take the α quantile of F to be any value Q^α for which

$$E(1_{X \le Q^\alpha} - \alpha) = 0. \tag{3.13}$$

If F has a positive density in an open interval containing Q^α then Q^α is unique.

For quantiles, it is not necessary to restrict attention to distributions supported on the sample. The distribution maximizing $\prod_{i=1}^n F(\{X_i\})$ subject to the constraint $\int (1_{x \le Q^\alpha} - \alpha) dF(x) = 0$ puts all its probability on the sample if $X_{(1)} \le Q^\alpha < X_{(n)}$. For $Q^\alpha < X_{(1)}$ the maximizer is not unique, but the maximum is achieved by a distribution that puts some probability on a value $X_0 \le Q^\alpha$. For $Q^\alpha \ge X_{(n)}$, one maximizer puts some probability on a value $X_{n+1} > Q^\alpha$. The likelihood does not depend on the exact values of X_0 and X_{n+1}. For convenience, define $X_0 = -\infty$ and $X_{n+1} = \infty$.

For $-\infty < q < \infty$ and $0 < p < 1$, let $Z_i(p, q) = 1_{X_i \le q} - p$. Then define

$$\mathcal{R}(p, q) = \max \left\{ \prod_{i=1}^n n w_i \mid \sum_{i=0}^{n+1} w_i Z_i(p, q) = 0, w_i \ge 0, \sum_{i=0}^{n+1} w_i = 1 \right\}.$$

Notice that the points X_0 and X_{n+1} count in the weighted mean of Z_i but do not contribute to the likelihood. For a given α, the value Q^α is rejected if $\mathcal{R}(\alpha, Q^\alpha)$ is too small. The confidence region for Q_0^α consists of all values q for which $\mathcal{R}(\alpha, q)$ is sufficiently large.

An easy calculation shows that $w_0 = w_{n+1} = 0$ for $X_{(1)} < Q^\alpha \le X_{(n)}$. It follows that from some n on, $\mathcal{R}(\alpha, Q^\alpha)$ reduces to the usual empirical profile likelihood ratio for the mean of $Z_i(\alpha, Q^\alpha)$, and so has a $\chi^2_{(1)}$ calibration by the univariate ELT, Theorem 2.2. The likelihood ratio function is

$$\mathcal{R}(p, q) = \left(\frac{p}{\hat p}\right)^{n\hat p} \left(\frac{1-p}{1-\hat p}\right)^{n(1-\hat p)}, \tag{3.14}$$

where $\hat p = \hat p(q) = \#\{X_i \le q\}/n$.

For $q < X_{(1)}$, the maximizing F has weight $w_0 = p$ below $X_{(1)}$, at X_0 say, weight $w_i = (1-p)/n$ for $i = 1, \ldots, n$, and $w_{n+1} = 0$. In this case

$$\mathcal{R}(p, q) = (1-p)^n.$$

Similarly for $q \ge X_{(n)}$ the maximizing F has weight $w_0 = 0$, $w_{n+1} = 1-p$, and $w_i = p/n$ for $i = 1, \ldots, n$. The empirical log likelihood ratio is

$$\mathcal{R}(p, q) = p^n.$$

Cow	Milk	Days	Cow	Milk	Days
Belle	4628	248	Jessie	2793	283
Bessie	3914	312	Ladybird	1932	256
Beauty	2851	278	Marigold	2992	254
Bertha	2765	267	Peggy	4444	268
Blossom	2067	235	Pansy	3944	333
Blackie	1817	211	Queen Ann	5448	299
Buttercup	4248	327	Queeny	4512	298
Dolly	3037	262	Rosemary	5435	317
Dandy	2627	235	Rosy	4792	415
Irene	3527	255	Sally	2635	264
Iona	3257	260	Wildy	3346	271

Table 3.2 *Shown are the pounds of milk produced and the number of days milked, in 1936, for 22 dairy cows. It is not known whether Rosy's number of days milked is in error, or if it includes some days from 1935.*

In all of these cases,

$$-\log \mathcal{R}(p, q) = n\left[\hat{p}\log(\hat{p}/p) + (1-\hat{p})\log((1-\hat{p})/(1-p))\right] \quad (3.15)$$

where $\hat{p} = \#\{X_i \leq q\}/n$ and $0\log 0$ is taken to be zero.

Table 3.2 shows the milk production data for the year 1936 for 22 cows from a family farm. We will use these milk production values, to illustrate the shape of the empirical likelihood function for quantiles and tail probabilities.

Figure 3.7 shows the empirical likelihood ratio function for the median amount of milk produced. As is clear from the definition of $Z_i(p, q)$, the empirical likelihood function is piecewise constant, taking steps only at observed data values. As a consequence, empirical likelihood confidence intervals for quantiles have endpoints equal to sample values.

Suppose that the interval for quantile q takes the form $[X_{(a)}, X_{(b)}]$. When X has a continuous distribution F, the value of $\Pr(X_{(a)} \leq q \leq X_{(b)})$ is the same for all F and can be found by a combinatorial argument. Thus χ^2 or F or bootstrap calibrations can be replaced by an exact calibration. If $[X_{(a)}, X_{(b)}]$ has nearly a 95% confidence level for a quantile, then the effect of changing a or b by one typically changes the coverage level by $O(n^{-1/2})$. It follows that the discreteness in the data limits the coverage accuracy of confidence intervals based on sample quantiles to $O(n^{-1/2})$. In order to get more precise confidence intervals, it is necessary to have a method of selecting end points between observed values. See Chapter 3.17.

Our test of Q_0^α as the α quantile of X is equivalent to a test of α as a tail

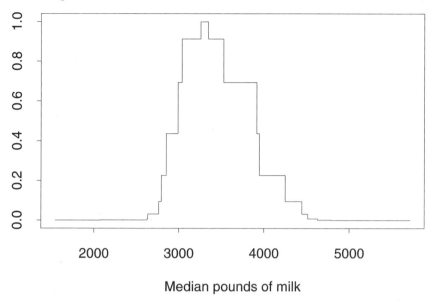

Figure 3.7 *The empirical likelihood function for the median is shown for the milk production numbers of Table 3.2.*

probability $\Pr(X \leq Q_0^\alpha)$. But a confidence interval for Q^α behaves differently from a confidence interval for $F_0((-\infty, Q^\alpha])$. The root cause is that $m(X, p, q) \equiv 1_{X \leq q} - p$ is a smooth function of p but not of q. Figure 3.8 shows the empirical likelihood ratio function for $\Pr(X \leq 3301.5)$ for the cow data. The value 3301.5 is the median of the cow data, and so the curve in Figure 3.8 has its peak at 0.5. The likelihood for a tail probability is very smooth. Central confidence intervals found by thresholding the likelihood do not have to have their endpoints at sample quantiles, and they get coverage errors of order n^{-1}.

Figure 3.9 shows a perspective plot of $\mathcal{R}(p, q)$ for the milk production data on the plane with $X_{(1)} \leq q \leq X_{(n)}$ and $0 \leq p \leq 1$. Slices through this function for a fixed quantile are smooth, while those for a fixed probability are step functions.

3.7 Ties and quantiles

In Chapter 3.6 we simplified the definition of the α quantile Q^α from

$$\Pr(X \leq Q^\alpha) \geq \alpha, \quad \text{and} \quad \Pr(X \geq Q^\alpha) \geq 1 - \alpha \tag{3.16}$$

to the estimating equation (3.13). If X has a continuous distribution, these definitions are equivalent, but for some other distributions they differ. This divergence

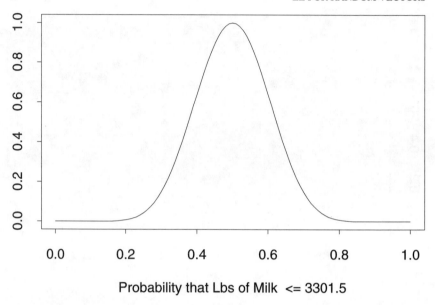

Figure 3.8 *The empirical likelihood function for $Pr(X \le 3301.5)$ is shown for the milk production numbers of Table 3.2. The value 3301.5 is the median milk production.*

can cause a problem when there are lots of ties in the data, but that problem is easily fixed as described here.

Suppose that m is the median of X, and that $\Pr(X = m) > 0$. Now X has an atom of probability on m, and it may then happen that $E(1_{X \le m} - 1/2) \ne 0$. Indeed the empirical likelihood confidence region for m satisfying $E(1_{X \le m} - 1/2) = 0$ may be empty for large enough n.

Suppose that there are a lot of tied values for the sample α quantile, and that the empirical likelihood confidence region for Q^α from equation (3.13) is empty. Then a simple remedy is to jitter the data slightly. We can add to each X_i a very small amount, such as a $U(-A, A)$ random variable where A is 10^{-6} times the smallest nonzero gap in the ordered data values. The resulting likelihood ratio curve will have a narrow spike at or near the sample α quantile.

We may take the limit $A \downarrow 0$ of small jitters, as follows. Compute the empirical likelihood function for a quantile of Y for the n values $Y_{(i)} = i, i = 1, \ldots, n$ and then transform it to the $X_{(i)}$ scale. The likelihood for $X_{(i)}$ as the α quantile of X is the maximum of that found for $Y_{(j)}$ as the α quantile of Y, over all j with $X_{(j)} = X_{(i)}$.

Thus, the pragmatic approach to testing whether q is the p quantile is to use

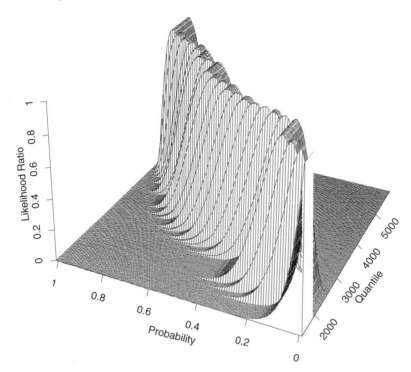

Figure 3.9 *The vertical axis depicts the empirical likelihood that* $\Pr(X \le q) = p$, *using the cow data of Table 3.2, as probabilities p and q vary. Slices through this surface, for fixed p or q, define the empirical likelihood function for quantiles and tail probabilities, respectively.*

equation (3.14) for any candidate Q that is not a sample value, and use

$$\mathcal{R}(p, X_{(i)}) = \max_{\hat{p}} \left(\frac{p}{\hat{p}}\right)^{n\hat{p}} \left(\frac{1-p}{1-\hat{p}}\right)^{n(1-\hat{p})} \quad (3.17)$$

at sample points. The maximum in (3.17) is taken over integer values of $n\hat{p}$ from $\sum_{j=1}^{n} 1_{X_j < X_{(i)}}$ to $\sum_{j=1}^{n} 1_{X_j \le X_{(i)}}$ inclusive.

By jittering we reduce the size of the atoms in the NPMLE to at most $1/n$. The exact definition of a quantile, namely $\Pr(X \le Q^\alpha) \ge \alpha$ and $\Pr(X \ge Q^\alpha) \ge 1 - \alpha$, cannot quite be attained through tests of $E(1_{X \le Q^\alpha}) = \alpha$, even when we split ties by jittering the data. For example, suppose that $n = 2k + 1$, and we put equal weight $1/n$ on all data points. Then weight $(k+1)/n \ge 1/2$ is on $[X_{(k+1)}, \infty)$ and the same weight is on $(-\infty, X_{(k+1)}]$. The sample median

$X_{(k+1)}$, should have an empirical likelihood ratio of 1.0. From equation (3.14) the likelihood ratio for the median at this value is

$$\left(\frac{n}{2k}\right)^k \left(\frac{n}{2(k+1)}\right)^{k+1} \doteq 1 - \frac{1}{2n}.$$

The only point on the likelihood ratio curve for the median that is incorrect is the MLE and it is incorrect by a small amount that can usually be safely ignored.

We can do better than equation (3.13) by employing the definition

$$E(1_{X<Q^\alpha} + \alpha 1_{X=Q^\alpha}) - \alpha = 0. \tag{3.18}$$

Equation (3.18) leads to an empirical likelihood ratio of 1.0 for the sample median with odd sample sizes, but for $\alpha \ne 1/2$ it is not necessarily exact, though likely to be better than equation (3.13).

That we have to worry about ties at all is perhaps surprising given that Chapter 2.3 shows ties do not affect the empirical likelihood confidence region for any statistical function. Indeed ties do not affect the confidence region for Q^α as defined by equation (3.16) nor do they affect the region for Q^α as defined by equation (3.13). But ties do affect whether the computationally convenient definition (3.13) agrees with the intended definition (3.16).

For the distribution F that is uniform on $[0,1] \cup [2,3]$, any value $1 \le m \le 2$ is a median. An empirical likelihood confidence region that contains any point from $(1, 2)$ will contain the whole interval $(1, 2)$.

3.8 Likelihood-based estimating equations

Perhaps the most widely used way of defining estimating equations is through parametric likelihoods. Suppose that

$$m(X, \theta) = \frac{\partial}{\partial \theta} \log f(X; \theta) = \frac{g(X, \theta)}{f(X; \theta)},$$

for a probability density function $f(X; \theta)$, with parameter θ, and gradient $g(X, \theta)$ with respect to θ. Then the estimating equation (3.7) is called the score equation and m is the score function. Solving the score equation (3.7) is the usual way that the parametric MLE $\hat\theta$ is found. Sometimes there is a unique solution $\hat\theta$, other times there are several. Parametric likelihood inferences are then centered around a consistent, though not necessarily unique, root $\hat\theta$ of the score equation.

Theorem 3.4 applies to the parameter θ under very weak conditions. In particular it is not necessary for X_i to have a probability density of the form $f(X; \theta)$ for any θ. The quantity that $\hat\theta$ estimates may retain some interpretability, despite small departures (and sometimes despite large departures) from the parametric model motivating it. As a case in point, consider the coefficients in multiple regression which we study in Chapter 4. These are often derived under strong model assumptions, such as normally distributed errors with mean zero and constant variance, but have useful least squares interpretations without those assumptions. We adopt

LIKELIHOOD-BASED ESTIMATING EQUATIONS

the model $f(X;\theta)$ as a "working likelihood" and use it to derive an estimate. If the parametric model holds, then $\hat{\theta}$ will have good asymptotic properties. Whether the parametric model holds or not, empirical likelihood confidence regions and tests on θ will have an asymptotic chisquared calibration.

We would expect that a true parametric likelihood ratio function should produce better confidence regions and tests around the parametric MLE than the empirical likelihood does. Surprisingly, this is not necessarily so, at least in large samples. Empirical likelihood tests have power very close to the parametric likelihood tests, and in some cases have better power. See Chapter 3.17 for details and a reference.

Estimating equations may also be generated from marginal and conditional likelihoods. Suppose that the data are (X_i, Y_i) pairs, for $i = 1, \ldots, n$, and that a parametric model gives the joint density or mass function $f_{XY}(x, y; \theta, \nu)$. We can factor f_{XY} into $f_X \times f_{Y|X}$, where f_X is the marginal distribution of X and $f_{Y|X}$ is the conditional distribution of Y given X. We may write the likelihood as

$$L(\theta, \nu) = \prod_{i=1}^{n} f_{XY}(X_i, Y_i; \theta, \nu)$$
$$= \prod_{i=1}^{n} f_X(X_i; \theta, \nu) \prod_{i=1}^{n} f_{Y|X}(Y_i \mid X_i; \theta, \nu)$$
$$\equiv L_m(\theta, \nu) L_c(\theta, \nu),$$

the product of a marginal and a conditional likelihood. The marginal likelihood is the likelihood from the marginal distribution of the X_i as if we had not observed Y_i. The conditional likelihood is the likelihood from the conditional distribution of Y_i given X_i as if the X_i were not random.

We may prefer to use marginal likelihood estimating equations

$$E\left(\frac{\partial}{\partial \theta} \log f_X(X_i; \theta, \nu)\right) = 0$$

or conditional ones

$$E\left(\frac{\partial}{\partial \theta} \log f_{Y|X}(Y_i \mid X_i; \theta, \nu)\right) = 0.$$

For example, if θ only appears in one of the likelihood factors L_m and L_c, then there is no information lost by using that factor as the likelihood, and there may be a computational advantage. Conversely, if one factor does not involve the nuisance parameter ν then there may be a computational advantage to using that factor. If the factor not involving ν contains all or most of the information on θ, then only a small information loss arises from using the factor instead of the full likelihood. Another consideration in some problems is that there may be much greater doubt about the appropriateness of one of the two parametric models f_X and $f_{Y|X}$. Then a likelihood based on the more credible model may be preferable.

A partial likelihood is one in which a series of events $E_1, \ldots, E_n$ defines the data, and instead of using $\prod_{i=1}^{n} \Pr(E_i \mid E_1, \ldots, E_{i-1})$, one uses a convenient subset of the factors. The original example comes from survival analysis, and is described in Chapter 6.7.

Theorem 3.4 applies under very weak conditions to maximum likelihood estimates based on either full, marginal, conditional, or partial log likelihoods.

A difficult problem arises when the likelihood that motivates a statistic comes from an unrealistic model and the lack of fit is such that the resulting estimator is no longer relevant or interpretable. In such a case, we may have to return to square one, describe what we hoped to learn about the distribution of the data, and devise a statistic that captures this property.

3.9 Transformation invariance of EL

Parametric likelihoods are invariant under one to one transformations of the parameter. For example, suppose that $\phi = \tau(\theta)$, and the relationship is invertible, $\theta = \tau^{-1}(\phi)$. Let C_θ be a parametric likelihood ratio confidence region for θ. Then the parametric confidence region for ϕ is $C_\phi = \{\tau(\theta) \mid \theta \in C_\theta\}$.

Empirical likelihood confidence regions are similarly invariant under parameter changes: If $\sum_{i=1}^{n} w_i m(X_i, \theta, \nu) = 0$, then $\sum_{i=1}^{n} w_i m(X_i, \tau^{-1}(\phi), \nu) = 0$. It follows that the profile empirical log likelihood ratio for $\theta = \theta_0$ is the same as that for $\phi = \phi_0 = \tau(\theta_0)$. Similarly, invertible transformations of the nuisance parameter ν do not affect empirical likelihood confidence regions for θ or ϕ.

Transformations are often usefully applied to data, too. Let $Y = \tau(X)$ be a one-to-one invertible function. Then $\sum_{i=1}^{n} w_i m(X_i, \theta) = 0$, if and only if $\sum_{i=1}^{n} w_i m(\tau^{-1}(Y_i), \theta) = 0$. So empirical likelihood inferences are also invariant under invertible transformations of the data.

Some insight is gained by a more detailed look at reparameterization. Suppose that

$$m(X, \theta) = \frac{\partial}{\partial \theta} \log f(X; \theta) = \frac{g(X, \theta)}{f(X; \theta)},$$

for a density or probability function $f(X; \theta)$, with parameter θ, and gradient $g(X, \theta)$ with respect to θ, as described in Chapter 3.8. The estimating equations for ϕ are found by alternative estimating equations

$$\widetilde{m}(X, \phi) = \frac{\partial}{\partial \phi} \log f(X; \tau^{-1}(\phi)) = \frac{g(X, \tau^{-1}(\phi))}{f(X; \tau^{-1}(\phi))} \frac{\partial \tau^{-1}(\phi)}{\partial \phi}.$$

Now $\widetilde{m}(X, \phi) = m(X, \tau^{-1}(\phi)) J(\phi)$, where J is a p by p Jacobian matrix. For a one-to-one transformation, J is of full rank. It follows that

$$\sum_{i=1}^{n} w_i \widetilde{m}(X_i, \phi) = J(\tau(\theta)) \sum_{i=1}^{n} w_i m(X_i, \theta),$$

so that $\sum_{i=1}^{n} w_i m(X_i, \theta) = 0$ if and only if $\sum_{i=1}^{n} w_i \widetilde{m}(X_i, \tau(\theta)) = 0$.

Now suppose that $Y = \tau(X)$ is a one-to-one invertible transformation of the

SIDE INFORMATION

data. For estimating equations arising as MLE's, if $X \sim f(X;\theta)$, then $Y \sim f(\tau^{-1}(Y);\theta) \times |\det(\partial \tau^{-1}(Y)/\partial Y)|$, where $\det(A)$ denotes the determinant of the matrix A. Now

$$\frac{\partial}{\partial \theta} \log \left(f(\tau^{-1}(Y);\theta) \times \left| \det \left(\frac{\partial \tau^{-1}(Y)}{\partial Y} \right) \right| \right) = \frac{\partial}{\partial \theta} \log f(X;\theta),$$

because the determinant does not involve θ, and so the same estimating equations arise whether one works with X_i or Y_i.

3.10 Side information

Suppose that we observe $(X,Y) \in \mathbb{R}^2$, where we know the mean of X and we want an estimate and confidence region for the mean of Y. This is a common circumstance in finite population settings such as survey sampling, where the ratio and regression estimators are used to get more accurate estimates of $E(Y)$ by exploiting the known value of $E(X)$. These techniques are also used in Monte Carlo simulation where X is called a control variate.

We suppose that $X \in \mathbb{R}^p$, that $Y \in \mathbb{R}^q$, that $E(X) = \mu_{x0}$ is known and that we are interested in drawing inferences on $E(Y) = \mu_{y0}$. Let

$$\operatorname{Var}\begin{pmatrix} X \\ Y \end{pmatrix} = \begin{pmatrix} V_{xx} & V_{xy} \\ V_{yx} & V_{yy} \end{pmatrix}$$

denote the variances and covariances of X and Y, using the natural matrix partition.

Since we know $E(X) = \mu_{x0}$, it is natural to constrain the weights w_i to satisfy $\sum_{i=1}^n w_i X_i = \mu_{x0}$. Define $\mathcal{R}_{XY}(\mu_x, \mu_y)$ by

$$\max \left\{ \prod_{i=1}^n n w_i \mid \sum_{i=1}^n w_i X_i = \mu_x, \sum_{i=1}^n w_i Y_i = \mu_y, w_i \geq 0, \sum_{i=1}^n w_i = 1 \right\},$$

set

$$\mathcal{R}_X(\mu_x) = \max \left\{ \prod_{i=1}^n n w_i \mid \sum_{i=1}^n w_i X_i = \mu_x, w_i \geq 0, \sum_{i=1}^n w_i = 1 \right\},$$

and define

$$\mathcal{R}_{Y|X}(\mu_y \mid \mu_x) = \frac{\mathcal{R}_{XY}(\mu_x, \mu_y)}{\mathcal{R}_X(\mu_x)}. \tag{3.19}$$

Theorem 3.5 *Let $(X_i, Y_i) \in \mathbb{R}^{p+q}$ be independent random vectors from a distribution F_0 with mean (μ_{x0}, μ_{y0}) and variance matrix of full rank $p + q$. Then $-2 \log \mathcal{R}_{Y|X}(\mu_{y0} \mid \mu_{x0}) \to \chi^2_{(q)}$ in distribution as $n \to \infty$.*

Proof. From the proof of Theorem 3.2, we may find that

$$-2\log \mathcal{R}_X(\mu_{x0}) \doteq n(\bar{X} - \mu_{x0})'V_{xx}^{-1}(\bar{X} - \mu_{x0}),$$

with an error of $o_p(1)$, and that to a similar accuracy

$$-2\log \mathcal{R}_{XY}(\mu_{x0}, \mu_{y0}) \doteq n \begin{pmatrix} \bar{X} - \mu_{x0} \\ \bar{Y} - \mu_{y0} \end{pmatrix}' \begin{pmatrix} V_{xx} & V_{xy} \\ V_{yx} & V_{yy} \end{pmatrix}^{-1} \begin{pmatrix} \bar{X} - \mu_{x0} \\ \bar{Y} - \mu_{y0} \end{pmatrix}.$$

Introduce the variable $Z = Y - V_{yx}V_{xx}^{-1}X$, with mean $\mu_{z0} = \mu_{y0} - V_{yx}V_{xx}^{-1}\mu_{x0}$ and variance matrix $V_{zz} = V_{yy} - V_{yx}V_{xx}^{-1}V_{xy}$. It is easy to show that Z and X are uncorrelated. Theorem 3.2 applies to the vector $(X', Z')'$ as well, so

$$\begin{aligned}
&-2\log \mathcal{R}_{XY}(\mu_{x0}, \mu_{y0}) \\
&= -2\log \mathcal{R}_{XZ}(\mu_{x0}, \mu_{z0}) \\
&= n(\bar{X} - \mu_{x0})'V_{xx}^{-1}(\bar{X} - \mu_{x0}) + n(\bar{Z} - \mu_{z0})'V_{zz}^{-1}(\bar{Z} - \mu_{z0}) + o_p(1),
\end{aligned}$$

and then

$$-2\log \mathcal{R}_{Y|X}(\mu_{y0} \mid \mu_{x0}) = n(\bar{Z} - \mu_{z0})'V_{zz}^{-1}(\bar{Z} - \mu_{z0}) + o_p(1) \to \chi^2_{(q)}$$

in distribution as $n \to \infty$. □

The approximations used in the proof of Theorem 3.5 are valid for μ_x, μ_y, and μ_z in $O(n^{-1/2})$ neighborhoods of μ_{x0}, μ_{y0}, and μ_{z0}, respectively, and so the curvature in $\mathcal{R}_{Y|X}(\mu_y \mid \mu_x)$ as a function of μ_y, at the MELE is $nV_{y|x}^{-1}$, where

$$V_{y|x} \equiv V_{yy} - V_{yx}V_{xx}^{-1}V_{xy}.$$

The variance $V_{y|x}$ is no larger than V_{yy} in that $u'V_{y|x}u \le u'V_{yy}u$, for any $u \in \mathbb{R}^q$, with equality holding for all u, if $V_{xy} = 0$. The constrained confidence region for μ_y is asymptotically smaller than the unconstrained one, except when every component of Y is uncorrelated with every component of X. In that case the regions are asymptotically equivalent. The empirical likelihood confidence regions for $E(Y)$ using the known value of $E(X)$, are not in general centered at $\bar{Y}$, but are instead centered at

$$\tilde{\mu}_y = \arg\max_{\mu_y} \mathcal{R}_{Y|X}(\mu_y \mid \mu_{x0}) \doteq \bar{Y} - V_{yx}V_{xx}^{-1}(\bar{X} - \mu_{x0}).$$

The asymptotic variance of $\tilde{\mu}_y$ is at least as small as that of $\bar{Y}$ and is strictly smaller unless Y is uncorrelated with X. The estimate $\tilde{\mu}_y$ is called the maximum empirical likelihood estimate (MELE) to distinguish it from the NPMLE $\bar{Y}$.

The notation $\mathcal{R}_{Y|X}$ suggests that some conditioning is taking place. Constraining the reweighted mean of X_i to match μ_{x0} is similar to conditioning on the observed value of $\bar{X} - \mu_{x0}$. The connection is that for large n the distribution of $\sqrt{n}((\bar{X} - \mu_{x0})', (\bar{Y} - \mu_{y0})')'$ is approximately normal with mean 0 and variance $\mathrm{Var}((X', Y')')$. In this asymptotic normal distribution, the conditional variance of $\sqrt{n}(\bar{Y} - \mu_{y0})$ given $\sqrt{n}(\bar{X} - \mu_{x0})$ is precisely $V_{y|x}$.

It is also possible to test whether the assumed mean μ_{x0} is really equal to $E(X)$

using empirical likelihood. If the hypothesis that $E(X) = \mu_{x0}$ is rejected, then the assumption that $E(X) = \mu_{x0}$ should be questioned. Failure to reject the hypothesis that $E(X) = \mu_{x0}$ might simply mean that n was not large enough, and does not suffice to justify imposing the constraint $\sum_{i=1}^{n} w_i X_i = \mu_{x0}$. Instead we should be fairly confident that μ_{x0} is the true value, or at least that $E(X)$ is close enough to μ_{x0} that only a small error is introduced through the constraint.

Knowledge of μ_{x0} is sometimes called side information, or auxiliary information. Much more general forms of side information can be expressed through estimating equations, as the following examples show. If (X, Y) pairs are known to have a common but unknown mean μ, this can be expressed through estimating equations

$$E(X - \mu) = E(Y - \mu) = 0.$$

These equations are equivalent to

$$E(X - \mu) = E(X - Y) = 0,$$

the second of which does not involve the parameter. If we know that the mean and median of X coincide, as for example with symmetric distributions (see Chapter 10), then

$$E(X - \mu) = E\left(1_{X \leq \mu} - 1/2\right) = 0,$$

for some unknown μ. If we know a quantile Q^α and are interested in the mean, then

$$E(X - \mu) = 0,$$
$$E\left(1_{X \leq Q^\alpha} - \alpha\right) = 0.$$

Quantities defined conditionally can be expressed through estimating equations involving indicator variables. Thus for events A and B

$$E((1_A - \rho)1_B) = 0$$

defines ρ as the conditional probability of A given B, and similarly

$$E((Y - \mu)1_{X > 32}) = 0$$

defines μ as the conditional mean of Y given that $X > 32$. In Chapter 4, regression of Y on a scalar X through the origin is expressed through

$$E(Y - X\beta) = 0,$$
$$E((Y - X\beta)X) = 0.$$

If we believe that the variance is functionally related to the mean, $\mathrm{Var}(X) = h(E(X))$, for some function h, as is commonly assumed in quasi-likelihood inference, then

$$E(X - \mu) = 0$$
$$E\left((X - \mu)^2 - h(\mu)\right) = 0.$$

The common feature of these problems is that there are more estimating equations than there are parameters. Such overdetermined problems can pose a difficulty for some parametric inferences. The sample mean is not likely to equal the sample median, and a regression through the origin does not usually have a zero mean residual. For (X, Y) pairs known to have a common mean, we could choose a weighted combination $\lambda \bar{X} + (1 - \lambda)\bar{Y}$, but the optimal value of λ depends on the variances of X and Y and on their covariance.

In parametric settings, with $p + q$ estimating equations in q unknown parameters, we could choose to work with q of the estimating equations and ignore the other p equations. More generally, we could use q linear combinations of the $p + q$ estimating equations

$$E(m(X, \theta)A(\theta)) = 0, \tag{3.20}$$

where $A(\theta)$ is a $p + q$ by p matrix that is of rank p for all values of θ.

Empirical likelihood uses an $n - 1$ parameter family of distributions. We expect to be able to impose $p + q$ constraints, when $p + q < n - 1$. Usually $p + q \ll n - 1$. It turns out that empirical likelihood effectively makes a data-determined choice of $A(\theta)$. For estimating equations where m is sufficiently smooth in θ, the asymptotic variance of the empirical likelihood estimator is at least as small as that obtained by any set of linear combinations of $A(\theta)$.

Theorem 3.6 *Let $X_i \in \mathbb{R}^d$ be IID random vectors, and suppose that $\theta_0 \in \mathbb{R}^p$ is uniquely determined via $E(m(X, \theta)) = 0$, where $m(X, \theta)$ takes values in $\mathbb{R}^{p+q}$, for $q \geq 0$. Let $\widetilde{\theta} = \arg\max_\theta \mathcal{R}(\theta)$, where*

$$\mathcal{R}(\theta) = \max\left\{\prod_{i=1}^n nw_i \mid \sum_{i=1}^n w_i m(X_i, \theta) = 0, w_i \geq 0, \sum_{i=1}^n w_i = 1\right\}.$$

Assume that there is a neighborhood Θ of θ_0 and a function $M(x)$ with $E(M(X)) < \infty$, for which:

1: $E(\partial m(X, \theta_0)/\partial \theta)$ *has rank p,*
2: $E(m(X, \theta_0)m(X, \theta_0)')$ *is positive definite,*
3: $\partial m(x, \theta)/\partial \theta$ *is continuous for $\theta \in \Theta$,*
4: $\partial^2 m(x, \theta)/\partial \theta \partial \theta'$ *is continuous in θ, for θ in Θ,*
5: $\|m(x, \theta)\|^3 \leq M(x)$, *for $\theta \in \Theta$,*
6: $\|\partial m(x, \theta)/\partial \theta\| \leq M(x)$, *for $\theta \in \Theta$, and*
7: $\|\partial^2 m(x, \theta)/\partial \theta \partial \theta'\| \leq M(x)$, *for $\theta \in \Theta$.*

Then

$$\lim_{n \to \infty} n\mathrm{Var}(\widetilde{\theta}) = \left[E\left(\frac{\partial m}{\partial \theta}\right)' (E(mm'))^{-1} E\left(\frac{\partial m}{\partial \theta}\right)\right]^{-1},$$

and this asymptotic variance is at least as small as that of any estimator of the form (3.20). Furthermore,

$$-2\log\bigl(\mathcal{R}(\theta_0)/\mathcal{R}(\tilde{\theta})\bigr) \to \chi^2_{(p)},$$

and

$$-2\log\mathcal{R}(\tilde{\theta}) \to \chi^2_{(q)},$$

both converging in distribution as $n \to \infty$.

Proof. This theorem restates several results proved in Qin & Lawless (1994). □

A corollary is that for smooth estimating equations on p parameters defined by introducing q nuisance parameters, maximizing out the nuisance parameters gives the anticipated $\chi^2_{(p)}$ limit for the parameters of interest.

Theorem 3.6 makes smoothness assumptions that are strong enough to rule out the estimating equations used to define quantiles, causing difficulties for functions, smooth or otherwise, of quantiles. Consider the interquartile range (IQR) of $X \in \mathbb{R}$. The equations

$$0 = E(1_{X \le Q^{0.25}} - 0.25), \quad \text{and} \tag{3.21}$$
$$0 = E(1_{X \le Q^{0.25} + \text{IQR}} - 0.75) \tag{3.22}$$

define IQR and the lower quartile $Q^{0.25}$. A joint region for $Q^{0.25}$ and IQR can be calibrated using Theorem 3.4 but neither Theorem 3.4 nor Theorem 3.6 allows us to use a $\chi^2_{(1)}$ calibration for IQR after maximizing over $Q^{0.25}$. Chapter 4.9 presents a more involved example of this problem arising in calibration and prediction of regression models. Chapter 10.6 presents some specialized results from the literature describing cases where a parameter can be profiled out despite entering the estimation equation in a nonsmooth way.

3.11 Sandwich estimator

In the just determined case, and subject to mild regularity conditions, the variance of the solution $\hat{\theta}$ to $\sum_{i=1}^n m(X_i, \theta) = 0$ satisfies

$$n\text{Var}(\hat{\theta}) \to I^{-1}C(I')^{-1}, \tag{3.23}$$

where each row of

$$I = \frac{\partial}{\partial \theta} \int m(X, \theta) dF \Big|_{\theta = \theta_0}$$

has the partial derivatives from one component of m, and

$$C = \int m(X, \theta_0) m(X, \theta_0)' dF(X).$$

The sandwich estimator of the variance of $\hat{\theta}$ is

$$\widehat{\text{Var}}_{\text{Sand}}(\hat{\theta}) = \frac{1}{n}\hat{I}^{-1}\hat{C}\hat{I}'^{-1} = \frac{1}{n}(\hat{I}'\hat{C}^{-1}\hat{I})^{-1}, \tag{3.24}$$

with $\hat{I} = (1/n)\sum_{i=1}^{n} \partial m(X_i, \hat{\theta})/\partial\theta$ and $\hat{C} = (1/n)\sum_{i=1}^{n} m(X_i, \hat{\theta})m(X_i, \hat{\theta})'$.

When m is obtained from a parametric likelihood, then $-I = -I'$ is the Fisher information in X_i for θ. When the data are from the motivating parametric model, then $\hat{C}$ and $-\hat{I}$ both estimate the Fisher information and so $\hat{I}^{-1}/n$ or $\hat{C}^{-1}/n$ may be used to estimate $\text{Var}(\hat{\theta})$. The sandwich estimator is widely used to generate variance estimates that are reliable without assuming a parametric form for the distribution of the data.

In the just determined case, the matrix of second derivatives of the empirical likelihood function at $\hat{\theta}$ is equal to the inverse of the sandwich estimator of variance. This relationship can be established through the partial derivatives given in Chapter 12.4.

3.12 Robust estimators

Empirical likelihood confidence intervals for a vector mean tend to extend further in directions that the data are skewed than they do in the opposite directions. If the data contain one or more outliers, then empirical likelihood confidence regions may be greatly lengthened in the direction of the outlier.

We consider two main approaches to robustness. The first is to replace non-robust statistics like the mean by more robust alternatives such as the median, or Huber's M-estimate. The second, robustifying the likelihood, is discussed in Chapter 3.13.

The mean of a univariate random variable may be derived as a parametric MLE using the normal distribution, and some others such as the Poisson or the exponential. The mean is notoriously non-robust. A single outlier can change it by an arbitrarily large amount.

The median is considered in Chapter 3.6. Although the median is robust and the mean is not, we often hesitate to replace means by medians. The median does not estimate the same quantity as the mean, except in special circumstances, such as sampling from a symmetric distribution. Even there we may lose some efficiency in replacing the mean by the median. Robust statistics, such as Huber's M-estimate below, have been designed to capture most of the efficiency of the mean, but with some robustness.

Huber's M-estimate is defined by solving the estimating equation

$$\frac{1}{n}\sum_{i=1}^{n} \psi\left(\frac{X_i - \mu}{\hat{\sigma}}\right) = 0 \qquad (3.25)$$

for μ, where $\hat{\sigma}$ is a robust estimate of scale, and

$$\psi(z) = \begin{cases} -c, & z \leq -c \\ z, & |z| \leq c \\ c, & z \geq c. \end{cases} \qquad (3.26)$$

ROBUST ESTIMATORS

A common choice for $\hat\sigma$ is MAD/0.674 where

$$\text{MAD} = \text{median}(|X_i - \text{median}(X_i)|)$$

is the median absolute deviation of X_i from their median. The factor 0.674 scales the MAD so that it estimates the standard deviation if the X_i are normally distributed.

The function ψ is a compromise between the mean and the median. These can be defined through functions $\psi(z)$ proportional to z and $\text{sign}(z)$, respectively. Smaller deviations z are handled as for the mean, while large ones are handled as for the median. By choosing $c = 1.35$ the resulting M estimate has 95% efficiency relative to the mean should the data be normally distributed. The 5% efficiency loss is then a small insurance premium to pay in order to get a statistic robust to outliers.

The simplest approach to empirical likelihood for Huber's statistic is to compute $\hat\sigma$ by the MAD and then for each candidate value μ, reweight $Z_i = \psi((X_i - \mu)/\hat\sigma)$ to have mean zero. But as the data are reweighted, the scale estimate $\hat\sigma$ should also change. For this reason we prefer to estimate the scale σ simultaneously with μ, through:

$$0 = \frac{1}{n}\sum_{i=1}^{n} \psi\left(\frac{X_i - \mu}{\sigma}\right), \quad \text{and} \tag{3.27}$$

$$0 = \frac{1}{n}\sum_{i=1}^{n} \left[\psi\left(\frac{X_i - \mu}{\sigma}\right)\right]^2 - 1. \tag{3.28}$$

These equations give robust estimates for both μ and σ. Moreover, it can be shown that solving these equations is equivalent to minimizing

$$\frac{1}{n}\sum_{i=1}^{n} \rho\left(\frac{X_i - \mu}{\sigma}\right)\sigma \tag{3.29}$$

with respect to μ and σ, where

$$\rho(z) = \begin{cases} \frac{1}{2}(z^2 + 1), & |z| \leq c \\ \frac{1}{2}(2c|z| - c^2 + 1), & |z| \geq c. \end{cases}$$

Because (3.29) can be shown to be jointly strictly convex in μ and σ, there is a unique global optimum $(\hat\mu, \hat\sigma)$.

Now we may define the empirical likelihood $\mathcal{R}(\mu, \sigma)$ for estimating equations (3.27) and (3.28), and then set $\mathcal{R}(\mu) = \max_\sigma \mathcal{R}(\mu, \sigma)$. Maximization over σ or μ with the other fixed is more convenient with a differentiable function ψ. Little changes in the estimate of (μ, σ) if the function ψ is replaced by another one that is twice continuously differentiable everywhere and only differs from ψ where $||z| - c| \leq \epsilon$. The examples below use $\epsilon = 0.01$.

Figure 3.10 shows 20 of Newcomb's measurements of the passage time of light. Figure 3.11 shows empirical likelihood ratio curves for both the mean and Huber's

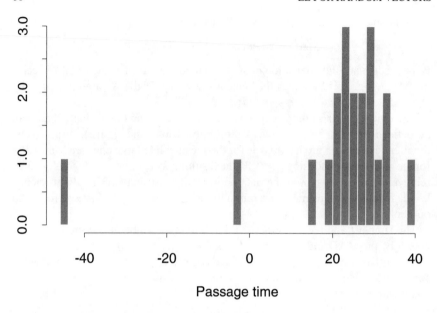

Figure 3.10 *Shown are 20 of Newcomb's measurements of the passage time of light. They represent a shift and rescaling of the time for light to travel a given distance. The presently accepted value of the speed of light corresponds to 33.02. Source: Stigler (1977).*

location estimate, applied to Newcomb's data. Huber's estimator is much less sensitive to outliers and it has a narrower confidence region. The curve for the mean is centered at the sample mean of 21.75. Huber's estimate for this data set is $\hat\mu = 25.64$, with $\hat\sigma = 4.16$.

In 1000 bootstrap simulations of this data, the 0.95 quantile of $-2\log\mathcal{R}(\bar{X})$ was 19.26 (using $\mathcal{R}$ for the mean). This is very far from 3.84, the 0.95 quantile of the $\chi^2_{(1)}$ distribution. For Huber's location estimator, the corresponding resampled 0.95 quantile was 4.19, closely matching the $\chi^2_{(1)}$ value. The discrepancy is so large for the mean because one or more outliers has made the passage time data very strongly skewed. By contrast, Huber's estimator is defined through the mean of a much less skewed function of the data.

There were 7 times when the sample mean was outside of the convex hull of the resampled data, for which $-\infty$ is the resampled log likelihood. Because 15 of the 20 data values are larger than μ, we would have expected the number of infinite resampled log likelihoods to have a Poisson distribution with mean $1000 \times (0.75^{20} + 0.25^{20}) = 3.17$. (Getting 7 occurrences of $\mathcal{R}(21.75) = 0$ is unusual. As a check, the next 9000 bootstraps have only 25 occurrences, closely matching the expected number 28.53.)

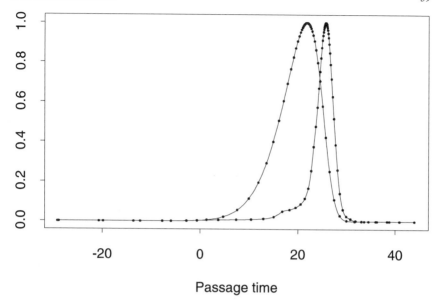

Figure 3.11 *Shown are empirical likelihood curves for the mean and for the value of Huber's location M-estimator, for the passage time of light data in Figure 3.10. The likelihood function for the mean is broader and shifted to the left compared to that of Huber's estimate. The cause is an outlier in the data at -40.*

A bootstrap-calibrated 95% confidence interval for the mean is from -9.17 to 32.08. The corresponding interval for Huber's M-estimate is from 21.57 to 28.68.

A general robustification technique is to replace the estimating equation of a non-robust statistic by a similar equation that defines a robust statistic. To control against the effects of outliers, we typically choose a bounded estimating function $m(x, \theta)$, perhaps by clipping a likelihood-derived estimating function.

3.13 Robust likelihood

Chapter 3.12 considered constructing empirical likelihood confidence regions for robust statistics defined through estimating equations.

Another approach to robustness (see Chapter 3.17) is to modify the likelihood function. Suppose that the data are thought to be from a discrete distribution with probability function $\Pr(X = x; \theta) = f(x; \theta)$, contaminated by outliers. That is with probability $1 - \epsilon$ where $\epsilon > 0$ is small, each X_i is drawn from $f(x; \theta)$, and with probability ϵ it is drawn from an unknown distribution G_i. We suppose further that the observations are independent, in particular that the contamination process is independent. Then the likelihood is

$$L(\theta, G_1, \ldots, G_n) = \prod_{i=1}^{n} \Big((1-\epsilon) f(X_i; \theta) + \epsilon G_i(X_i) \Big).$$

We can maximize over all n distributions G_i by taking $G_i(X_i) = 1$, leaving

$$L(\theta) = \prod_{i=1}^{n} \Big((1-\epsilon) f(X_i; \theta) + \epsilon \Big) \qquad (3.30)$$

$$= (1-\epsilon)^n \prod_{i=1}^{n} \big(f(X_i; \theta) + \eta \big),$$

where $\eta = \epsilon/(1-\epsilon)$.

Although this argument begins by robustifying the likelihood, it ends with estimating equations

$$m(X_i, \theta) = \frac{\frac{\partial}{\partial \theta} f(X_i; \theta)}{f(X_i; \theta) + \eta}, \qquad (3.31)$$

where η bounds the denominator away from 0.

Robustifying the likelihood this way is more difficult for models with continuous distributions for X_i. Then the likelihood is a product of terms

$$(1-\epsilon) f(X_i; \theta) dX_i + \epsilon G_i(X_i) \propto f(X_i; \theta) + \frac{\eta G_i(X_i)}{dX_i}.$$

The dX_i is there because we do not observe X_i to infinite precision, but only to within a small set of volume dX_i. If we suppose that each observation has the same dX_i, and maximize over discrete G_i, then once again we arrive at an estimating equation of the form (3.31), with η replaced by $\tau = \eta/dX$. Taking every $G_i(X_i) = 1$ and dX corresponding to the data precision could easily lead to unreasonably large τ. In practice we might prefer an ad hoc choice of a smaller τ for a particular data set.

3.14 Computation and convex duality

A convex dual problem arises in the computation of the profile empirical likelihood function for a vector mean, and leads to the development of an iterated least squares algorithm. This dual problem is also useful for a theoretical understanding of empirical likelihood.

The set $\mathcal{H} = \mathcal{H}(X_1, \ldots, X_n) = \{\sum_{i=1}^{n} w_i X_i \mid w_i \geq 0, \sum_{i=1}^{n} w_i = 1\}$ is the convex hull of X_i. It generalizes the interval $[X_{(1)}, X_{(n)}]$ from the one-dimensional case. If $\mu \notin \mathcal{H}$, then we take $\mathcal{R}(\mu) = 0$. If μ is on the boundary of $\mathcal{H}$ then we also take $\mathcal{R}(\mu) = 0$, unless the X_i all lie in a q-dimensional hyperplane with $1 \leq q < d$. In that case $\mathcal{R}(\mu)$ is defined by selecting q components of the X_i having full rank and taking the q-dimensional empirical likelihood of the corresponding components of μ.

COMPUTATION AND CONVEX DUALITY

The nontrivial cases of interest are those in which μ is an interior point of $\mathcal{H}$. As in the one-dimensional case, we introduce the Lagrangian

$$G = \sum_{i=1}^{n} \log(nw_i) - n\lambda'\left(\sum_{i=1}^{n} w_i(X_i - \mu)\right) + \gamma\left(\sum_{i=1}^{n} w_i - 1\right), \quad (3.32)$$

where $\lambda \in \mathbb{R}^d$ is a vector of Lagrange multipliers. As in the univariate case we can find that $\gamma = -n$. Then

$$w_i = \frac{1}{n}\frac{1}{1 + \lambda'(X_i - \mu)},$$

where $\lambda = \lambda(\mu)$ satisfies d equations given by

$$0 = \frac{1}{n}\sum_{i=1}^{n}\frac{X_i - \mu}{1 + \lambda'(X_i - \mu)}. \quad (3.33)$$

Substituting the expression for w_i into $\log R(F)$ yields

$$\log R(F) = -\sum_{i=1}^{n} \log(1 + \lambda'(X_i - \mu)) \equiv \mathbb{L}(\lambda).$$

We now seek the *minimum* of $\mathbb{L}$ over λ. Setting the gradient of $\mathbb{L}$ with respect to λ to zero is equivalent to solving the d equations (3.33) above. This is an example of convex duality, because a maximization over n variables w_i subject to $d+1$ equality constraints has become a minimization over d variables λ. It would have been a $d+1$ dimensional minimization, except that we were able to eliminate the multiplier γ.

The domain of $\mathbb{L}$ must exclude any λ for which some $w_i \leq 0$. Therefore the inequality constraints

$$1 + \lambda'(X_i - \mu) > 0, \quad i = 1, \ldots, n \quad (3.34)$$

must be imposed. The set of λ for which (3.34) holds is an intersection of n half spaces. It contains the origin $\lambda = 0$, and so it is a nonempty and convex subset of $\mathbb{R}^d$.

The Hessian of $\mathbb{L}$ with respect to λ is

$$\sum_{i=1}^{n} \frac{(X_i - \mu)(X_i - \mu)'}{[1 + \lambda'(X_i - \mu)]^2}.$$

This is a positive semidefinite function of λ and is in fact positive definite unless the X_i lie in a linear subspace of dimension $q < d$. Assuming that the X_i are not contained in a subspace of dimension smaller than d, the search for λ is the minimization of a strictly convex function over a convex domain.

We may redefine $\mathbb{L}$ in such a way that it is convex over all of $\mathbb{R}^d$ without changing its value near the solution. Then we need not explicitly impose the n

constraints in (3.34). Define a pseudo-logarithm function

$$\log_\star(z) = \begin{cases} \log(z), & \text{if } z \geq 1/n \\ \log(1/n) - 1.5 + 2nz - (nz)^2/2, & \text{if } z \leq 1/n. \end{cases}$$

The function $\log_\star$ matches $\log$ for arguments greater than $1/n$, is a quadratic for $z \leq 1/n$, and has two continuous derivatives. Arguments smaller than $1/n$ correspond to w_i larger than 1, and so they are not part of a valid solution.

Now instead of minimizing $\mathbb{L}(\lambda)$ subject to n inequality constraints, we minimize the function

$$\mathbb{L}_\star(\lambda) = -\sum_{i=1}^n \log_\star\left(1 + \lambda'(X_i - \mu)\right)$$

over $\lambda \in \mathbb{R}^d$ without any constraints. The value of $\mathbb{L}_\star$ matches $\mathbb{L}$ at any λ satisfying the inequality constraints, including the solution $\lambda(\mu)$ if it exists.

After changing $\mathbb{L}$ to $\mathbb{L}_\star$, setting the gradient to zero amounts to solving

$$0 = \frac{1}{n}\sum_{i=1}^n \log'_\star\left(1 + \lambda'(X_i - \mu)\right)(X_i - \mu),$$

for λ, where

$$\log'_\star(z) = \begin{cases} 1/z, & \text{if } z \geq 1/n \\ 2n - n^2 z, & \text{if } z \leq 1/n, \end{cases}$$

thus avoiding the problem of a vanishing denominator in (3.33).

The dual problem is one of minimizing a convex function over $\mathbb{R}^d$. There is a unique global minimizer, in the nontrivial cases where the X_i are not contained in a $d - 1$ dimensional hyperplane, and μ is interior to the convex hull $\mathcal{H}$ of the X_i. This global minimum can be found at a superlinear rate by modified versions of Newton's method, and by the Davidon-Fletcher-Powell algorithm. This fact justifies the claim in Chapter 2 that computing the empirical likelihood is not particularly difficult, though real difficulties may arise in maximizing over nuisance parameters. The same two facts are generally true of parametric likelihoods.

If $\mu \in \mathcal{H}$ but the X_i lie in a lower dimensional hyperplane, then there is not a unique solution λ. If λ is a solution, then so is $\lambda + \tau$ for any τ orthogonal to all of the $X_i - \mu$. There is however a unique set of weights w_i common to all solutions.

If μ is not an interior point of $\mathcal{H}$, then iterative algorithms based on Newton's method produce a sequence of vectors λ with length $\|\lambda\|$ diverging to infinity. As $\|\lambda\|$ increases, the size of the gradient decreases, until it goes below the value used to detect convergence.

In practice it is convenient to minimize $\mathbb{L}_\star(\lambda)$ over λ without checking whether $\mu \in \mathcal{H}$. One can inspect the resulting solution to determine whether μ is in the convex hull of the data. If $\mu \notin \mathcal{H}$, then the weights $w_i = n^{-1}(1 + \lambda'(X_i - \mu))$ will not sum to one.

The value of $\mathbb{L}_\star$ corresponding to a point μ, that is not in the convex hull of

EUCLIDEAN LIKELIHOOD

the data, is typically so small that μ will not be in any reasonable confidence set. There is little practical advantage in drawing a distinction between the true value $\mathbb{L} = -\infty$ and large negative values of $\mathbb{L}_\star$ such as -300 or -1000.

The trick of parameterizing away the constraints does have a cost. It can slow down the algorithm, turning superlinear convergence into linear convergence. But this only happens for μ that are not interior points of the convex hull of the data.

The Newton step in computing $\lambda(\mu)$ can be expressed as a least squares computation. We begin by writing the gradient and Hessian as

$$\frac{\partial}{\partial \lambda} \mathbb{L}_\star = -\sum_{i=1}^{n} \log'_\star\!\left(1 + \lambda'(X_i - \mu)\right)(X_i - \mu)$$

$$\frac{\partial^2}{\partial \lambda \partial \lambda'} \mathbb{L}_\star = -\sum_{i=1}^{n} \log''_\star\!\left(1 + \lambda'(X_i - \mu)\right)(X_i - \mu)(X_i - \mu)',$$

where $\log''_\star(z)$ is $-z^{-2}$ if $z \geq 1/n$ and is $-n^2$ otherwise. These derivatives can be expressed as

$$\frac{\partial}{\partial \lambda} \mathbb{L}_\star = -J'y, \quad \text{and} \quad \frac{\partial^2}{\partial \lambda \partial \lambda'} \mathbb{L}_\star = J'J, \qquad (3.35)$$

where J is the $n \times d$ matrix with i'th row

$$J_i = \left[-\log''_\star(1 + \lambda'(X_i - \mu))\right]^{1/2} (X_i - \mu)',$$

and y is the column vector of n components with i'th component

$$y_i = \frac{\log'_\star(1 + \lambda'(X_i - \mu))}{\left[-\log''_\star(1 + \lambda'(X_i - \mu))\right]^{1/2}}.$$

The Newton step is

$$\lambda \to \lambda + (J'J)^{-1} J'y,$$

where the increment $(J'J)^{-1} J'y$ may be found by least squares regression of y on J. Notice that when $1 + \lambda(X_i - \mu) \geq 1/n$ then $y_i = 1$.

3.15 Euclidean likelihood

The log likelihood statistic $-\sum_{i=1}^{n} \log(nw_i)$ can be viewed as a measure of the distance of $(w_1, \ldots, w_n)$ from the center $(n^{-1}, \ldots, n^{-1})$ of the simplex (2.6). Contours of this function are shown in Figure 2.3 for the case $n = 3$. The reweighted mean $\sum_{i=1}^{n} w_i X_i$ has linear contours, and a test value μ is rejected if the hyperplane with $\sum_{i=1}^{n} w_i X_i = \mu$ does not come close to the center.

Results similar to empirical likelihood can be obtained by replacing the empirical log likelihood by other distance functions. A Euclidean log likelihood ratio may be obtained by taking

$$\ell_E = -\frac{1}{2} \sum_{i=1}^{n} (nw_i - 1)^2.$$

This function is maximized over $\sum_{i=1}^{n} w_i = 1$ by taking $w_i = 1/n$ for all i. Notice that ℓ_E is well defined even if some $w_i < 0$; we return to this point below. Strictly speaking, this is not a genuine log likelihood because it is not the logarithm of the probability of the data. But ℓ_E behaves like a log likelihood in some ways.

The Euclidean likelihood can be used to derive new statistical tests, and it also recovers some well known ones such as Hotelling's T^2 for inferences on the multivariate mean, Neyman's χ^2 for multinomial inferences (Chapter 3.16), White's heteroscedasticity robust regression covariance estimator (Chapter 4.10), the sandwich estimator of variance (Chapter 3.11), and the continuous updating version of the generalized method of moments in econometrics.

The Euclidean log likelihood is a quadratic, so it has the same curvature everywhere. Therefore it equals its own quadratic approximation, and so Taylor approximation methods, like those used in Wald tests and confidence intervals, leave it unchanged. Because the Euclidean and empirical log likelihoods have the same curvature at the NPMLE, Euclidean likelihood ratio tests are analogous to Wald tests for the empirical likelihood.

To maximize ℓ_E subject to $\sum_{i=1}^{n} w_i = 1$ and $\sum_{i=1}^{n} w_i X_i = \mu \in \mathbb{R}^d$, introduce the Lagrangian

$$G = -\frac{1}{2} \sum_{i=1}^{n} (nw_i - 1)^2 - n\lambda' \sum_{i=1}^{n} w_i (X_i - \mu) + \gamma \left(\sum_{i=1}^{n} w_i - 1 \right).$$

Here $\lambda \in \mathbb{R}^d$ and $\gamma \in \mathbb{R}$ are Lagrange multipliers for the equality constraints on w_i, and the factor n is the same scale factor used in empirical likelihood. By setting $\partial G / \partial w_i = 0$, we obtain

$$0 = -n(nw_i - 1) - n\lambda'(X_i - \mu) + \gamma. \tag{3.36}$$

Averaging (3.36) over i we find $\gamma = n\lambda'(\bar{X} - \mu)$, and then

$$w_i = \frac{1}{n} \left[1 - \lambda'(X_i - \bar{X}) \right].$$

EUCLIDEAN LIKELIHOOD

Now by some standard manipulations

$$\sum_{i=1}^{n} w_i (X_i - \mu) = \bar{X} - \mu - \frac{1}{n}\lambda' \sum_{i=1}^{n} (X_i - \bar{X})(X_i - \mu)$$

$$= \bar{X} - \mu - \frac{1}{n}\lambda' \sum_{i=1}^{n} (X_i - \bar{X})(X_i - \bar{X})$$

$$= \bar{X} - \mu - \frac{1}{n} \sum_{i=1}^{n} (X_i - \bar{X})(X_i - \bar{X})' \lambda$$

$$= \bar{X} - \mu - S\lambda,$$

where

$$S = \frac{1}{n} \sum_{i=1}^{n} (X_i - \bar{X})(X_i - \bar{X})' \quad (3.37)$$

is $(n-1)/n$ times the usual sample covariance matrix. It follows that $\lambda = S^{-1}(\bar{X} - \mu)$.

Substituting this λ into w_i and then into ℓ_E we find

$$-2\ell_E = \sum_{i=1}^{n} (nw_i - 1)^2$$

$$= \sum_{i=1}^{n} \left(\lambda'(X_i - \bar{X})\right)^2$$

$$= n(\bar{X} - \mu)' S^{-1} (\bar{X} - \mu).$$

Thus $-2\ell_E = (1 - n^{-1})T^2$ where T^2 is Hotelling's T^2. The factor of $1 - n^{-1}$ arises because the normalization of S in equation (3.37) is $1/n$ not $1/(n-1)$. When X has a finite nonsingular variance matrix, T^2 has an asymptotic $\chi^2_{(d)}$ distribution.

The main difference between Euclidean and empirical likelihoods is that the former allows $w_i < 0$ while the latter does not. By allowing negative w_i it is possible for the Euclidean likelihood to produce confidence intervals for a variance that include negative values $\sum_{i=1}^{n} w_i (X_i - \sum_{j=1}^{n} w_j X_j)^2 < 0$. By keeping all $w_i \geq 0$, empirical likelihood guarantees that the reweighted data form a genuine distribution which therefore has a nonnegative variance. Similarly, empirical likelihood keeps probabilities in $[0, 1]$ and correlations in $[-1, 1]$, while Euclidean likelihood need not.

While allowing negative w_i can cause Euclidean likelihood to violate some range restrictions, it does yield confidence regions for the mean that extend beyond the convex hull of the data. This can be an advantage when d is large or when n is small. The empirical likelihood-t method in Chapter 10.4 yields a form of empirical likelihood confidence region for the mean that can extend beyond the convex hull of the data.

Euclidean likelihood for the mean of d-dimensional vectors requires solving a set of d linear equations. Empirical likelihood requires a d-dimensional unconstrained convex optimization in d variables. The empirical likelihood computations reduce to iterated d-dimensional least squares problems. Practically speaking this means that Euclidean likelihood for a vector mean might cost as much as one iteration of empirical likelihood for the mean. Empirical likelihood commonly takes four or five iterations for a vector mean.

Just as with parametric and empirical likelihoods, the challenge with a Euclidean likelihood is in maximizing it subject to nonlinear equality constraints. If the equality constraints are linear in w_i then Euclidean likelihood and empirical likelihood become, respectively, linearly constrained quadratic and linearly constrained convex optimization.

3.16 Other nonparametric likelihoods

Other distances can be constructed in the simplex to give a χ^2 limit for inferences on a vector mean, perhaps after suitable scaling. Widely studied examples include the Kullback-Liebler distance

$$\mathrm{KL} = \sum_{i=1}^{n} w_i \log(nw_i) \qquad (3.38)$$

and the Hellinger distance

$$H = \sum_{i=1}^{n} \left(w_i^{1/2} - n^{-1/2} \right)^2.$$

The empirical and Euclidean likelihoods have the advantage that the Lagrange multiplier for $\sum_{i=1}^{n} w_i = 1$ can be eliminated. The Euclidean version has the further advantage that the multiplier for $\sum_{i=1}^{n} w_i(X_i - \mu) = 0$ can also be eliminated.

The Cressie-Read power divergence statistic for a multinomial with O_i observations where E_i observations were expected, for $i = 1, \ldots, k$, takes the form

$$\mathrm{CR}(\lambda) = \frac{2}{\lambda(\lambda+1)} \sum_{i=1}^{k} O_i \left[\left(\frac{O_i}{E_i} \right)^{\lambda} - 1 \right]$$

where $-\infty < \lambda < \infty$. The degenerate cases $\lambda \in \{-1, 0\}$ are handled by taking limits.

The empirical likelihood setup has n distinct data values observed once each. Thus $k = n$, and $O_i = 1$ and we write $E_i = nw_i$. Then

$$\mathrm{CR}(\lambda) = \frac{2}{\lambda(\lambda+1)} \sum_{i=1}^{n} \left[(nw_i)^{-\lambda} - 1 \right].$$

OTHER NONPARAMETRIC LIKELIHOODS

Also, after taking the required limits,

$$\text{CR}(0) = -2\sum_{i=1}^{n} \log(nw_i), \quad \text{and}$$

$$\text{CR}(-1) = 2\sum_{i=1}^{n} nw_i \log(nw_i).$$

Figure 3.12 shows contours of CR(λ) for $n = 3$ and six values of λ. The same contour levels are shown for each λ. These correspond to the contours in Figure 2.3.

The quantity CR(0) is minus twice the empirical log likelihood ratio, and CR(-1) is equal to $2n$KL from equation (3.38). The value $\lambda = -2$ corresponds to the Euclidean log likelihood, through CR(-2) = $-2\ell_E$, under the usual condition that $\sum_i w_i = 1$. In this setting, Euclidean log likelihood is also known as Neyman's χ^2. The value $\lambda = 2/3$ is known to give an especially good fit for χ^2 calibration in multinomial families on a fixed number of values. The value $\lambda = 1$ corresponds to Pearson's χ^2, and the Freeman-Tukey statistic has $\lambda = -1/2$.

All members of the Cressie-Read family give rise to empirical divergence analogues of the empirical likelihood in which an asymptotic χ^2 calibration holds for the mean. They all give convex confidence regions for the mean.

The Kullback-Liebler distance is often singled out because of its interpretation as an entropy. We use the term empirical entropy for this method, also known as exponential empirical likelihood. The empirical entropy is between empirical likelihood and Euclidean likelihood in the Cressie-Read family. It is also between them in that empirical likelihood uses positive weights, empirical entropy uses nonnegative weights, and Euclidean likelihood uses real-valued weights.

By convention $w_i \log(nw_i) = 0$ for $w_i = 0$, and so the empirical entropy is bounded over the simplex of weights. The largest possible value of $CR(-1)$ is $2n\log(n)$, attained when some $w_i = 1$ and the others are all 0. By contrast, empirical likelihood is unbounded as $\min w_i \to 0$. Thus if the true mean is near the convex hull of the data, empirical likelihood discrepancy $CR(0)$ will be much larger than $CR(-1)$. Where empirical entropy allows $w_i = 0$, the Euclidean likelihood allows $w_i < 0$, and so $CR(-2)$ might be still smaller for a true mean just barely within the convex hull of the data. These features do not appear to give a faster χ^2 approximation. Figure 3.13 shows resampled values of empirical entropy and Euclidean likelihood using the same set of resampled data sets as Figure 3.2 uses for empirical likelihood. Although the resampling generated some infinite $CR(-1)$ values, the empirical log likelihood was not more extreme than the other discrepancies, except in the very deep tail of the distribution. Surprisingly, the distribution of $CR(-2)$ appears to be farthest from the nominal χ^2 distributions.

Chapter 3.14 shows how empirical likelihood for a vector mean can be solved

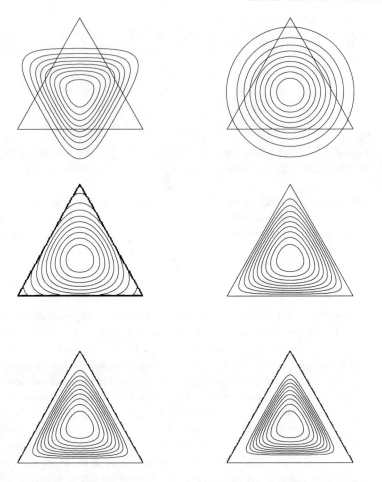

Figure 3.12 *Shown are contours for Cressie-Read divergences for $n = 3$, as described in the text. The values of λ are, reading left to right within rows from top to bottom, -5, -2, -1, 0, $2/3$, and $3/2$. Euclidean likelihood is in the upper right corner. The middle row has empirical entropy on the left and empirical likelihood on the right.*

by iterated least squares regression. There is also a simple algorithm for computing empirical entropy, using Poisson regression methods. See Chapter 3.17.

One way to compare empirical likelihood and empirical entropy is to introduce

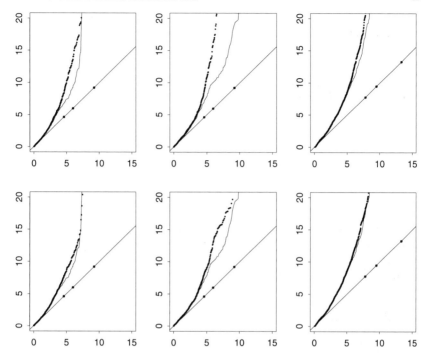

Figure 3.13 *The top row of plots repeats the QQ plots of Figure 3.2 replacing the empirical likelihood by the Euclidean likelihood. The bottom row uses the empirical entropy. In all 6 plots the empirical likelihood QQ plot is also shown as an additional reference line.*

the weights $v_i = 1/n$, for $i = 1, \ldots, n$. Then

$$\mathrm{KL} = \sum_{i=1}^{n} w_i \log\left(\frac{w_i}{v_i}\right), \quad \text{and}$$

$$\log R(F) = \sum_{i=1}^{n} v_i \log\left(\frac{v_i}{w_i}\right),$$

making empirical entropy appear to be a backward log likelihood—with the roles of model and data reversed. Of course, this also makes the log likelihood a backward entropy. A practical consequence is that the difficulties empirical likelihood has with $w_i = 0$ would be expected for empirical entropy with $v_i = 0$. A value of $v_i = 0$ corresponds to a potential observation for which the observed count is zero, as was used in defining quantiles in Chapter 3.6. Such "undata" could also be used to good effect when upper and lower limits are known for $m(X, \theta)$.

Compared to other nonparametric discrepancy measures, empirical likelihood extends more directly to problems with biased data, censoring, truncation, and prior distributions. The extensions involve modifying the parametric likelihood treatments of those issues.

Of the distances above, empirical likelihood is the only one that allows a Bartlett correction. For another nonparametric likelihood to be Bartlett correctable it must match empirical likelihood to sufficiently high order. A distance function

$$-2\sum_{i=1}^{n}\widetilde{\log}(nw_i)$$

with

$$\widetilde{\log}(1+z) = z - \frac{1}{2}z^2 + \frac{1}{3}z^3 - \frac{1}{4}z^4 \qquad (3.39)$$

gives rise to a Bartlett correctable nonparametric likelihood. The key is to match four derivatives of the log function at 1. A bounded function that, like $\widetilde{\log}$, matches log near 1, may generate a nonparametric likelihood with the best features of empirical likelihood and empirical entropy.

3.17 Bibliographic notes

Coverage and calibration

Theorem 3.2 was proved by Owen (1990*b*), who also shows that the coverage error in empirical likelihood is $O(n^{-1/2})$, if $E(\|X\|^4) < \infty$. The coverage error also attains the much better rate $O(n^{-1})$, under mild extra conditions, as shown by DiCiccio, Hall & Romano (1991) for smooth functions of means, and by Zhang (1996*b*) for estimating equations (of a scalar parameter). These extra conditions include Cramér's condition (Equation (13.1) on page 249) and stronger moment conditions.

Box (1949) considers approximations to the distribution of the parametric log likelihood ratio using the F distribution. Most of the bibliographic notes for Bartlett correction of empirical likelihood appear in Chapter 13.6 (which begins on page 257). Tsao (2001) proposes a calibration method in which the exact critical value of $-2\log\mathcal{R}$ is obtained for the mean, assuming a normal distribution, and then used for possibly non-normal data. In some univariate simulations, this proposal improves the coverage level, although the resulting method still tends to undercover. La Rocca (1995*b*) proposes a calibration method based on replacing the $\chi_{(p)}^{2,1-\alpha}$ quantile by $\chi_{(p)}^{2,1-\alpha+\Delta}$ where $\Delta(\alpha)$ takes a simple functional form. Bootstrap calibration of Δ should then require fewer resamples than bootstrap calibration of a percentile.

Bootstrap calibration of empirical likelihood was proposed by Owen (1988*b*), Hall & La Scala (1990), and Wood, Do & Broom (1996). Bootstrap calibration in general settings has been studied by Beran (1987), Beran (1988), and Loh (1991). The limiting distribution of an empirical likelihood test is typically the same χ^2

distribution regardless of the true value of the quantity being tested. In such settings bootstrap calibration typically reduces the order of magnitude of the level error in the test. The result is that bootstrap calibration of empirical likelihood has the same order of coverage error, typically $O(1/n^2)$ as obtained in bootstrapping the bootstrap (Hall & Martin 1988).

Hall & LePage (1996, Remark 3.5) describe how the bootstrap can be used to calibrate empirical likelihood confidence regions for the mean in some settings with infinite variance. The data have to come from a distribution satisfying a tail condition, and the bootstraps are based on $m < n$ resampled values.

Fisher, Hall, Jing & Wood (1996) consider constructing confidence regions for the mean of directional data on the surface of the sphere. They find that the combination of empirical likelihood to choose the shape of the region and the bootstrap to choose the size performs best.

Power

Chen (1994a) compares the asymptotic power of empirical likelihood and a bootstrap T^2 method, for tests on the mean of a vector. For alternatives that are $O(n^{-1/2})$ from the true mean, both tests have the same noncentral chisquared limit. Bootstrap T^2 confidence regions are symmetric about the sample mean while empirical likelihood regions are elongated in directions of skewness. Not surprisingly, the next term in the power expansion shows that the bootstrap T^2 tests have better power against alternatives in directions of positive skewness and empirical likelihood tests have better power in directions of negative skewness.

Lazar & Mykland (1998) compare the asymptotic power of empirical and parametric likelihoods. The data are assumed to be sampled from a member of the parametric family, and the empirical likelihoods are assumed to use the same estimating equation as the likelihood. The power of empirical likelihood tests matches that of the parametric likelihood to second order. At third order the empirical likelihood can have either greater or lesser power than the parametric likelihood.

Estimating equations

Godambe (1960) introduces the subject of estimating equations, and shows that in parametric models, the score equation is the optimal estimating equation. Godambe & Thompson (1974) consider nuisance parameters. Statistics defined via estimating equations are also known as M-estimates. Huber (1981) describes their use in constructing robust extensions of maximum likelihood estimates. Theorem 3.6 with results on smooth estimating equations for empirical likelihood are from Qin & Lawless (1994).

Marginal and conditional likelihoods are described by Cox & Hinkley (1974). The partial likelihood was introduced by Cox (1975). Lindsay (1980) shows that for a vector X consistent estimators can be put together by multiplying together likelihood factors, which may be conditional distributions of some components of X given others or marginal distributions of some components. The resulting product need not have a probability interpretation.

Many problems in econometrics involve testing a model in which the number of estimating equations is larger than the number of parameters in them. The overidentifying restrictions test is based on the generalized method of moments. See Hansen (1982). Kitamura (2001) points out that the Euclidean likelihood tests are equivalent to the continuous-updating GMM approach to overidentifying restrictions, due to Hansen, Heaton & Yaron (1996).

Qin & Lawless (1994) provide the examples of side information in which two co-observed random variables have common mean. Owen (1991) proves Theorem 3.4 and illustrates it with regression through the origin. The example of a common mean and median is discussed at length in Qu (1995); see also Chapter 10.8.

Quantiles

Exact confidence levels can be obtained for quantiles of continuous distributions. Useful references include David (1981) and Hahn & Meeker (1991).

Chen & Hall (1993) introduce a kernel smoothing technique to construct confidence regions for quantiles. Smoothing the CDF allows them to obtain confidence interval endpoints between observations. Coverage error can be reduced from $O(n^{-1/2})$ to $O(n^{-1})$ by smoothing, and then to $O(n^{-2})$ by Bartlett correction. Much less smoothing is required for quantiles than for density estimation. When the kernel is the density function of a symmetric bounded random variable, then a bandwidth of $h = o(n^{-1/2})$ provides a χ^2 calibration, including $h = 0$ for unsmoothed empirical likelihood. To get coverage error $O(n^{-1})$ it must also hold that $nh/\log(n) \to \infty$. Bartlett correctability requires additionally that $h = O(n^{-3/4})$. When $h = O(n^{-3/4})$ and $nh/\log(n) \to \infty$, the Bartlett constant is approximately $p^{-1}(1-p)^{-1}(1-p+p^2)^{-1}/6$ for the quantile Q^p. Using this approximation the coverage error is $O(n^{-1}h)$. This can be nearly as good as the rate $O(n^{-2})$ attainable by exact Bartlett correction. Geoff Eagleson raised the question of the interquartile range at a seminar.

Several authors have considered the estimation of an EDF subject to side information on moments or quantiles. Haberman (1984) introduced an approach based on Kullback-Liebler distance. Sheehy (1987) considered several other information distances. Qin & Lawless (1994) consider estimation of the distribution function subject to known means of smooth estimating equations. Sheehy (1987) considers Kullback-Liebler and Hellinger distances as do DiCiccio & Romano (1990).

Robustness

Huber (1981) describes robust methods in statistics, including M-estimators and alternatives. The joint location-scale M-estimate presented is based on a proposal for regression problems in Chapter 7.7 of Huber (1981). Owen (1988*b*) computes an empirical likelihood ratio function for Huber's M-estimate, applied to 20 of Newcomb's measurements of the passage time of light (data set 9 of Stigler (1977)). Owen (1988b) uses a scale value estimated from the data, and fixed throughout the likelihood profiling. Tsao & Zhou (2001) study the robustness properties of the endpoints of empirical likelihood confidence intervals. They find that the endpoints of the interval for a univariate mean are not robust to outliers, but that the endpoints of an interval for a robust M-estimate are robust to outliers.

Hall & Presnell (1999a) describe an approach to robustness, through the empirical entropy. They maximize the empirical entropy subject to an equality constraint on a location statistic and another constraint on a concomitant scale statistic. The constraint on the scale parameter forces the weight on outliers to be small. Unlike empirical likelihood, empirical entropy remains finite when some observation weights are set to zero. This allows robust confidence regions to be constructed for nonrobust statistics like the mean.

Colin Mallows described robust likelihood by contamination in a conversation with the author in the mid-1980s. Mallows computed some examples for data with nearly a Poisson distribution, apart from one or more gross outliers. Small values of ϵ were enough to effectively trim away the outliers, and leave an estimate of the mean of the uncontaminated part of the distribution. The method did not appear to be very sensitive to ϵ, though of course ϵ could not be zero. Modifying the likelihood this way introduces a small systematic error, because some of the Poisson probabilities are not small compared to ϵ.

Empirical discrepancies

Cover & Thomas (1991) is a background reference on entropy methods. Read & Cressie (1988) describe power divergence discrepancies. They find that the value $\lambda = 2/3$ has the most accurate χ^2 calibration in parametric settings. The power divergence discrepancies were also studied by Rényi (1961). Mittelhammer, Judge & Miller (2000) describe empirical entropy and empirical likelihood in econometric settings. The focus is on methods such as generalized method of moments and instrumental variables.

Baggerly (1998) makes the connection between empirical likelihood and Cressie-Read statistics, and proves that all members of the Cressie-Read family have a χ^2 calibration. Baggerly (1998) proves that empirical likelihood is the only Bartlett correctable member of the family. Baggerly (1998) notes that for $\lambda \geq 0$, the confidence regions lie within the convex hull of the data. Baggerly (1998) shows that for $\lambda \geq -1$ the divergence minimizing weights for a candidate value of the mean are nonnegative whenever that candidate value is within the convex hull of the data. For $\lambda < -1$, it is possible that some of the optimal weights are negative even if the candidate for the mean is within the convex hull of the data. For $\lambda = -2$, confidence regions for the mean are ellipsoidal. Baggerly (1998) shows that for $\lambda > -2$, the regions tend to be lengthened in the direction of outlying observations, while for $\lambda < -2$ they tend to be shorter in those directions. Baggerly (1998) notes a sufficiency property for the first k moments when $\lambda = -k/(k-1)$, with Euclidean likelihood corresponding to $k = -2$. Letting $k \to \infty$ provides a sense in which empirical entropy ($\lambda = -1$) is the limit of a sequence of moment-based inference methods.

Corcoran, Davison & Spady (1995) show some an example where the distribution of $-2\log \mathcal{R}(\mu)$ is not close to χ^2, and for which the QQ plot bends upwards like those in Figures 3.2 and 3.13, making Bartlett correction ineffective. Their data were sampled from a χ^2 distribution and a parametric log likelihood ratio statistic had a QQ plot very close to the 45° line. Bartlett-correctable nonparametric likelihoods including (3.39) were published posthumously by Corcoran (1998).

Other bibliographic notes

The dipper survey data come from Iles (1993). The milk production data belonged to my father, George Owen. It is from the farm on which he grew up.

The sandwich variance formula (3.23) is due to Huber (1967). Sample versions of the sandwich formula have been advocated by White (1980) for regression and by Liang & Zeger (1986) for generalized estimating equations.

Some background on financial returns may be found in Hull (2000), Duffie (1996), and Campbell, Lo & MacKinlay (1996), who describe ways of testing whether returns are independent from one time period to the next. The combined actions of investors, speculators, and arbitragers are likely to make for very small correlations in returns. There is some empirical evidence that the variance of returns changes over time and that nearby time periods have more similar volatilities than do separated periods.

Standard references on numerical optimization include Gill, Murray & Wright (1981) and Fletcher (1987). Fletcher describes a convex duality problem very similar to empirical likelihood, except that the entropy distance is used. Davison & Hinkley (1997, Chapter 10) describe an algorithm for empirical entropy based on Poisson regression.

3.18 Exercises

Exercise 3.1 In $d = 1$ dimension, a distribution F can have roughly $1.92/n$ probability off of the sample, and still contribute to the asymptotic 95% confidence region. Modify the argument following Lemma 2.1 for higher dimensions. For dimensions $d = 2, \ldots, 100$ find how much probability can be placed off of the sample while keeping $-2 \log \mathcal{R}(F) \leq \chi^{2,0.95}_{(d)}$. Describe the pattern with respect to d.

Exercise 3.2 For the Euclidean log likelihood, suppose that k of the weights w_i are 0 or less, for $1 < k < n$. What is the largest possible value of ℓ_E subject

EXERCISES

to these constraints? For $d = 1, 2, 5, 10$, roughly what proportion of the n observations can a distribution give nonpositive weight, while still contributing to the 95% confidence interval? Assume that w_i sum to 1.

Exercise 3.3 Suppose that the data consist of pairs (X_i, Y_i) of real values, and that interest centers on the ratio $\tau = \sigma_Y^2/\sigma_X^2$ of the variance of Y to that of X. Write a set of estimating equations, for a parameter vector $\theta = (\tau, \nu')'$. Here τ is one component of θ and ν is a vector of nuisance parameters. The number of estimating equations should equal the dimension of θ.

Exercise 3.4 Suppose that (X_i, Y_i) are real data pairs, with a continuous distribution. We are interested in the hypothesis H_0 which stipulates that, conditionally on $X \le 100$, the median of Y is equal to 50. Construct an estimating function whose expectation is zero if and only if H_0 holds.

Exercise 3.5 For the setting in Exercise 3.4, suppose that $n = 100$, that exactly 70 of the X_i are less than or equal to 100, and that of these 70 observations exactly 30 have $Y_i < 50$ while 40 have $Y_i > 50$. What is the maximum empirical likelihood ratio under H_0? What weight do the 30 observations with $X_i > 100$ get under the maximizing distribution?

Exercise 3.6 Bootstrap calibration may also be done using the 950th largest of 999 resampled log likelihood ratios. Suppose that $X_1, \ldots, X_B \in \mathbb{R}$ are independent random variables with distribution F. Let $X_{(1)} < X_{(2)} < \cdots < X_{(B)}$ be their order statistics. Show that $\Pr(X < X_{(A)}) = A/(B+1)$ where $1 \le A \le B$ and $X \sim F$ is independent of X_i.

Exercise 3.7 Explain why restricting the data $X_i \in \mathbb{R}^d$ to have a variance matrix of full rank d results in no loss of generality when formulating statistics as smooth functions of means.

Exercise 3.8 Suppose that m is the unique median of F, but $\Pr(X = m) > 0$. Suppose that among $X_1, \ldots, X_n$, there are n_- observations $X_i < m$, and n_0 observations $X_i = m$, and n_+ observations $X_i > m$.

1. Assume that $n_- > 0$ and $n_+ > 0$. Let w_i maximize $\prod_{i=1}^n nw_i$ subject to $w_i > 0$, $\sum_{i=1}^n w_i = 1$, $\sum_{i=1}^n w_i 1_{X_i \ge m} \ge 1/2$ and $\sum_{i=1}^n w_i 1_{X_i \le m} \ge 1/2$. Show that at most 3 different values of w_i arise, one for $X_i < m$, one for $X_i > m$, and one for $X_i = m$.

2. Let w_+, w_- and w_0 be the weights described above. Let $p_+ = n_+ w_+$, $p_0 = n_0 w_0$ and $p_- = n_- w_-$. The maximum empirical log likelihood for the median value m is the maximum of

$$n_+ \log(nw_+) + n_0 \log(nw_0) + n_- \log(nw_-)$$
$$= n_+ \log(np_+/n_+) + n_0 \log(np_0/n_0) + n_- \log(np_-/n_-)$$

over nonnegative p's satisfying $p_+ + p_0 + p_- = 1$, $p_+ + p_0 \geq 1/2$, and $p_0 + p_- \geq 1/2$. Find these p's for $n_+ = 30$, $n_0 = 1$, $n_- = 30$.

3. Find these p's for $n_+ = 30$, $n_0 = 1$, $n_- = 20$.
4. Find these p's for $n_+ = 30$, $n_0 = 100$, $n_- = 10$.
5. Find these p's for $n_+ = 40$, $n_0 = 10$, $n_- = 10$.

Exercise 3.9 Suppose that the median is defined through $E(1_{X \leq m} - 1/2) = 0$, and that there are no ties among $X_1, \ldots, X_n$. When $n = 2k$ is even, then any value of $m \in [X_{(k)}, X_{(k+1)})$ is a sample median.

1. Show that no value of m has $n^{-1} \sum_{i=1}^{n} (1_{X_i \leq m} - 1/2) = 0$ when $n = 2k+1$ for integer $k \geq 1$.
2. For even n, it is possible to attain $-2\log(\prod_{i=1}^{n} nw_i) = 0$ subject to $w_i \geq 0$, $\sum_{i=1}^{n} w_i = 1$, and $\sum_{i=1}^{n} w_i(1_{X_i \leq m} - 1/2) = 0$. Show that for odd n, it is possible to attain $-2\log(\prod_{i=1}^{n} nw_i) = n^{-2} + O(n^{-3})$.
3. Suppose that X_i are IID from a continuous distribution F. Describe how to construct Z_i as a function of X_i and m, so that $E(Z_i) = 0$ when m is any median of F, and so that $n^{-1} \sum_{i=1}^{n} Z_i = 0$ when m is any sample median of the X_i.

Exercise 3.10 Suppose that $(X, Y) \in \mathbb{R}^2$ are independent. The bivariate median of X and Y may be defined as any point (m_x, m_y) such that

$$\min\{\Pr(X \leq m_x), \Pr(X \geq m_x), \Pr(Y \leq m_y), \Pr(Y \geq m_y)\} \geq 1/2$$

Suppose that the data are

X: 9, 8, 7, 6, 5, 4, 3, 2, 1, 0, 0, 0, 0, 0, 0, 0, 0, 0, 0
Y: 0, 0, 0, 0, 0, 0, 0, 0, 0, 0, 1, 2, 3, 4, 5, 6, 7, 8, 9

1. Does the empirical distribution on this set of data have a bivariate median?
2. Show that if $m_x > 0$ and $m_y > 0$ then any distribution formed by attaching nonnegative weights summing to one to the (X, Y) pairs above, and having (m_x, m_y) as its bivariate median, has empirical likelihood 0.
3. Describe the set of points (m_x, m_y) arising as bivariate medians of distributions formed by applying positive weights summing to one to the data values above.

Exercise 3.11 Determine whether the score equations for the bivariate normal parameters $\mu_x, \mu_y, \sigma_x, \sigma_y$ and ρ give rise to the estimating equations (3.8) through (3.12) used to define the correlation in the text. Do they give the same estimates?

Exercise 3.12 Suppose that $X \in \mathbb{R}$ has the distribution F. For $\theta \in \mathbb{R}$, let $m(X, \theta) = \sin(X - \theta)$. Show that for any distribution F, there is a θ_0 with $E(m(X, \theta_0)) = 0$. Obviously $\theta_0 + 2k\pi$ is also a solution for any integer k. Find a distribution F for which there is more than one solution θ_0 in the interval $[0, 2\pi)$.

Exercise 3.13 Suppose that $X \sim N(0,1)$. Let $m(x, \theta) = (x-\theta)^2 - 0.5$ for real-valued θ. Show that there is no value θ_0 with $E(m(X, \theta_0)) = 0$. Draw a sample of 100 observations X_i. Let $\mathcal{R}_n(\theta)$ be the empirical likelihood ratio function for θ based on $X_1, \ldots, X_n$. Plot $\log \mathcal{R}_n(\theta)$ versus θ for $n = 10k$, $k = 1, \ldots, 10$. What happens to the confidence set for θ as $n \to \infty$?

Exercise 3.14 Consider a setting in which we do not know whether there is any solution θ_0 to $E(m(X, \theta)) = 0$. If there is exactly one solution, then the empirical likelihood confidence region is nonempty with asymptotic probability at least $1 - \alpha$. Now suppose that there are two distinct solutions θ_{0A}, and θ_{0B}. Is the asymptotic probability of an empty confidence region still smaller than α?

Exercise 3.15 Suppose that $Y \in \{0, 1\}$, and that $X \in \mathbb{R}^d$. Let $\pi_0 = \Pr(Y = 1) \in (0, 1)$, and $\pi_1 = 1 - \pi_0$. When $Y = y$, the distribution of X is $N(\mu_y, \Sigma)$, where Σ is finite and of full rank. Show that

$$\Pr(Y = 1 \mid X = x) = [1 + \exp(-\alpha - \beta'x)]^{-1}$$

for a scalar α and a vector $\beta \in \mathbb{R}^d$. This derives logistic regression as a conditional likelihood method.

Exercise 3.16 Suppose $X_1, \ldots, X_n \in \mathbb{R}$ are IID from F_0 with mean μ_0, and finite variance $\sigma_0^2 > 0$. Suppose that $n = 2m$ is even. Then it also follows that $Y_i = (X_{2i-1}, X_{2i})'$, for $i = 1, \ldots, m$ are IID with common mean $(\mu_0, \mu_0)'$. Suppose that $\hat{\mu}$ is defined as the MELE for the estimating equations

$$\sum_{i=1}^{m} w_i \begin{pmatrix} X_{2i-1} - \mu \\ X_{2i} - \mu \end{pmatrix} = 0.$$

These are equivalent to the equations

$$\sum_{i=1}^{m} w_i \begin{pmatrix} X_{2i-1} - \mu \\ X_{2i} - X_{2i-1} \end{pmatrix} = 0.$$

Find $\lim_{m \to \infty} m \text{Var}(\hat{\mu})$. The natural estimator in this setting is

$$\bar{X} = \frac{1}{n} \sum_{i=1}^{n} X_i.$$

It is hard to imagine that $\hat{\mu}$ is better than $\bar{X}$. Perhaps it is asymptotically as good, or perhaps because $m = n/2$ it roughly doubles the variance. Settle the issue by finding $\lim_{m \to \infty} \text{Var}(\hat{\mu})/\text{Var}(\bar{X})$. Describe an applied setting in which doubt about the joint distribution of X_i would lead you to prefer $\hat{\mu}$ to $\bar{X}$.

Exercise 3.17 Suppose that (X, Y) pairs are jointly observed, with $X \in \mathbb{R}^p$ and $Y \in \mathbb{R}^q$. The mean of X is known to be μ_{x0} and a confidence region for the mean μ_{y0} is to be constructed using the side information on X. The theorem in the text assumes that the variance of $(X', Y')'$ has full rank $p + q$. When the rank is less than $P+1$, a $\chi^2_{(r)}$ calibration might still hold for some r. Suppose that U, V, W, Z are independent random variables and that $\mathrm{Var}((U, V, W, Z)')$ has rank 4. Decide whether an appropriate value of r exists in the following settings, and if one does, say what it is:

1. $X = (U, V, U + V)'$ and $Y = (W, Z)'$.
2. $X = (U, V)'$ and $Y = (W, Z, W + Z)'$.
3. $X = (U, V, U + V)'$ and $Y = (U, W, Z)'$
4. $X = (U, V, W)'$ and $Y = (U + V, V + W)'$.

Exercise 3.18 Find an expression for $m(x, \theta)$ for Mallows's robust estimating equation (3.30), in the case where $f(x; \theta)$ is the Poisson probability function with mean $\theta \in [0, \infty)$ and observations $x \in \{0, 1, \dots\}$. Decide among:

- **A.** $m(x, \theta)$ is bounded uniformly in x and θ,
- **B.** $m(x, \theta)$ is bounded in x for each θ, but A fails,
- **C.** $m(x, \theta)$ is bounded in θ for each x, but A fails,
- **D.** both B and C hold,
- **E.** $m(x, \theta)$ is unbounded in x for some θ and unbounded in θ for some x.

Exercise 3.19 Consider the estimating function

$$m(x, \mu) = \frac{\frac{\partial}{\partial \mu} \varphi(x - \mu)}{\varphi(x - \mu) + \tau}$$

where $\varphi(z)$ is the $N(0, 1)$ density. Plot $m(x, \mu)$ versus $x - \mu$ for several values of $\tau > 0$. Describe qualitatively how the curves look. Find a value of τ so that the resulting curve is nearly linear for $|x - \mu| \leq 1.35$.

Exercise 3.20 Determine whether ties can be ignored for other empirical divergences in the Cressie-Read family.

Exercise 3.21 For a power $\lambda \neq -1$, show that the weights w_i that minimize the empirical divergence subject to $\sum_{i=1}^n w_i = 1$ and $\sum_{i=1}^n w_i X_i = \mu$ are

$$w_i = \frac{1}{n} (1 + a + b'(X_i - \mu))^{-1/(\lambda+1)}$$
$$= \frac{c}{n} (1 + d'(X_i - \mu))^{-1/(\lambda+1)},$$

where a and c are scalars and b and d are vectors of the same dimension as X_i.

Exercise 3.22 Show that for $\lambda = -1$, the weights are

$$w_i = c \exp\left(d'(X_i - \mu)\right).$$

CHAPTER 4

Regression and modeling

Linear and multiple regression are among the most widely used statistical methods. This chapter considers empirical likelihood inferences for linear regression and other models with covariates, such as generalized linear models. The standard setting for linear regression has fixed predictors and a random response. Empirical likelihood was developed in Chapter 3.4 for smooth functions of means and for estimating equations, but assuming independent identically distributed data. Some new techniques are required to extend empirical likelihood to settings with fixed regressors.

Figure 4.1 shows a measure of breast cancer mortality versus population size for a set of counties in the southern U.S.A. A linear regression fits this data set well, though it is clear from the data, and obvious scientifically, that the variance of mortality increases with the population size. It is also reasonable to consider a regression through the origin for this set of data.

In simple linear regression, we observe pairs (X_i, Y_i), $i = 1, \ldots, n$ and it is thought that for a generic pair (X, Y),

$$E(Y \mid X = x) \doteq \beta_0 + \beta_1 x,$$

where the approximate inequality allows for some practically insignificant lack of fit. There are two widely used sampling models for linear regression. In one case (X_i, Y_i) are independent random vectors from a joint distribution $F_{X,Y}$ on $\mathbb{R}^2$, while in the other, $X_i = x_i$ are fixed and Y_i are then sampled independently from the conditional distributions $F_{Y|X=x_i}$. It is also a common practice to sample random pairs, but to analyze the data as if the X_i had been fixed at their observed values. These two sampling models extend to multiple regression in an obvious way.

This chapter first considers independent sampling of X, Y pairs because the results from Chapter 3 apply directly. Then sampling with fixed X_i is handled using a more general ELT. Then extensions are made to generalized linear models, nonlinear least squares, and the analysis of variance.

4.1 Random predictors

Suppose that $X \in \mathbb{R}^p$ and $Y \in \mathbb{R}$ are the generic predictor vector and response. Let (X_i, Y_i) be independent random observations from a common distribution. To make the notation simpler, suppose that X_i includes any necessary functions

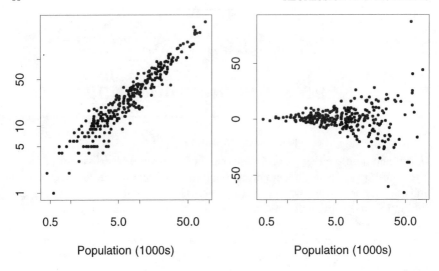

Figure 4.1 *The left plot shows some cancer mortality counts (plus one) versus the population sizes for some counties in the southern U.S.A. The right plot shows residuals for these data from a linear regression of cancer on population. Source: Rice (1988).*

of the available measurements. Thus $X = (1, W)'$ for simple linear regression on W, and $X = (1, U, W, U^2, W^2, UW)'$ for a quadratic response surface in U and W. We will suppose that $E(X'X)$ has full rank p. It may be necessary to remove one or more redundant predictor variables to arrive at X with $E(X'X)$ of full rank.

A linear model takes the form $X'\beta$ for some vector $\beta \in \mathbb{R}^p$. The value β_{LS} that minimizes $E((Y - X'\beta)^2)$ is

$$\beta_{\mathrm{LS}} = E(X'X)^{-1} E(X'Y),$$

and the sample least squares estimate of β_{LS} is

$$\hat\beta_{\mathrm{LS}} = \left(\frac{1}{n}\sum_{i=1}^n X_i'X_i\right)^{-1}\left(\frac{1}{n}\sum_{i=1}^n X_i'Y_i\right),$$

the NPMLE of β_{LS} in the sense of Chapter 2. We use β_{LS} instead of β_0 to designate the population value of β, because in regression problems β_0 has a firmly established use as an intercept coefficient.

Because β_{LS} is a smooth function of means, empirical likelihood inferences for it follow by the theory in Chapter 3.4, under mild moment conditions: that $E(\|X\|^4) < \infty$, $E(\|X\|^2 Y^2) < \infty$, and $E(X'X)$ is nonsingular. Invertibility of $E(X'X)$ is needed in order to make β_{LS} a smooth function of means. There is

RANDOM PREDICTORS

no need to assume that either X or Y has a normal distribution, or that $\sigma^2(x) = \mathrm{Var}(Y \mid X = x)$ is constant with respect to x.

There is also no need to assume that $\mu(x) = E(Y \mid X = x)$ is of the form $x'\beta$. In this case the interpretation is that empirical likelihood confidence regions for the best value β_{LS} are properly calibrated, despite the lack of fit that might be inherent in that best value. A very large lack of fit could make the linear model irrelevant, and perhaps require the addition of some components to X. But a small lack of fit might be acceptable. In practice, some lack of fit is inevitable for many applications, and empirical likelihood tests are not sensitive to it.

The mean square prediction error $E((Y - X'\beta_{\mathrm{LS}})^2)$ is also a smooth function of means and so empirical likelihood inferences apply to it. This squared error can be written $E(\sigma^2(X)) + E((\mu(X) - X'\beta_{\mathrm{LS}})^2)$, combining variance and lack of fit terms.

The regression model can also be approached through estimating equations. The definition of β_{LS} is equivalent to

$$E\left(X\left(Y - X'\beta_{\mathrm{LS}}\right)\right) = 0, \tag{4.1}$$

and the definition of $\hat{\beta}_{\mathrm{LS}}$ is equivalent to the normal equations

$$\frac{1}{n}\sum_{i=1}^{n} X_i\left(Y_i - X'_i\hat{\beta}_{\mathrm{LS}}\right) = 0.$$

That is, the errors $Y - X'\beta_{\mathrm{LS}}$ are uncorrelated with X and the residuals $Y_i - X'_i\hat{\beta}_{\mathrm{LS}}$ are orthogonal to the sample X_is. This formulation allows us to weaker the moment conditions for empirical likelihood regression inferences. The conditions $E(\|X\|^4) < \infty$, and $E(\|X\|^2 Y^2) < \infty$ can be replaced by $E(\|X\|^2(Y - X'\beta_{\mathrm{LS}})^2) < \infty$. It is still necessary to have $E(X'X)$ invertible so that β_{LS} is determined by (4.1).

Define the auxiliary variables $Z_i = Z_i(\beta) = X_i(Y_i - X'_i\beta)$. The empirical likelihood ratio function for β is defined by

$$\mathcal{R}(\beta) = \max\left\{\prod_{i=1}^{n} nw_i \mid \sum_{i=1}^{n} w_i Z_i(\beta) = 0, w_i \geq 0, \sum_{i=1}^{n} w_i = 1\right\}.$$

For any vector β this empirical likelihood ratio may be computed using the algorithm for a vector mean, applied to the Z_i values. When a single component of β is of interest, then we maximize $\mathcal{R}$ over the other components to obtain the profile empirical likelihood ratio function for the component of interest.

For the cancer data, we take $X_i = (1, P_i)'$ where P_i is the population of the i'th county, and $Y_i = C_i$ where C_i is the number of cancer deaths in the i'th county. Then $\beta = (\beta_0, \beta_1)'$. For these data $\hat{\beta}_1 = 3.58$, corresponding to a rate of 3.58 cancer deaths per 1000 population. Because deaths were counted over 20 years, the annualized rate is $3.58/20 = 0.18$ per thousand. Also the intercept $\hat{\beta}_0 = -0.53$ is quite close to zero, as we would expect.

Figure 4.2 shows the profile empirical log likelihood ratio functions for β_0 and

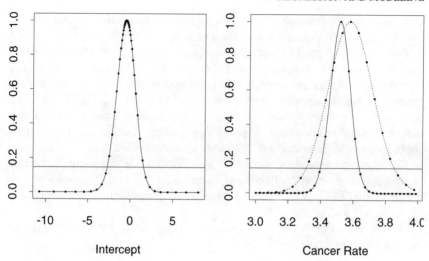

Figure 4.2 *The left plot shows the empirical likelihood ratio function for the intercept in a simple linear regression model relating cancer mortality to county population for the data shown in Figure 4.1. A value of zero is reasonable for the intercept. The right plot shows the empirical likelihood ratio function for the slope. The solid curve is for regression through the origin, the dotted curve is for ordinary regression not through the origin.*

β_1. Horizontal reference lines mark asymptotic confidence thresholds. These data are clearly non-normal, and have nonconstant variance, but empirical likelihood inferences still have good coverage properties in this setting.

It is natural with data such as these to consider a regression through the origin. It is clear that a county with near zero population must have near zero cancer deaths. Furthermore, linearity appears to hold over the observed range of data, including some very small counties. To obtain a regression through the origin with slope β_1 we insert $\beta_0 = 0$ into the reweighted normal equations, obtaining

$$\sum_{i=1}^{n} w_i(C_i - P_i\beta_1) = 0, \quad \text{and} \qquad (4.2)$$

$$\sum_{i=1}^{n} w_i P_i(C_i - P_i\beta_1) = 0. \qquad (4.3)$$

This almost appears to be solving two equations in one unknown β_1, one to make the residuals have reweighted mean 0 and the other to make them uncorrelated with the population size. This is very different from what we would do with ordinary regression through the origin. In ordinary regression through the origin, the slope is estimated by $\sum_{i=1}^{n} X_i Y_i / \sum_{i=1}^{n} X_i^2$, for scalar X_i, corresponding to the second equation above. Then the resulting residuals do not sum to zero.

The profile empirical likelihood ratio $\mathcal{R}(\beta_1)$ is found by maximizing $\prod_{i=1}^{n} nw_i$ subject to $w_i \ge 0$, $\sum_{i=1}^{n} w_i = 1$, and the estimating equations (4.2) and (4.3) above, for a fixed value of β_1. The maximizing value is taken to be $\hat{\beta}_1$. Under mild conditions $-2\log(\mathcal{R}(\beta_{\mathrm{LS},1})/\mathcal{R}(\hat{\beta}_1)) \to \chi^2_{(1)}$ in distribution as $n \to \infty$, when $\beta_{\mathrm{LS},1}$ is the true value of β_1 and the regression is truly through the origin.

Figure 4.2 also shows the profile empirical log likelihood ratio curve for the slope using regression through the origin. Because there is a correlation between the estimated intercept and slope, fixing the intercept at its known value makes a difference. This constraint shifts the MLE of the cancer rate down slightly to 3.52. It substantially narrows the likelihood ratio curve. As Figure 4.2 shows the confidence interval for $\beta_{\mathrm{LS},1}$ using regression through the origin is less than half as wide as that without the constraint. Halving the length of a confidence interval usually requires a quadrupling of the sample size. Thus, one might interpret Figure 4.2 to mean that choosing regression through the origin more than quadruples the effective sample size n. We will revisit this issue, with another interpretation, in Chapter 4.5.

4.2 Nonrandom predictors

The usual model for linear regression has deterministic values $X_i = x_i$. Sometimes these values have been fixed by an experimental design. Sometimes they are fixed by conditioning. To fix X by conditioning, in a parametric model, proceed as follows. Factor the joint density or probability of X and Y as a product with one factor for the X distribution and another for the distribution of Y given X. Commonly the X distribution does not involve the regression parameters, and so all the information about them is in the conditional distribution of Y given X. Then it makes sense to use the conditional likelihood of $Y_1, \ldots, Y_n$ given $X_1 = x_1, \ldots, X_n = x_n$, as in Chapter 3.8.

With X fixed, either by design or conditioning, the data are analyzed conditionally on the observed values of X. Now we suppose that $E(Y_i) = \mu_i$ and that $\mathrm{Var}(Y_i) = \sigma_i^2$.

With fixed regressors, $Z_i(\beta) = x_i(Y_i - x_i'\beta)$ has mean $x_i(\mu_i - x_i'\beta)$ and variance $x_i x_i' \sigma_i^2$. A linear model assumption $\mu_i = x_i'\beta_{\mathrm{LS}}$ makes the $Z_i(\beta_{\mathrm{LS}})$ independent with mean zero and nonconstant variance. They would still have nonconstant variance even if σ_i^2 were constant, because of the factor $x_i x_i'$. Chapter 4.3 presents a triangular array ELT, for independent but not necessarily identically distributed random vectors, that applies to regression with fixed X_i when $\mu_i = x_i'\beta$ holds for some β. As with random sampling of (X, Y) pairs, the main requirements are on moments.

Lack of fit is more serious in the fixed regressor setting, because it makes it problematic to define a true value β_{LS} upon which to draw inferences. A natural default value to consider for β_{LS} is

$$\left(\sum_{i=1}^n x_i x_i'\right)^{-1}\left(\sum_{i=1}^n x_i \mu_i\right),$$

corresponding to random regressors sampled from the empirical distribution on X. Empirical likelihood inferences in this setting tend to be conservative. Asymptotically they cover the true value more often than the nominal level indicates.

Theorem 3.5 shows that imposing the constraint $\sum w_i U_i = E(U)$ for sample values U_i has asymptotically the same effect as conditioning on the observed value of $(1/n)\sum_{i=1}^n U_i$. So this suggests that we might be able to capture the effects of conditioning on x_i by employing the constraints

$$\sum_{i=1}^n w_i x_i = \frac{1}{n}\sum_{i=1}^n x_i$$

and

$$\sum_{i=1}^n w_i x_i x_i' = \frac{1}{n}\sum_{i=1}^n x_i x_i'$$

while computing $\mathcal{R}(\beta)$. Then our analysis will, in the sense of Theorem 3.5, be conditional on the first and second moments of x.

To match even more features of the observed x values, let $Q(x)$ be an s-dimensional function of x, and consider imposing the constraint $\sum_{i=1}^n w_i Q(x_i) = (1/n)\sum_{i=1}^n Q(x_i) = \overline{Q}$ analogous to conditioning on the observed value of $\overline{Q}$. These s constraints cannot widen the confidence region for β. They would narrow it asymptotically if there were any correlation between $Q(X_i)$ and $Z_i(\beta)$.

When β_{LS} is the true regression vector,

$$E\left(Z_i(\beta_{\text{LS}})Q(X_i)\right) = E\left(E\left(Z_i(\beta_{\text{LS}}) \mid X_i\right)Q\left(X_i\right)\right) = 0.$$

There appears to be no advantage to conditioning on any finite set of moments of X in the limit as $n \to \infty$. For a scalar function of β_{LS}, the ratio of constrained to unconstrained confidence interval lengths approaches 1.0 as $n \to \infty$. Perhaps there are benefits to imposing constraints on the reweighted x_i, but if so they tend to disappear as $n \to \infty$.

There are two common reasons for fixing regressors by conditioning. The first is that the conditional variance of the least squares estimator $\hat{\beta}_{\text{LS}}$ is simpler than the unconditional one, especially when the errors are assumed to have constant variance.

The second, and more subtle, reason is based on the statistical idea of ancillarity. Suppose that observing the X_i does not give us any information about the value of β_{LS} but does give us information about the variance of $\hat{\beta}_{\text{LS}}$. For instance, with constant error variance, $\text{Var}(\hat{\beta}_{\text{LS}}) = (\sum_{i=1}^n x_i x_i')^{-1}\sigma^2$. A statistic like $\sum_{i=1}^n x_i x_i'$, which tells us nothing about β_{LS} but something about $\text{Var}(\hat{\beta}_{\text{LS}})$, is called ancillary. The point of conditioning on an ancillary statistic is to use the known variance of our estimate rather than considering other x_i and corresponding other $\text{Var}(\hat{\beta}_{\text{LS}})$'s that we might have had instead. In particular, we should expect confidence regions for β_{LS} to be large when $\sum_{i=1}^{n} x_i x_i'$ is small and vice versa. This indeed holds for empirical likelihood, and most if not all other widely used confidence interval methods for regression.

4.3 Triangular array ELT

In applications such as linear regression with nonrandom predictors, and kernel smoothing (Chapter 5), empirical likelihood is applied to the mean of random variables that are not necessarily independent and identically distributed. Instead a triangular array structure $Z_{in} \in \mathbb{R}^p$, for $i = 1, \ldots, n$ is appropriate. The vectors can be arranged into a triangular array,

$$
\begin{array}{cccc}
Z_{11} & & & \\
Z_{12} & Z_{22} & & \\
Z_{13} & Z_{23} & Z_{33} & \\
\vdots & \vdots & \vdots & \ddots \\
Z_{1n} & Z_{2n} & Z_{3n} & \cdots & Z_{nn}.
\end{array}
$$

For each n, we will assume that $Z_{1n}, \ldots, Z_{nn}$ are independent, but not necessarily that they are identically distributed. For regression with fixed regressors, $Z_{in}(\beta) = x_i(Y_i - x_i'\beta)$. In regression $Z_{in} = Z_{im}$ for $i \leq n < m$, though this does not hold in other settings. We assume that $E(Z_{in})$ is the same for $i = 1, \ldots, n$. In regression, this common value is 0 for all n, but in other settings the common mean can depend on n. The variances of the Z_{in} have to be of roughly the same order of magnitude for a central limit theorem to hold.

Introduce $V_{in} = \text{Var}(Z_{in})$ and $V_n = (1/n) \sum_{i=1}^{n} V_{in}$, and for a real symmetric matrix A, let $\text{maxeig}(A)$ and $\text{mineig}(A)$ denote the largest and smallest eigenvalues of A, respectively.

Theorem 4.1 (Triangular array ELT) *Let $Z_{in} \in \mathbb{R}^p$ for $1 \leq i \leq n$ and $n \geq n_{\min}$ be a triangular array of random vectors. Suppose that for each n, that $Z_{1n}, \ldots, Z_{nn}$ are independent and have common mean μ_n. Let $\mathcal{H}_n$ denote the convex hull of $Z_{1n}, \ldots, Z_{nn}$, and put $\sigma_{1n} = \text{maxeig}(V_n)$, and $\sigma_{pn} = \text{mineig}(V_n)$. Assume that as $n \to \infty$*

$$\Pr(\mu_n \in \mathcal{H}_n) \to 1 \tag{4.4}$$

and

$$\frac{1}{n^2} \sum_{i=1}^{n} E\left(\|Z_{in} - \mu_n\|^4 \sigma_{1n}^{-2}\right) \to 0 \tag{4.5}$$

and that for some $c > 0$ and all $n \geq n_{\min}$,

$$\frac{\sigma_{pn}}{\sigma_{1n}} \geq c. \tag{4.6}$$

Then $-2\log \mathcal{R}(\mu_n) \to \chi^2_{(p)}$ in distribution as $n \to \infty$, where

$$\mathcal{R}(\mu_n) = \max\left\{\prod_{i=1}^n nw_i \mid \sum_{i=1}^n w_i(Z_{in} - \mu_n) = 0, w_i \geq 0, \sum_{i=1}^n w_i = 1\right\}.$$

Proof. This theorem is proved in Chapter 11.3. □

For regression problems with fixed predictors x_i and independent responses Y_i, the vectors $Z_{in}(\beta_0) = x_i(Y_i - x_i'\beta_0)$ are independent. They all have mean 0 if β_0 is the true common regression vector. That is, if $E(Y \mid X = x_i) = x_i'\beta_0$, for $i = 1, \ldots, n$.

The asymptotic results do not depend on which $n_{\min}$ is used. For regression problems $n_{\min} \geq p$, for otherwise $\sigma_{pn} = 0$. In applications there are usually many more data points than regression coefficients, so the value $n_{\min}$ has only a minor role.

The convex hull condition (4.4), which was easily satisfied for the mean of IID random vectors, must be investigated separately for each use of the triangular array ELT. If all the errors e_i are positive then it is certain that $0 \notin \mathcal{H}_n$. Similarly, for simple linear regression with $x_i = (1, t_i)'$, if the regression $\beta_0 + \beta_1 t$ is increasing, with $\beta_1 > 0$, but the sample data are decreasing, with $Y_i < Y_j$ whenever $t_i > t_j$, then $0 \notin \mathcal{H}_n$. We cannot reweight a decreasing sample to get an increasing regression line.

Under normal circumstances, we expect however that (4.4) will be satisfied rapidly unless p is large or n is small. Let $e_i = Y_i - x_i'\beta_0$ be the error for observation i, so that $Z_{in} = x_i e_i$. We require that some vector of weights $w_i \geq 0$ exists with $\sum_{i=1}^n w_i = 1$ and $\sum_{i=1}^n w_i x_i e_i = 0$. Here is a simple sufficient condition.

Lemma 4.1 *Let $\mathcal{H}_n^+$ be the convex hull of the set $\{x_i \mid e_i > 0, 1 \leq i \leq n\}$ and let $\mathcal{H}_n^-$ be the convex hull of the set $\{x_i \mid e_i < 0, 1 \leq i \leq n\}$. If $\mathcal{H}_n^+ \cap \mathcal{H}_n^- \neq \emptyset$, then $0 \in \mathcal{H}_n$.*

Proof. If $x \in \mathcal{H}_n^+ \cap \mathcal{H}_n^-$, then we may write $x = \sum_{i=1}^n w_i^+ x_i = \sum_{i=1}^n w_i^- x_i$ with all $w_i^\pm \geq 0$, $\sum_{i=1}^n w_i^\pm = 1$, $w_i^+ = 0$ if $e_i \leq 0$, and $w_i^- = 0$ if $e_i \geq 0$. Then $\sum_{i=1}^n w_i^\pm Z_{in} = x\tilde{e}_\pm$ where $\tilde{e}_\pm = \sum_{i=1}^n w_i^\pm e_i$. We have $\tilde{e}_- < 0 < \tilde{e}_+$, and by taking $w_i = (w_i^+|\tilde{e}_-| + w_i^-|\tilde{e}_+|)/(|\tilde{e}_+| + |\tilde{e}_-|)$, we find

$$\sum_{i=1}^n w_i Z_{in} = \frac{|\tilde{e}_-|x\tilde{e}_+ + |\tilde{e}_+|x\tilde{e}_-}{|\tilde{e}_+| + |\tilde{e}_-|} = 0,$$

and so $0 \in \mathcal{H}_n$. □

For simple linear regression with $Y_i = \beta_0 + \beta_1 t_i + e_i$ and $x_i = (1, t_i)'$, the convex hull condition simplifies further. Let I_n^+ be the interval from the smallest t_i with $e_i > 0$ to the largest t_i with $e_i > 0$, and let I_n^- be similarly defined using t_i with $e_i < 0$. If these intervals overlap then $0 \in \mathcal{H}_n$. We simply require a triple $t_i < t_j < t_k$ where the sign of e_j differs from those of e_i and e_k. If on the other hand there is some value t such that $e_i > 0$ whenever $t_i > t$ and $e_i < 0$ whenever $t_i < t$, then 0 might not be in $\mathcal{H}_n$.

The vectors Z_{in} have variance $V_{in} = x_i x_i' \sigma_i^2$, where $\sigma_i^2 = \mathrm{Var}(Y_i)$, so

$$V_n = \frac{1}{n} \sum_{i=1}^{n} x_i x_i' \sigma_i^2.$$

Under mild conditions on x_i and σ_i^2, both $\sigma_{1n} = \mathrm{maxeig}(V_n)$, and $\sigma_{pn} = \mathrm{mineig}(V_n)$ have finite nonzero limits, $\sigma_{1\infty}$ and $\sigma_{p\infty}$, respectively, as $n \to \infty$.

For regression, condition (4.5) becomes

$$\frac{1}{n^2} \sum_{i=1}^{n} \|x_i\|^4 E\left(e_i^4\right) \sigma_{1n}^{-2} \to 0.$$

If σ_{1n} tends to a finite nonzero limit, then a sufficient condition for (4.5) is that $\max_{1 \le i \le n} E(e_i^4)$ and $\max_{1 \le i \le n} \|x_i\|^4$ both have a finite upper bound holding for all n. Still weaker conditions are sufficient. These quantities are allowed to diverge slowly to infinity, and even the average of $\|x_i\|^4 E(e_i^4)$ can diverge to infinity as long as it remains $o(n)$.

For condition (4.6), the constant c can be taken to be slightly smaller than $\sigma_{p\infty}/\sigma_{1\infty}$ when these exist. To violate condition (4.6) would require either unbounded ratios of observation variances σ_i^2 or unbounded ratios of extreme eigenvalues of $n^{-1} \sum_{i=1}^{n} x_i x_i'$.

4.4 Analysis of variance

The one way analysis of variance (ANOVA) is widely used to compare the means of different populations. Suppose that we observe independent random variables $Y_{ij} \in \mathbb{R}$, for $i = 1, \ldots, k$ and $j = 1, \ldots, n_i$. The groups i are considered to have possibly different means, and usually an identical variance. Then the null hypothesis of identical means is rejected when the ratio

$$\frac{\frac{1}{k-1} \sum_{i=1}^{k} n_i (\bar{Y}_{i\bullet} - \bar{Y}_{\bullet\bullet})^2}{\frac{1}{N-k} \sum_{i=1}^{k} \sum_{j=1}^{n_i} (Y_{ij} - \bar{Y}_{i\bullet})^2}$$

exceeds the $1 - \alpha$ quantile $F_{k-1, N-k}^{1-\alpha}$ of the $F_{k-1, N-k}$ distribution. Here $N = \sum_{i=1}^{k} n_i$, and $\bar{Y}_{i\bullet} = (1/n_i) \sum_{j=1}^{n_i} Y_{ij}$, and $\bar{Y}_{\bullet\bullet} = (1/N) \sum_{i=1}^{k} \sum_{j=1}^{n_i} Y_{ij}$. The F distribution holds if the Y_{ij} are normally distributed, and it holds asymptotically for non-normal data. The assumption of a common variance is critical for asymptotic validity, unless the n_i are equal or nearly so.

For an empirical likelihood approach to ANOVA, we suppose that $Y_{ij} \in \mathbb{R}^d$ are independent and have the distribution F_{i0}, for $d \geq 1$. We do not need to distinguish ANOVA ($d = 1$) from multivariate ANOVA (MANOVA) with $d > 1$. A natural approach to empirical likelihood for this setting is to define the likelihood function

$$L_k(F_1, \ldots, F_k) = \prod_{i=1}^{k} \prod_{j=1}^{n_i} v_{ij},$$

where $v_{ij} = F_i(\{Y_{ij}\})$. The empirical likelihood ratio function is then

$$R_k(F_1, \ldots, F_k) = \prod_{i=1}^{k} \prod_{j=1}^{n_i} n_i v_{ij}.$$

An alternative formulation is to encode the data as N pairs (I, Y) where $I \in \{1, \ldots, k\}$ and $Y \in \mathbb{R}^d$. The observation Y_{ij} is represented by a pair with $I = i$ and $Y = Y_{ij}$. Let F be a distribution on (I, Y) pairs. The data are not an IID sample from any such distribution F_0, in the usual setting where n_i are nonrandom. Instead the variable I behaves more like a nonrandom categorical predictor. Define the likelihood

$$L(F) = \prod_{i=1}^{k} \prod_{j=1}^{n_i} w_{ij},$$

where $w_{ij} = F(\{(i, Y_{ij})\}) \geq 0$ and $\sum_{i=1}^{k} \sum_{j=1}^{n_i} w_{ij} = 1$.

The weights w_{ij} can be factored into $w_{j|i} w_{i\bullet}$ where $w_{i\bullet} = \sum_{j=1}^{n_i} w_{ij}$ and $w_{j|i} = w_{ij}/w_{i\bullet}$. The $w_{i\bullet}$ factor describes the probability attached by F to group $I = i$, while the factor $w_{j|i}$ describes the distribution of Y_{Ij} given that $I = i$. In ANOVA problems, we are usually interested in F only through $w_{j|i}$. This is necessarily true when the n_i have been fixed by the experimental design. The empirical likelihood ratio function on data pairs may be written

$$R(F) = \prod_{i=1}^{k} \prod_{j=1}^{n_i} N w_{i\bullet} w_{j|i}$$

$$= \left(\prod_{i=1}^{k} \left(\frac{N w_{i\bullet}}{n_i} \right)^{n_i} \right) \left(\prod_{i=1}^{k} \prod_{j=1}^{n_i} n_i w_{j|i} \right).$$

If we maximize $R(F)$ subject to constraints that only involve $w_{j|i}$, then the result will have $w_{i\bullet} = n_i/N$ and so

$$R(F) = R_k(F_1, \ldots, F_k)$$

where $F_i(\{Y_{ij}\}) = w_{j|i}$. Maximizing the likelihood ratio for such constraints automatically keeps the group weights $w_{i\bullet}$ proportional to the actual sample sizes n_i.

Empirical likelihood confidence regions and tests for the one way ANOVA will

ANALYSIS OF VARIANCE

be identical whether they are constructed through $R(F)$ or through the more natural $R_k(F_1, \ldots, F_k)$, as long as the statistic $T(F)$ depends only on the conditional distributions of Y_{Ij} given $I = i$ and not on the marginal distribution of the group variable I. The triangular array ELT is then available to justify the method based on sampling (I, Y) pairs.

Suppose that $\mu_{i0} = \int y \, dF_{i0}(y) \in \mathbb{R}^d$ and define

$$\mathcal{R}(\mu_1, \ldots, \mu_k) = \max\left\{\prod_{i=1}^{k}\prod_{j=1}^{n_i} N w_{ij} \mid w_{ij} \geq 0, \sum_{i=1}^{k}\sum_{j=1}^{n_i} w_{ij} = 1 \right.$$

$$\left. \sum_{j=1}^{n_i} w_{ij}(Y_{ij} - \mu_i) = 0, \ j = 1, \ldots, k, \right\}.$$

To apply the triangular array ELT, define the auxiliary variables $Z_{ijN} \in \mathbb{R}^D$, where $D = kd$. Taking Y_{ij} to be column vectors, we write

$$Z_{ijN} = (0, \ldots, 0, Y'_{ij} - \mu'_i, 0, \ldots, 0)',$$

where $Y'_{ij} - \mu'_i$ is preceded by $(i-1)d$ zeros and followed by $(k-i)d$ zeros. We could rewrite Z_{ijN} as Z_{lN} for $1 \leq l \leq N$ in order to make the notation more closely match that of the triangular array ELT, but this is not necessary. The key quantity in applying that theorem is the matrix V_N given by

$$V_N = \frac{1}{N}\begin{pmatrix} n_1 \text{Var}(Y_{11}) & 0 & \cdots & 0 \\ 0 & n_2 \text{Var}(Y_{21}) & \cdots & 0 \\ \vdots & \vdots & \ddots & \vdots \\ 0 & 0 & \cdots & n_k \text{Var}(Y_{k1}) \end{pmatrix}.$$

If each $\text{Var}(Y_{i1})$ is finite and nonsingular, then the condition on the eigenvalues of V_N is satisfied so long as the sample sizes grow subject to

$$\lim_{N\to\infty} \frac{\min_i n_i}{\max_i n_i} > 0. \tag{4.7}$$

The convex hull condition for Z_{ijN} becomes k convex hull conditions: for each $i = 1, \ldots, k$, the convex hull of Y_{ij} needs to contain μ_{i0}. Thus under very mild conditions,

$$-2\log \mathcal{R}(\mu_{10}, \ldots, \mu_{k0}) \to \chi^2_{(D)}$$

in distribution as $N \to \infty$.

Returning to ANOVA, suppose that $d = 1$. Then

$$-2\log \mathcal{R}(\mu_{10}, \ldots, \mu_{k0}) \to \chi^2_{(k)}$$

in distribution as $N \to \infty$. The most common hypothesis is that $\mu_{i0} = \mu_0$ all take the same (unknown) value. This hypothesis corresponds to $k - 1$ constraints on the mean of Z_{ijN} instead of k constraints and so

$$-2 \max_{\mu} \log \mathcal{R}(\mu, \ldots, \mu) \to \chi^2_{(k-1)}$$

in distribution as $N \to \infty$ when $\mu_{10} = \cdots = \mu_{k0}$. The observations do not need to be normally distributed, or to have a common variance, and the sample sizes need not be equal. Each group needs a finite nonzero variance.

For $d > 1$, empirical likelihood produces MANOVA tests for equality of μ_i requiring very weak assumptions and having an asymptotic $\chi^2_{(d(k-1))}$ calibration. The formulation above can be further generalized to a setting where the data have possibly different dimensions d_i in each population. An asymptotic χ^2_D result holds for $-2 \log \mathcal{R}(\mu_1, \ldots, \mu_k)$ under conditions including a bound on the ratio of eigenvalues of the matrix V_N. Now the matrix V_N is D by D where $D = \sum_{i=1}^{k} d_i$. When $d_1 \neq d_2$, it is not natural to compare μ_{10} and μ_{20}, though comparisons of functions of $\mu_1, \ldots, \mu_k$ may be of interest.

More general multi-sample statistics are considered in Chapter 11.4. For example, $\Pr(Y_{1j} > Y_{2j})$ is covered there, but not by the ANOVA formulation described above.

4.5 Variance modeling

Least squares inferences can be inefficient when the response Y_i has nonconstant variance σ_i^2, for fixed regressors, or when $\sigma^2(x) = \text{Var}(Y \mid X = x)$ is nonconstant in x, for random regressors. Greater accuracy can be obtained by weighting observation i in inverse proportion to the variance of Y_i. In some cases the inefficiency of unweighted least squares is mild and may be tolerated. In others, the inefficiency may be large enough that we seek to put more weight on the less variable observations. This can be done by introducing a model for the variance of Y. Sometimes we may introduce such a model because the variance is interesting in its own right.

Suppose that we observe (X, Z, Y) triples, where X and Z are thought to be related to the mean and variance of Y, respectively. It may be that Z is X, a subvector of X, or a transformation of X. Perhaps (X_i, Z_i, Y_i) are IID vectors, or alternatively x_i and z_i are fixed. Because the variance of Y cannot be negative it is natural to model the logarithm of the variance of Y using Z. The model $Y \mid (X, Z) \sim N(X'\beta, \exp(2Z'\gamma))$, leads to the estimating equations

$$0 = \frac{1}{n} \sum_{i=1}^{n} \exp(-2z_i'\gamma) x_i (Y_i - x_i'\beta) \tag{4.8}$$

$$0 = \frac{1}{n} \sum_{i=1}^{n} z_i \Big(1 - \exp(-2z_i'\gamma)(Y_i - x_i'\beta)^2\Big). \tag{4.9}$$

For the breast cancer data, take $x_i = (1, P_i)'$, and $z_i = (1, \log P_i)'$, where P_i is the population of the ith county, and take $Y_i = C_i$, the number of cancer deaths in that county. The model for the expected value of C_i is now $\beta_0 + \beta_1 P_i$, but we will run the regression through the origin, by constraining $\beta_0 = 0$. The model for the conditional standard deviation of C_i is a power model $\exp(\gamma_0 + \gamma_1 \log(P_i)) = \exp(\gamma_0) P_i^{\gamma_1}$. A normal distribution clearly cannot hold for discrete data, but the parameters β_1, γ_0, and γ_1 retain their interpretations as parameters of the mean and variance of C given P. Specifically, $E(C|P) = \beta_1 P$, and $\text{Var}(Y|P) = \exp(2\gamma_0) P^{2\gamma_1}$.

The simplest parametric model one might believe to hold for cancer incidence is Poisson with intensity proportional to the population size. This model would give a regression through the origin. It would also give $\gamma_1 = 1/2$ and $\gamma_0 = \log(\beta_1)/2$, because for a Poisson model the variance equals the mean. It is often found that a Poisson model fails to fit epidemiological data. A Poisson model may be derived by assuming that different people get cancer independently of each other, and with the same small probability. Commonly there is overdispersion, wherein the variance of observations with different means increases faster than does the mean. This can happen with clustering (as for families within counties) or because there is variation from county to county in factors such as industrial exposure, age, or smoking. A commonly used model for data overdispersed relative to the Poisson is the Gamma distribution. For Gamma distributions, $\gamma_1 = 1$.

The MLE's for the cancer data are $\hat{\beta}_1 = 3.57$, $\hat{\gamma}_0 = 0.602$ and $\hat{\gamma}_1 = 0.731$. The value for γ_1 describes overdispersion relative to a Poisson model, but less overdispersion than would hold in a Gamma model.

The left panel of Figure 4.3 shows the profile empirical likelihood ratio function for β_1. For comparison purposes, the curve for β_1 from a regression through the origin without variance modeling is also shown. The two curves have roughly the same width. The approximate four-fold efficiency gain from regression through the origin can also be obtained by variance modeling which puts more weight on observations from small counties. Also, using both techniques is not much more accurate than using just one of them.

The right panel of Figure 4.3 plots the profile empirical likelihood function for γ_1 using both regression through the origin and unconstrained regression. The constraint narrows the empirical likelihood function peak slightly. In particular, constraining the regression to go through the origin raises the lower confidence limit for γ_1 somewhat. By either curve, we can infer that both the overdispersion relative to the Poisson model and the underdispersion relative to the Gamma model are statistically significant. The value 0.75 is near the center of the confidence interval for γ_1, and this corresponds to weighting the observations proportionally to $P^{1.5}$ instead of P or P^2 as in Poisson and Gamma models, respectively.

4.6 Nonlinear least squares

In some applications there is a specific functional form $f(x, \theta)$ that is known or suspected to give $E(Y|X = x)$. Then a reasonable way to estimate θ is to

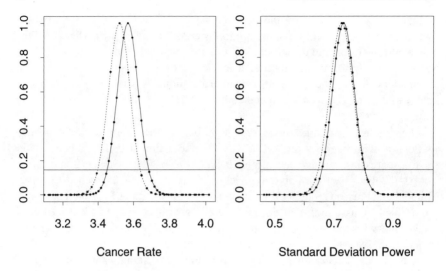

Figure 4.3 *The left plot shows the empirical likelihood function for the intercept in a linear regression through the origin, relating cancer mortality to county population for the data shown in Figure 4.1. The solid curve is from a heteroscedastic regression model, the dotted curve is from a regression model using constant variance. The right plot shows the empirical likelihood function for the exponent in a power law relating the standard deviation of the cancer mortality level to the population size. Values 1/2 and 1, corresponding to Poisson and Gamma models, respectively, do not fit the data. The solid curve is for regression through the origin, the dotted curve is for regression not contrained to pass through the origin.*

minimize

$$S(\theta) = \sum_{i=1}^{n} (Y_i - f(x_i, \theta))^2 \quad (4.10)$$

with respect to θ. Minimizing equation (4.10) gives the maximum likelihood estimate of θ under a model with independent $Y_i \sim N(f(x_i, \theta), \sigma^2)$. We suppose that $f(x, \theta)$ is not linear in θ, for otherwise linear regression methods could be used.

The least squares estimate is still valuable, even if Y_i are not normally distributed or do not have equal variance. For random X_i, the population quantity estimated is that value of θ minimizing $E((Y - f(X, \theta))^2)$. In particular, if $E(Y|X = x)$ were really of the form $f(x, \theta_0)$, then θ_0 minimizes $E((Y - f(X, \theta))^2)$.

Finding the nonlinear least squares estimate $\hat{\theta}$ can be a challenge. Success can depend on using good starting values, and it may be necessary to rescale the data or parameter values to avoid loss of numerical accuracy. The same estimate $\hat{\theta}$ is used for empirical likelihood and parametric inferences. The methods differ when it comes to constructing confidence sets.

Normal theory confidence regions for θ may be obtained either by linearizing the model or by profiling the log likelihood function. Linearization inferences begin with the approximation

$$f(x_i, \theta) \doteq f(x_i, \theta_0) + (\theta - \theta_0)' g(x_i, \theta_0),$$

where

$$g(x_i, \theta) = \frac{\partial}{\partial \theta} f(x_i, \theta).$$

The vector $\theta - \theta_0$ may be approximated by a linear regression with an n by p predictor matrix J and a response vector Z given by

$$J = J(\theta_0) = \begin{pmatrix} g(x_1, \theta_0)' \\ g(x_2, \theta_0)' \\ \vdots \\ g(x_n, \theta_0)' \end{pmatrix} \quad \text{and} \quad Z = Z(\theta_0) = \begin{pmatrix} Y_1 - f(x_1, \theta_0) \\ Y_2 - f(x_2, \theta_0) \\ \vdots \\ Y_n - f(x_n, \theta_0) \end{pmatrix}.$$

The value θ_0 is unknown. The Gauss-Newton iteration estimates θ_0 by iterated least squares, replacing $\hat{\theta}$ by $\hat{\theta} + (\hat{J}'\hat{J})^{-1}\hat{J}'\hat{Z}$, where $\hat{J} = J(\hat{\theta})$ and $\hat{Z} = Z(\hat{\theta})$.

Under standard assumptions, the asymptotic distribution of $n^{1/2}(\hat{\theta} - \theta_0)$ is $N(0, \sigma^2(J'J)^{-1})$. This matches our expectations under a linear regression model on J, and justifies ellipsoidal confidence regions for θ of the form

$$\left\{ \theta \mid (\theta - \hat{\theta})' \hat{J}' \hat{J} (\theta - \hat{\theta}) \leq s^2 p F_{p,n-p}^{1-\alpha} \right\} \tag{4.11}$$

where $s^2 = S(\hat{\theta})/(n-p)$.

Confidence regions for θ_0 formed by thresholding the normal theory likelihood reduce to thresholding the function $S(\theta)$. The asymptotic distribution of $(S(\theta_0) - S(\hat{\theta}))/ps^2$ is $F_{p,n-p}$ justifying confidence regions of the form

$$\left\{ \theta \mid S(\theta) \leq S(\hat{\theta}) \left[1 + \frac{p}{n-p} F_{p,n-p}^{1-\alpha} \right] \right\} \tag{4.12}$$

It can be very difficult to get accurate inferences in nonlinear least squares problems. The source of this difficulty is curvature of the vector

$$f_n(\theta) = (f(x_1, \theta), \ldots, f(x_n, \theta))'$$

expressed as a function of θ. This mapping from $\mathbb{R}^p$ to $\mathbb{R}^n$ has two kinds of curvature: intrinsic curvature and parameter effects curvature. Intrinsic curvature arises because the p-dimensional surface $\{f_n(\theta) \mid \theta \in \mathbb{R}^p\}$ in n-dimensional space is not flat but curved. Parameter effects curvature arises because of the labeling through θ of the points in this surface. In a linear mapping, changing θ to $\theta + \Delta$ produces the same change in f_n for any value of θ. In a nonlinear mapping

$f_n(\theta+\Delta) - f_n(\theta)$ can depend strongly on θ for fixed Δ. Linearization inferences implicitly approximate $f_n(\theta+\Delta) - f_n(\theta)$ at every θ by $f_n(\hat\theta+\Delta) - f_n(\hat\theta)$, while methods based directly on the sum of squares do not make this approximation.

If we reparameterize, replacing θ by $\tau(\theta)$ the intrinsic curvature of the surface is unchanged, but the parameter effects curvature will usually have changed. It is common for nonlinear models to employ the exponential function, or even to have the exponential in an exponent. These models can have very high parameter effects curvature.

It has been found empirically that confidence regions for θ and functions of θ found by thresholding (profiling) the sum of squares are usually well calibrated but that confidence regions based on linearization can be very badly calibrated. The usual explanation is that intrinsic curvatures are typically small compared to parameter effects curvatures.

Another explanation for the success of methods based on the sum of squares runs as follows. For normally distributed Y_i with known and constant σ^2, the set $\{\theta \mid S(\theta) \leq \sigma^2 \chi_{(n)}^{2,1-\alpha}\}$ has exactly $1 - \alpha$ coverage probability, regardless of the form of f. Coverage error can enter when σ^2 is estimated. In a normal linear model, the average squared residual is $S(\hat\theta)/n = (n-p)s^2/n \sim \sigma^2 \chi^2_{(n-p)}/n$. This average squared residual tends to be smaller than σ^2, but in a way that is well understood and easily corrected. For a pathological nonlinear model, in which the p-dimensional hyper-surface $\{f_n(\theta) \mid \theta \in \mathbb{R}^p\} \subset \mathbb{R}^n$ nearly fills the n-dimensional space $S(\hat\theta)$ can be far smaller than σ^2, and no practical correction is available. But for reasonable models, the correction of $S(\hat\theta)$ implicit in (4.12) does not go far wrong.

Empirical likelihood inferences for nonlinear least squares are based on

$$\theta(w_1, \ldots, w_n) = \arg\min_\theta \sum_{i=1}^n w_i(Y_i - f(x_i, \theta))^2.$$

If the sum of squares takes a unique minimum at a point where its gradient with respect to θ vanishes, then the estimating equations

$$\sum_{i=1}^n w_i(Y_i - f(x_i,\theta))g(x_i,\theta) = 0$$

serve to define θ for given weights w_i. Because empirical likelihood is parameterization invariant, it is only affected by parameter effects curvature, in the same way that a parametric likelihood is. Empirical likelihood inferences have an advantage in not requiring constant error variance.

Figure 4.4 shows data measuring calcium uptake Y versus time X. A reasonable model for these data is

$$E(Y_i|X_i = x_i) = \theta_1\left(1 - e^{-\theta_2 x_i}\right). \tag{4.13}$$

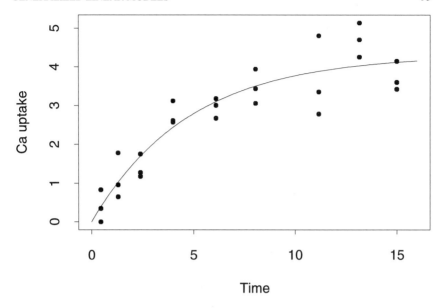

Figure 4.4 *Shown are the calcium uptake data (Rawlings 1988) and a model $E(Y|X = x) = \theta_1(1 - \exp(-\theta_2 x))$ fit to the data by nonlinear least squares.*

The least squares estimates are $\hat{\theta}_1 = 4.309$ and $\hat{\theta}_2 = 0.208$. The estimated curve $\hat{\theta}_1(1 - \exp(-\hat{\theta}_2 x))$ is shown as well.

Figure 4.5 shows the empirical likelihood ratio confidence regions for θ. These contours do not look very elliptical. Some of them are not even convex.

4.7 Generalized linear models

In a generalized linear model (GLM) we begin with a parametric model for the data $Y \sim f(y; \theta)$, where f may be either a probability density function, or a probability mass function, with a single real-valued parameter θ. This parameter is then written as $\theta = \tau(X'\beta)$ for predictors X, a coefficient vector β, and a known function τ. The same response surface models $X'\beta$ that are used in linear regression models may be used in GLM's. The function τ serves, at the least, to squash the real line into the natural domain for θ. GLM's are usually described in terms of the link function τ^{-1} for which $X'\beta = \tau^{-1}(\theta)$.

Apart from normal theory regression, the most widely used GLM is logistic regression. Here $Y_i \in \{0, 1\}$ are independent, with $\Pr(Y_i = 1 \mid X_i = x_i) =$

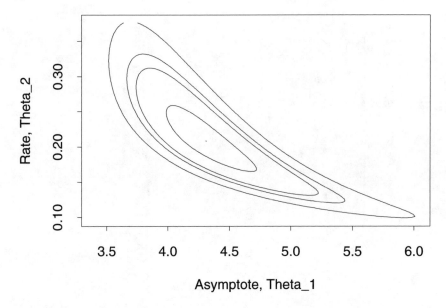

Figure 4.5 *Shown are empirical likelihood contours for the parameters θ_1, and θ_2 in the calcium uptake model. The confidence levels are 50%, 90%, 95%, and 99%.*

$\tau(x_i'\beta)$, for

$$\tau(z) = \frac{\exp(z)}{1+\exp(z)} = \left(1+\exp(-z)\right)^{-1}.$$

The parametric likelihood for β is

$$\prod_{i=1}^{n} [\tau(x_i'\beta)]^{Y_i} [1-\tau(x_i'\beta)]^{1-Y_i},$$

with log likelihood

$$\sum_{i=1}^{n} Y_i \log \tau(x_i'\beta) + (1-Y_i)\log(1-\tau(x_i'\beta)),$$

and estimating equations

$$\sum_{i=1}^{n} \left(\frac{Y_i}{\tau(x_i'\beta)} - \frac{1-Y_i}{1-\tau(x_i'\beta)}\right) \tau'(x_i'\beta)x_i = 0,$$

where $\tau'(z)$ denotes

$$\frac{d}{dz}\tau(z) = \frac{\exp(-z)}{[1+\exp(-z)]^2} = \tau(z)(1-\tau(z)).$$

GENERALIZED LINEAR MODELS

After some algebra, the estimating equations simplify to

$$\sum_{i=1}^{n} x_i(Y_i - \tau(x_i'\beta)) = 0. \tag{4.14}$$

Empirical likelihood inferences for logistic regression are based on

$$\mathcal{R}(\beta) = \max\left\{\prod_{i=1}^{n} nw_i \mid \sum_{i=1}^{n} w_i Z_i(\beta) = 0, w_i \geq 0, \sum_{i=1}^{n} w_i = 1\right\},$$

where $Z_i(\beta) = x_i(Y_i - \tau(x_i'\beta))$.

For a generalized linear model with $Y_i \sim f(y_i; \theta_i)$ and $\theta = \tau(x_i'\beta)$, the estimating equations are

$$\sum_{i=1}^{n} \frac{g(y_i, \tau(x_i'\beta))}{f(y_i; \tau(x_i'\beta))} \tau'(x_i'\beta) x_i = 0, \tag{4.15}$$

where as usual $g(y, \theta) = \partial f(y; \theta)/\partial \theta$. In practice it can pay to simplify (4.15) for the actual functions f, g, and τ of the model. For example, (4.14) provides insight into logistic regression that is not directly evident in (4.15).

A common source of difficulty with generalized linear models is overdispersion. The generalized linear model usually implies that the conditional variance of the response, given some predictors, is a known function of the conditional mean, given those same predictors. Overdispersed data has a conditional variance larger than what the model predicts for it. Overdispersion can invalidate statistical inferences based on parametric likelihood ratios. An empirical likelihood analysis treats the generalized linear model as a "working likelihood", using the same maximum likelihood estimate, but substituting a more generally applicable likelihood ratio for the parametric one.

Giant cell (temporal) arteritis (GCA) is a form of vasculitis – inflammation of blood or lymph vessels. A set of data on vasculitis cases was collected in order to investigate statistical methods of separating GCA from other forms of vasculitis. There were 585 cases, with the 8 binary features recorded in Table 4.1. The results of a logistic regression are shown in Table 4.2 and in Figure 4.6.

The coefficients β_j are all strongly significant except the one for scalp tenderness. For a patient with symptoms $z_j \in \{0, 1\}$, interest centers on the function $\theta = \beta_0 + \sum_{j=1}^{8} z_j \beta_j$ of the β_j. The primary interest may be in $(1 + \exp(-\theta))^{-1}$, the probability under the logistic regression model that this patient has GCA. Figure 4.7 shows the empirical likelihood ratio for this probability for nine hypothetical patients, the k'th one of which has the first $k-1$ symptoms and no others. For a patient with the first three or fewer of the symptoms, the likelihood concentrates around low probabilities of GCA. If six or more symptoms are present, then we may be similarly confident that the probability of GCA is very high. A GCA probability near $1/2$ would describe great uncertainty. For a patient with the first four or five symptoms and no others, not only is the outcome uncertain, even the amount of uncertainty is not well determined from the data.

Variable	Equals 1 if and only if
Headache	New onset of localized headache
Temporal artery	Tenderness or decreased pulsation
Polymyal rheumatism	Aching and morning stiffness
Artery biopsy	Histological changes on biopsy, showing destructive inflammatory process
ESR	Erythrocyte sedimentation rate $\geq$ 50 mm/hour
Claudication	Fatigue and discomfort while eating
Age	Disease onset after age 50
Scalp tenderness	Tender areas or nodules over scalp, away from arteries

Table 4.1 *Binary predictors of GCA. See the source (Bloch et al. 1990) for full definitions.*

Another quantity of interest here is the predictive accuracy of the logistic regression. Suppose that a threshold c is used so that if $Z\beta > c$ then Y is predicted to be 1, otherwise Y is predicted to be 0. The value of c might be 0, or it might be adjusted to take account of the prior odds that $Y = 1$, or the ratio $L(1,0)/L(0,1)$ where $L(j,k)$ is the loss from predicting $Y = j$ when $Y = k$.

Variable	Coefficient	Log likelihood
Intercept	-8.83	-272.33
Headache	1.54	-7.55
Temporal artery	2.45	-11.21
Polymyal rheumatism	1.07	-3.59
Artery biopsy	3.56	-43.96
ESR	1.69	-6.26
Claudication	2.06	-4.26
Age	3.50	-19.72
Scalp tenderness	-0.21	-0.59

Table 4.2 *Estimated logistic regression coefficients $\hat{\beta}_j$ for GCA predictors, with empirical log likelihood for $H_0 : \beta_j = 0$. Scalp tenderness is not a statistically significant predictor. The others are all strongly significant.*

GENERALIZED LINEAR MODELS

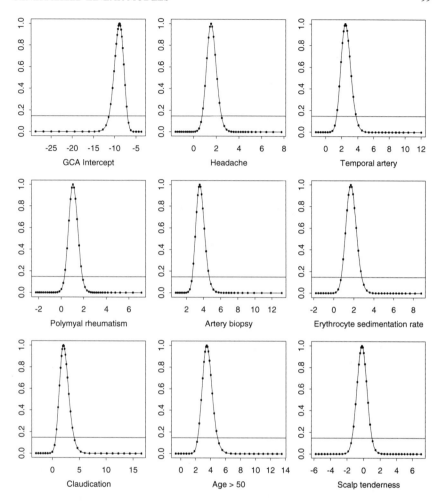

Figure 4.6 *Shown are empirical likelihood functions for the 9 parameters in the logistic regression of the GCA data. The plotting limits for β_j correspond to $\log \mathcal{R}(\beta_j)$ approximately equal to -25.0. The horizontal lines denote approximate 95% confidence levels, using a $\chi^2_{(1)}$ calibration of $-2 \log \mathcal{R}(\beta_j)$.*

Define θ_0 and θ_1 through estimating equations

$$0 = E(Y \times (1_{Z\beta \leq c} - \theta_1)), \quad \text{and}$$
$$0 = E((1-Y) \times (1_{Z\beta \geq c} - \theta_0))$$

Then θ_j is the probability of making a mistaken prediction, when $Y = j$. The

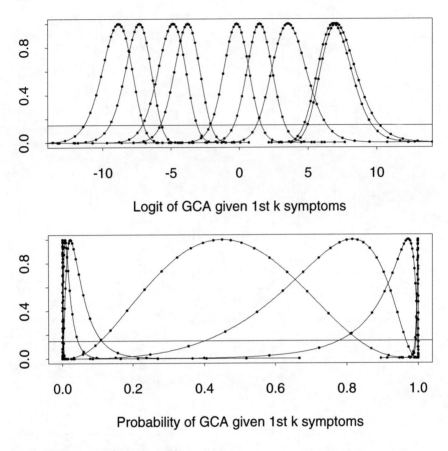

Figure 4.7 *Shown are empirical likelihood functions for nine hypothetical patients. Patient k presents the first k − 1 symptoms and only those, for k = 1, ..., 9. The top figure gives the empirical likelihood ratio function for $\sum_{j=0}^{k-1} \beta_j$. As the number of symptoms increases from zero to nine each curve is to the right of the previous one, except that the curve for all eight symptoms is just to the left of the one for the first seven symptoms. The lower figure plots the same likelihoods versus $(1 + \exp(-\sum_{j=0}^{k} \beta_j))^{-1}$, giving a likelihood for the estimated probability of GCA.*

quantity $1 - \theta_1$ is the sensitivity of a logistic regression classifier for GCA. We will work with $c = 0$ for simplicity.

Sample versions of θ_0 and θ_1 are subject to a bias, because the sample version of β has been fit to the data. In an example like this with a large number of observations and relatively few parameters, such bias is likely to be small. Thus we might consider estimating θ_j and forming confidence intervals for it by empiri-

POISSON REGRESSION

cal likelihood. A difficulty arises because the Heaviside function $H(u) = 1_{u \geq 0}$ is discontinuous, making both optimization and theory much harder. We replace H by the smooth function $G(u, \epsilon) = \Pr(t_{(4)} \leq u/\epsilon)$ where $t_{(4)}$ is a Student's t random variable on four degrees of freedom and $\epsilon > 0$. Taking $\epsilon = 0.05$, the estimating functions are:

$$0 = E\Big(Y \times \big(G(c - Z\beta, 0.05) - \theta_1\big)\Big)$$

$$0 = E\Big((1 - Y) \times \big(G(Z\beta - c, 0.05) - \theta_0\big)\Big).$$

The function $G(\cdot, 0.05)$, shown in Figure 4.8, is continuous and differentiable and is within 0.01 of H for values of u with $\exp(u)/(1 + \exp(u))$ outside the interval $[0.45, 0.55]$. This substitution makes θ_j more tractable at the cost of blurring the error count for near misses.

Figure 4.9 plots the empirical likelihood for the smoothed conditional error probabilities, when using a threshold of $c = 0$. The flatness at the top of these curves is not very common in profile log likelihoods. On inspection of the data, there are a number of observations with $Y = 0$ and $Z\hat{\beta}$ just barely less than 0. Small movements in the logistic regression parameters can produce modestly large positive $Z\hat{\beta}$ values for these observations.

4.8 Poisson regression

Poisson regression is a generalized linear model in which $Y_i \sim \text{Poi}(\tau(x_i'\beta))$. Here x_i is a vector of predictors, usually including a component always equal to 1. The most widely used model has $\tau(z) = \exp(z)$.

The number of home runs Y hit by a baseball player in one year may have approximately a Poisson distribution. It is reasonable to expect the number of home runs to depend on the number of times m the player came to bat, as well as the number of years t that the player has been playing. A natural model is that in year i, $Y_i \sim \text{Poi}(\lambda_i)$ where

$$\lambda_i = \exp(\beta_0 + \beta_1 t_i + \beta_2 \log(m_i))$$
$$= m_i^{\beta_2} \exp(\beta_0 + \beta_1 t_i).$$

It is natural to set $\beta_2 \equiv 1$ in order to study the number of home runs per at bat. Such a constraint is called an offset in generalized linear modeling. Thus we consider the model

$$Y_i \sim \text{Poi}(m_i \exp(\beta_0 + \beta_1 t_i)).$$

Baseball fans tend to study at bats per home run m_i/Y_i. Figure 4.10 shows this quantity plotted against the year for two singular home run hitters of the 20th century: Babe Ruth and Hank Aaron. It is well beyond the scope of this text to attempt to compare two players from different eras. However, the coefficient β_1 can be interpreted as a comparison for an individual player. A positive value

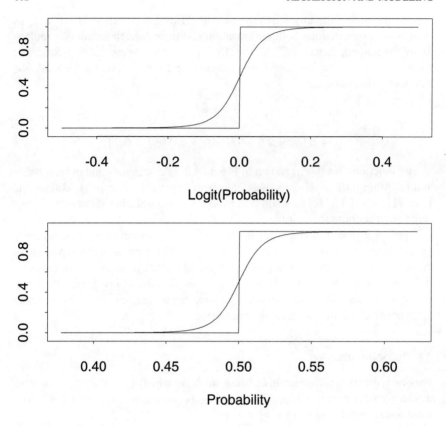

Figure 4.8 *The top plot shows $G(u, 0.05)$ and $H(u)$ for the logit, $u = log(\pi/(1-\pi))$. The bottom plot shows $G(u, 0.05)$ and $H(u)$ versus the probabiliity $\pi = exp(u)/(1+exp(u))$.*

describes a player whose home run production is increasing over time. A negative value has the opposite interpretation.

It appears from the plot that Ruth's home run production was fluctuating around a decreasing trend line. Aaron's was mostly increasing, except for his final two (or possibly three) seasons. A trend in home run production could be due to changes in a player's skill, or to many other factors, such as an opposite trend in the pitching talent, or changes in baseball manufacture, stadium size, and team strategy.

For Babe Ruth, $\hat{\beta}_1 = -.01841$, corresponding to a drop of about 1.84% per year in home run production per at bat. For Hank Aaron, $\hat{\beta}_1 = 0.01420$, corresponding to an increase of about 1.42% per year.

We can investigate these trends by constructing confidence intervals for β_1. A least squares regression confidence interval would ignore the fact that for Poisson data the variance is equal to the mean. A confidence interval based on the Poisson

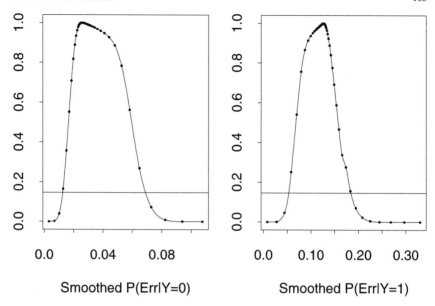

Figure 4.9 *The left curve shows the empirical likelihood function for the smoothed probability of error for a patient with GCA. The right curve shows the empirical likelihood function for the smoothed probability of error for a patient without GCA.*

likelihood could be in error because of overdispersion. Finally a confidence interval based on likelihood curvature at the MLE might be inaccurate because of the strong curvature in the exponential activation function.

Figure 4.11 shows profile empirical likelihood ratio curves for β_1, for these two players. The bootstrap threshold for $-2\log \mathcal{R}(\beta_1)$ is approximately 6.65 for Babe Ruth and 6.51 for Hank Aaron. These are indicated by horizontal calibration lines at $\exp(-6.65/2) = 0.0360$ and $\exp(-6.51/2) = 0.0385$, respectively. A short vertical line through $\beta_1 = 0$ shows that 0 is in the confidence interval for Aaron, and just barely inside the one for Ruth.

It is plausible that Ruth's home run rate only appears to decrease because of sampling fluctuations. Aaron's home run rate could more easily have been constant. Running the analysis on all of Aaron's seasons but the last two, one finds that 0 is clearly outside of the confidence interval. The interpretation of this is that Aaron might have been steadily increasing in home run output for the first 21 of his 23 seasons, but that the last two seasons do not fit the linear model.

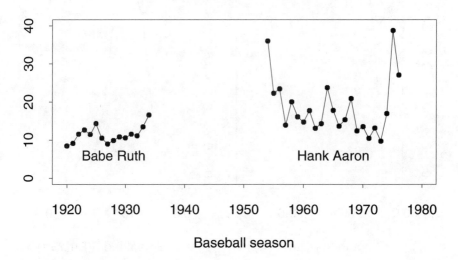

Figure 4.10 *Shown are the number of at bats per home run for Babe Ruth and Hank Aaron, in each year of their careers. The left curve is for Babe Ruth, the right is for Hank Aaron. The Aaron data can be found at* http://Baseball-Reference.com *and the Ruth data is from Stapleton (1995).*

4.9 Calibration, prediction, and tolerance regions

In linear regression, a common problem is to construct a confidence interval for $\beta_0 + \beta_1 x_0$, where x_0 is a specified value of the predictor variable. A related problem, called calibration or inverse regression, is to find a confidence region for the value x_0 with $\beta_0 + \beta_1 x_0 = y_0$, for a given response value y_0.

These problems are similar in that they may be handled by maximizing likelihood, parametric or empirical, subject to the additional constraint

$$\beta_0 + \beta_1 x_0 = y_0.$$

For regression, x_0 is fixed and the likelihood ratio is found for each y_0, after maximizing over β_0 and β_1. For calibration, x_0 varies for a given y_0. There is an important practical difference: when the slope β_1 is not well determined the confidence set for x_0 is not necessarily a finite interval. It can be the whole real line or even the set theoretic complement of a finite interval.

Prediction intervals extend easily to nonlinear and generalized linear models. Calibration intervals are more complicated. Specifying y_0 imposes only one constraint on x_0, so if the predictor is in a p-dimensional space, there will generally be a $p - 1$ dimensional space of x_0 values consistent with y_0 at the MLE of the

CALIBRATION, PREDICTION, AND TOLERANCE REGIONS

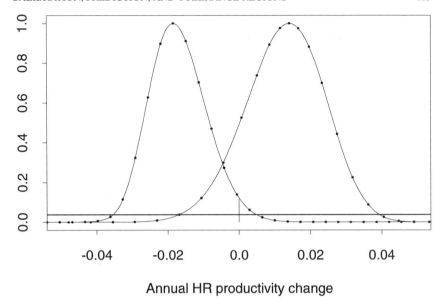

Annual HR productivity change

Figure 4.11 *Shown are empirical likelihood ratio curves for the time coefficient β_1 in the home run intensities of Babe Ruth and Hank Aaron. The coefficient β_1 measures a player's annual improvement in home run productivity, as described in the text. The curve for Ruth is to the left of that for Aaron. Horizontal reference lines and a short vertical reference line show that 0 is in the confidence interval for β_1 for both players, as described in the text.*

parameter. More generally, when the dimension of X is larger than that of Y, we can ordinarily expect that calibration estimating equations are underdetermined, while if Y has the larger dimension, we might anticipate the calibration problem to be overdetermined.

A related problem is that of tolerance intervals or regions. Given x_0, we seek a set that with confidence $1 - \alpha$ contains at least $1 - \gamma$ of the probability in the distribution of corresponding Y_0 values. Consider a linear regression problem with observations $(X_i, Y_i) \in \mathbb{R}^2$ independent and identically distributed for $i = 1, \ldots, n$. The estimating equations

$$0 = E(Y - \beta_0 - \beta_1 X) \tag{4.16}$$

$$0 = E(X(Y - \beta_0 - \beta_1 X)) \tag{4.17}$$

$$0 = E((Y - \beta_0 - \beta_1 X)^2 - \sigma^2) \tag{4.18}$$

$$0 = E(1_{Y < \beta_0 + \beta_1 X + \tau\sigma} - 0.95) \tag{4.19}$$

define the regression intercept, slope, and error standard deviation to be β_0, β_1, and σ, respectively. They also define τ such that 5% of the Y values lie τ or more standard deviations above the regression line. For a tolerance interval, $\beta_0 + \beta_1 x_0 + \tau\sigma$ is of interest, while β_0, β_1, τ, and σ themselves are nuisance parameters.

Equation (4.19) above is not differentiable with respect to the parameters, and so Theorem 3.6 does not apply. Theorem 3.4 does apply if we are satisfied with a joint confidence region $C^{1-\alpha}$ for $(\beta_0, \beta_1, \sigma, \tau)$. We could construct a region $C^{1-\alpha}$ using a $\chi^2_{(4)}$ calibration. The set $\{\beta_0 + \beta_1 x_0 + \tau\sigma \mid (\beta_0, \beta_1, \sigma, \tau) \in C^{1-\alpha}\}$ then has asymptotic coverage greater than or equal to the nominal level from the $\chi^2_{(4)}$ calibration. Intuitively we expect that the right coverage level should be obtained using one degree of freedom. Presently known results do not let us conclude this in general. See Chapter 10.6 for some related results.

4.10 Euclidean likelihood for regression and ANOVA

Applying the Euclidean log likelihood to the regression problem produces a test statistic

$$(\hat{\beta} - \beta_0)' \left[(\mathcal{X}'\mathcal{X})^{-1} \left(\sum_{i=1}^{n} (Y_i - X_i\hat{\beta})^2 X_i X_i' \right) (\mathcal{X}'\mathcal{X})^{-1} \right]^{-1} (\hat{\beta} - \beta_0)$$

where $\mathcal{X} = (X_1 \ X_2 \ \cdots \ X_n)'$.

This is equivalent to using White's heteroscedasticity robust estimate

$$(\mathcal{X}'\mathcal{X})^{-1} \left(\sum_{i=1}^{n} (Y_i - X_i'\hat{\beta})^2 X_i X_i' \right) (\mathcal{X}'\mathcal{X})^{-1} \quad (4.20)$$

of the covariance matrix of $\hat{\beta}$. The true covariance of $\hat{\beta}$ is

$$(\mathcal{X}'\mathcal{X})^{-1} \left(\sum_{i=1}^{n} \sigma_i^2 X_i X_i' \right) (\mathcal{X}'\mathcal{X})^{-1},$$

and so (4.20) can be thought of as using $(Y_i - X_i'\hat{\beta})^2$ as an estimate of σ_i^2. For each individual σ_i^2, this is a poor estimate, but the n estimation errors tend to cancel and the result gives properly calibrated inferences on β that do not require equal variance for the observations.

If there is a common value $\sigma_i^2 = \sigma^2$, the true covariance simplifies to the familiar form $\sigma^2(\mathcal{X}'\mathcal{X})^{-1}$, for which the usual estimate is of the form $\hat{\sigma}^2(\mathcal{X}'\mathcal{X})^{-1}$.

For the Euclidean log likelihood the two approaches to ANOVA in Chapter 4.4 do not in general provide the same answer. Taking the k distributions approach, with distance function $\sum_{i=1}^{k} \sum_{j=1}^{n_i} (n_i v_{ij} - 1)^2$ the resulting test statistic is

$$\sum_{i=1}^{k} n_i \frac{(\bar{Y}_{i\cdot} - \tilde{Y}_{\cdot\cdot})^2}{s_i^2},$$

where
$$s_i^2 = \frac{1}{n_i} \sum_{j=1}^{n_i} (Y_{ij} - \bar{Y}_{i\bullet})^2,$$
and
$$\tilde{Y}_{\bullet\bullet} = \frac{\sum_{i=1}^k n_i \bar{Y}_{i\bullet}/s_i^2}{\sum_{i=1}^k n_i/s_i^2}.$$

The test statistic for regression, and the test statistic for ANOVA (with minor modifications) have both been used in applications before, as described in the bibliographic notes in Chapter 4.11.

4.11 Bibliographic notes

Empirical likelihood for regression was considered by Owen (1991), who used the normal estimating equations to study first order coverage properties, including the case of nonrandom predictors. Song Chen investigated higher order terms, establishing Bartlett correctability and exhibiting the Bartlett factor, again for nonrandom predictors. Chen (1993) considers confidence regions for the regression coefficients and Chen (1994b) looks at confidence intervals for linear combinations of regression coefficients.

The formulation of empirical and Euclidean likelihoods for ANOVA is taken from Owen (1991). Jing (1995b) and Adimari (1995) independently considered the problem of comparing the means of two populations. Both find a chisquared limit for the difference in means, using a product of one sample empirical likelihoods. Adimari (1995) obtains a noncentral chisquared distribution under alternatives $O(n^{-1/2})$ from the true difference in means. Jing (1995b) shows Bartlett correctability.

The conservatism of empirical likelihood for fixed predictors and a misspecified regression model is explained in Owen (1991). Davidian & Carroll (1987) consider variance modeling in regression problems.

La Rocca (1998) studies the coverage error of empirical likelihood for linear regression models, including several error distributions, and both equal and unequal error variances. Bootstrap calibration of empirical likelihood proves to be most reliable.

The most basic form of model selection for regressions is conducted by fitting the model with and without one of the predictor variables and comparing the sum of squares. In empirical likelihood, the analogous procedure is to fit the model with and without constraining the corresponding parameter to be zero, and to compare the constrained and unconstrained empirical log likelihood ratios. Constraining a parameter to zero is different from dropping the corresponding predictor, because the constrained fit keeps the residuals uncorrelated with the missing predictor. Based on the results in Qin & Lawless (1994) we would expect this extra information to be helpful. Kolaczyk (1995) investigated empirical likelihood model selection, introducing an empirical information criterion (EIC) analogous to AIC (Akaike's information criterion) given by Akaike (1973).

Generalized linear models were introduced by Nelder & Wedderburn (1972). McCullagh & Nelder (1983) is the standard reference on GLM's. Empirical likelihood for GLM's was investigated by Kolaczyk (1994), who considered GLM's with a fixed amount of overdispersion. He also considered estimating an overdispersion constant and modeling the overdispersion via a link and a linear model. The usual inferences for GLM's do not require that the motivating parametric model hold. They can instead be defined through a quasi-likelihood. See Wedderburn (1974). This weakens the model to a requirement that the conditional variance of Y given $X = x$ be functionally related to the conditional mean, that is $\sigma^2(x) = h(\mu(x))$ for some known function x. There is no reason in general to expect data that depart from the parametric model to satisfy the quasi-likelihood condition, although the introduction of an overdispersion parameter or model can mitigate the problem. Nelder & Pregibon (1987) introduce extended quasi-likelihood to model overdispersion. See also Jorgensen (1987). Efron (1986) introduces double exponential families.

The GCA data is from Bloch et al. (1990). They investigate numerous prediction rules for GCA and find that logistic regression performs best. They do not use a zero threshold. Their logistic regression coefficients are not the same as the ones in Table 4.2, though they are qualitatively similar. There may have been slight differences in the data, or they might not have used maximum likelihood estimates. The coefficients in Table 4.2 agree with the ones computed by the glm function in S-PLUS.

The idea to study baseball home runs comes from Stapleton (1995), who provides the Babe Ruth data. Data for other players is easily available over the Internet at numerous sites. The site Baseball-Reference.com contains a lot of baseball data.

Background material on nonlinear least squares asymptotics and algorithms can be found in Bates & Watts (1988) and Seber & Wild (1989). The calcium uptake data is from Rawlings (1988). Davison & Hinkley (1997) present a bootstrap analysis of it.

The estimate (4.20) is due to White (1980).

For the analysis of variance (with $d = 1$), the Euclidean likelihood is essentially equivalent to the statistic used by James (1951) to test differences among group means when the variances are thought to be unequal. James (1951) used $n_i - 1$ instead of n_i in defining s_i^2. His critical value was not taken from a χ^2 distribution but instead took account of the differing values of s_i^2.

4.12 Exercises

Exercise 4.1 Consider regression with a predictor $X \in \mathbb{R}^d$ and a response $Y \in \mathbb{R}$. The data are IID (X, Y) pairs, including numerous ties among the X's. The data can then be labeled as $X_i \in \mathbb{R}^d$, $i = 1, \ldots, k$, with which we observe $Y_{ij} \in$

$\mathbb{R}$, $j = 1, \ldots, n_i$. Define $Z_{ij}(\beta) = X_i'(Y_{ij} - X_i'\beta)$. Let $\bar{Y}_i = (1/n_i)\sum_{j=1}^{n_i} Y_{ij}$, $\bar{Z}_i(\beta) = (1/n_i)\sum_{j=1}^{n_i} Z_{ij}(\beta)$, and

$$s_i^2 = \begin{cases} \frac{1}{n_i-1}\sum_{j=1}^{n_i}(Y_{ij} - \bar{Y}_i)^2, & n_i > 1 \\ 0, & n_i = 1. \end{cases}$$

Define pure error and lack of fit variances $\sigma_{PE}^2 = \iint (y - \mu(x))^2 dF_{X,Y}(x,y)$ and $\sigma_{LF}^2 = \int (\mu(x) - x'\beta)^2 dF_X(x)$, where $\mu(x) = \int y dF_{Y|X=x}(y)$. For $n_i > 1$, $E(s_i^2) = \sigma_{PE}^2$. Consider the estimating equations

$$0 = \sum_{i=1}^{k} w_i n_i \bar{Z}_i(\beta)$$

$$0 = \sum_{i=1}^{k} w_i \left[n_i \bar{Z}_i^2 - n_i \sigma_{LF}^2 - \sigma_{PE}^2 \right]$$

$$0 = \sum_{i=1}^{k} w_i (n_i - 1) \left[s_i^2 - \sigma_{PE}^2 \right].$$

Show that the quantities being averaged have expectation 0 if β is the least squares coefficient vector and σ_{PE}^2 and σ_{LF}^2 are the correct values. Do empirical likelihood confidence regions for these parameters have the correct calibration as $k \to \infty$? What conditions if any are required on n_i? Now suppose that k is fixed and that $n_i \to \infty$. Are empirical likelihood regions correctly calibrated? If not, suggest an alternative formulation of the problem.

Exercise 4.2 For the cancer data, construct the empirical likelihood ratio function for the slope β_1, using the estimating equation $E((C_i - \beta_1 P_i)P_i) = 0$. Compare it to the likelihood ratio curve for β_1 from ordinary linear regression and from regression through the origin with $N(0, \sigma^2)$ errors.

Exercise 4.3 Find a joint distribution for $X \in \mathbb{R}^d$ and $Y \in \mathbb{R}$, and a vector $\beta \in \mathbb{R}^d$ such that $E(\|X\|^2 (Y - X'\beta)^2) < \infty$, but at least one of $E(\|X\|^4) = \infty$ or $E(\|X\|^2 Y^2) = \infty$ holds.

Exercise 4.4 Consider a regression of Y_i on predictors $X_i = (1, U_i, V_i)'$, $i = 1, \ldots, n$. Suppose that $n = 100$ and that $U_{62}, U_{84}, V_{17}, V_{38}$, and V_{62} are missing, while all other observations are available. In addition to the regression parameter vector β introduce a parameter for each of the missing values. The estimating equations are still

$$\sum_{i=1}^{n} w_i X_i (Y_i - X_i' \beta) = 0,$$

except that there are now eight parameters $(\beta_1, \beta_2, \beta_3, U_{62}, U_{84}, V_{17}, V_{38}, V_{62})$ for these three estimating equations. Which, if any, of these parameters has a unique NPMLE? Assume that the matrix consisting of all completely observed X_i vectors has full rank.

Exercise 4.5 Consider the ratio

$$F = \frac{\frac{1}{k-1}\sum_{i=1}^{k} n_i (\bar{Y}_{i\cdot} - \bar{Y}_{\cdot\cdot})^2}{\frac{1}{N-k}\sum_{i=1}^{k}\sum_{j=1}^{n_i}(Y_{ij} - \bar{Y}_{i\cdot})^2}.$$

Suppose that $k = 2$, $n_1 = n$, and $n_2 = n^2$. Consider the limit as $n \to \infty$. Assume that independent $Y_{ij} \sim F_i$ have mean μ_i, variance σ_i^2, and $E(Y_{ij}^4) < \infty$. Show that F has a $\chi^2_{(1)}$ limit if $\mu_1 = \mu_2$ and $\sigma_1^2 = \sigma_2^2 > 0$. What happens if the means are equal but the variances are not?

Exercise 4.6 Verify that the heteroscedastic regression estimating equations (4.8) and (4.9) are the likelihood equations for the model $Y \sim N(X'\beta, \exp(2Z'\gamma))$.

Exercise 4.7 Consider these two ways of computing

$$\tau(z) = \frac{\exp(z)}{1 + \exp(z)}$$
$$= [1 + \exp(-z)]^{-1},$$

the squashing function for logistic regression. The IEEE floating point systems have numbers that represent $\pm\infty$. As one would expect, $1/\infty = 0$ and $1/0 = \infty$. There is even a reciprocal -0 for $-\infty$ with -0 equaling 0. These systems also have values NaN that designate "not a number". A floating point value of NaN arises from operations like dividing 0 by 0, subtracting ∞ from ∞, multiplying 0 by ∞ or dividing ∞ by ∞.

The first expression for τ above can produce NaN for finite z, the second one should not produce NaN for any $z \in [-\infty, \infty]$. Find two ways of expressing $\tau'(z) = d\tau(z)/dz$ for which one way produces NaNs and the other does not. Assume that $\exp(\infty) = \infty$ and $\exp(-\infty) = 0$. Find a way of computing $\tau''(z) = d^2\tau(z)/dz^2$ for which NaNs are not produced for any $z \in [-\infty, \infty]$.

Exercise 4.8 The specificity of a classifier for predicting that $Y = 1$ is the probability that $Y = 1$ given that the classifier predicts $Y = 1$. There is commonly a trade-off between sensitivity and specificity. For a logistic regression predicting $Y = 1$ when $X\beta > c$, the tradeoff is governed by the choice of c. Write the estimating equation for the specificity of a logistic regression rule. Is it smooth in the parameters?

CHAPTER 5

Empirical likelihood and smoothing

This chapter adapts empirical likelihood to curve estimation problems such as density estimation and nonparametric regression. Kernel methods are an attractive choice for this, because they lead easily to estimating equations. We also investigate some regression splines.

5.1 Kernel estimates

Figure 5.1 shows diastolic blood pressure, in millimeters of mercury, plotted against age in years, for some men in New Zealand. There are 7532 data points. The data come from two sources: the Auckland Heart & Health study (Jackson, Yee, Priest, Shaw & Beaglehole 1995) and another study called the Fletcher-Challenge study. The original data were integer valued; the plotted points have $U(-0.5, 0.5)$ random variables added to them to show them better. Blood pressures that are multiples of 10 are more common than others, due to rounding. The rounding stops at around age 60. This point is significant and we return to it later.

Superimposed on the data is a smooth curve taken as a local average of blood pressure. Letting X_i denote age and Y_i denote blood pressure, the curve is

$$\hat{\mu}(x) = \frac{\sum_{i=1}^{n} K_h(X_i - x) Y_i}{\sum_{i=1}^{n} K_h(X_i - x)}, \qquad (5.1)$$

the Nadaraya-Watson estimator, where

$$K_h(z) = \frac{1}{\sqrt{2\pi h}} \exp\left(-\frac{z^2}{2h^2}\right).$$

The bandwidth h used in Figure 5.1 is 5 years. This local average may be thought of as an estimate of $\mu(x) = E(Y \mid X = x)$. The denominator in (5.1) can be awkward, and so it is convenient to write $\hat{\mu}(x)$ in terms of the estimating equation

$$0 = \frac{1}{n} \sum_{i=1}^{n} K_h(X_i - x)(Y_i - \hat{\mu}(x)). \qquad (5.2)$$

More general kernel estimates may be formed through

$$K_h(z) = \frac{1}{h} K\left(\frac{z}{h}\right)$$

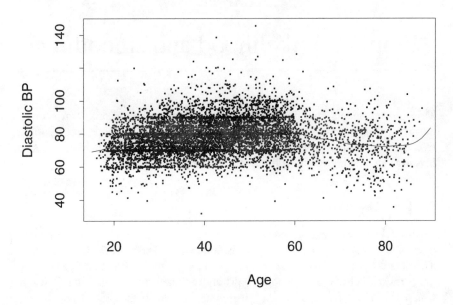

Figure 5.1 *Diastolic blood pressure is plotted versus age. There are 7532 points from men in New Zealand.*

where a kernel function K of integer order $r \geq 1$ satisfies

$$\int_{-\infty}^{\infty} K(z)dz = 1,$$
$$\int_{-\infty}^{\infty} z^j K(z)dz = 0, \quad 1 \leq j < r,$$
$$\int_{-\infty}^{\infty} z^r K(z)\, dz \neq 0.$$

A common choice is for K to be a symmetric probability density function, so that $r = 2$. Higher order kernels take some negative values, typically in "side lobes", and this can improve accuracy if μ is smooth enough. The benefits of higher order kernels can also be attained by replacing the local average estimator (5.1) by a local linear or local polynomial model. See Exercise 5.2.

Now suppose that $X_i \in \mathbb{R}^p$ and $Y_i \in \mathbb{R}^q$, and that independent identically distributed pairs (X_i, Y_i) are observed. An estimate $\hat{\mu}(x)$ of $\mu(x) = E(Y \mid X = x)$ may still be defined by (5.2), except that the kernel must obviously have a

p-dimensional argument, and the natural scaling is

$$K_h(z) = \frac{1}{h^p} K\left(\frac{z}{h}\right).$$

The order of K may be defined by analogy to the one-dimensional case. Due to a curse of dimensionality, the estimate $\hat{\mu}$ rapidly loses its effectiveness as p increases.

Kernel methods are also widely used to estimate probability density functions. Suppose $X_i \in \mathbb{R}^p$ are independent with a common density f. Then the kernel density estimate of f is

$$\hat{f}(x) = \frac{1}{n} \sum_{i=1}^{n} K_h(X_i - x).$$

As for kernel regression, kernel density estimation is really only practical for small p.

5.2 Bias and variance

This section illustrates the trade-off between bias and variance for kernel estimates, using a slightly simplified kernel regression estimator with $p = q = 1$.

For $p = 1$, kernel methods put relatively large weight only on the $O(nh)$ nearest observations to x_0. These neighbors are at distance $O(h)$ from x_0. If we choose a small h, then the observations are more nearly identically distributed, reducing bias, while for large h there are more of them, reducing variance. This sort of trade-off is not restricted to kernel methods, but is ubiquitous in curve estimation problems. As $n \to \infty$ we should have $h \to 0$ and $nh \to \infty$ to account for bias and variance, respectively.

Suppose for simplicity, that instead of using (5.2), we define

$$\hat{\mu}(x_0) = \frac{1}{n} \sum_{i=1}^{n} K_h(X_i - x) Y_i.$$

Then $E(\hat{\mu}(x_0)) = \tilde{\mu}(x_0)$, where

$$\tilde{\mu}(x_0) = E(K_h(X - x_0)Y) = \int_{-\infty}^{\infty} K_h(x - x_0) \mu(x) f_X(x) dx,$$

where f_X is the density of X. Assuming some smoothness of μ and small h we may write

$$\tilde{\mu}(x_0) \doteq \mu(x_0) + \frac{\mu^{(r)}(x_0)}{r!} \int_{-\infty}^{\infty} K_h(x - x_0)(x - x_0)^r f_X(x) dx$$

$$\doteq \mu(x_0) + \frac{h^r \mu^{(r)}(x_0)}{r!} f_X(x_0) \int K(z) z^r dz.$$

This bias of order h^r in kernel estimates raises difficulties for confidence interval construction below.

To study the variance of $\hat{\mu}(x_0)$, introduce $\sigma^2(x) = \text{Var}(Y \mid X = x)$. Now

$$\begin{aligned}
\text{Var}(\hat{\mu}(x_0)) &= \frac{1}{n} E\left((K_h(X - x_0)Y - \tilde{\mu}(x_0))^2\right) \\
&= \frac{1}{n} \int_{-\infty}^{\infty} \left[(K_h(x - x_0)\mu(x) - \tilde{\mu}(x_0))^2 \right. \\
&\quad \left. + K_h(x - x_0)^2 \sigma^2(x)\right] f_X(x) dx \\
&\doteq \frac{1}{nh} \int_{-\infty}^{\infty} K(z)^2 \left[\mu(x_0 + hz)^2 + \sigma^2(x_0 + hz)\right] f(x_0 + hz) dz \\
&\doteq \frac{1}{nh} \left(\mu(x_0)^2 + \sigma^2(x_0)\right) f(x_0) \int_{-\infty}^{\infty} K(z)^2 dz.
\end{aligned} \quad (5.3)$$

We see that the variance is of order $(nh)^{-1}$ instead of the rate n^{-1} familiar for finite dimensional parametric vectors. The appearance of $\mu^2(x_0)$ in (5.3) arises because the kernel weights are not made to sum to one. In practice, we fix this by constraining the reweighted mean of $K_h(X_i - x_0)$ to be 1, or by defining $\hat{\mu}$ through the estimating equation (5.2). Then, a more complicated derivation leads to the same asymptotic orders for bias and variance.

The above derivation shows that for $p = 1$, the bias is of order h^r and the variance is of order $(nh)^{-1}$, so the mean squared error is of order $h^{2r} + n^{-1}h^{-1}$. This order is minimized by taking $h \propto n^{-1/(2r+1)}$. The resulting mean squared error is of order $n^{-r/(2r+1)}$. Symmetric densities K have order $r = 2$. Then the optimal rate for h is $n^{-2/5}$, and therefore the mean squared error decreases as $n^{-4/5}$ compared to n^{-1} for vector parameters. The variance decreases as $n^{-4/5}$ so that the kernel method has an effective sample size of order $n^{4/5}$.

5.3 EL for kernel smooths

In practice, a kernel method requires a choice of h. There is a large literature on choosing h from the data. Some references are given in Chapter 5.8. We will study empirical likelihood for fixed sequences $h = h(n)$ in order to gain insight into its behavior. In practice, when h is determined from the data, bootstrap calibration of the profile empirical log likelihood should be used, with h being determined in each bootstrap replication. We begin by considering pointwise inferences at a single value x. Confidence bands with uniform coverage over a set of x values are considered in Chapter 5.6.

The profile empirical likelihood ratio function for kernel regressions based on (5.2) is

$$\mathcal{R}_x(\mu) = \max\left\{\prod_{i=1}^n nw_i \mid \sum_{i=1}^n w_i Z_{in}(x, \mu) = 0, w_i \geq 0, \sum_{i=1}^n w_i = 1\right\}, \quad (5.4)$$

where
$$Z_{in}(x, \mu) = K_h(X_i - x)(Y_i - \mu)$$
and $h = h(n)$. There are two difficulties in using this statistic for inferences. The first difficulty, which we return to below, is that due to bias in kernel estimation, the expected value of Z_{in} is zero not at $\mu = E(Y \mid X = x)$ but at some other value $\tilde{\mu}(x)$.

The second difficulty is in applying an ELT to Z_{in}. If we consider $X_i = x_i$ to be fixed predictors, then the Z_{in} for $i = 1, \ldots, n$ and a given value of n, are not identically distributed. For IID (X_i, Y_i) pairs, the distribution of Z_{in} changes with n and has a variance diverging to infinity. The difficulty in applying empirical likelihood may be resolved through the triangular array ELT, Theorem 4.1 of Chapter 4.3.

Theorem 4.1 requires a common mean for the Z_{in}, a convex hull condition, and two conditions on the variances of Z_{in}. Suppose that $(X_i, Y_i) \in \mathbb{R}^2$ are an IID sample. Then $Z_{1n}, \ldots, Z_{nn}$ have a common mean 0, but at a point $\tilde{\mu}(x)$ differing from μ by a bias of order h^r. The convex hull condition is satisfied as soon as at least two suitable data points appear: one has $Y_i > \tilde{\mu}(x)$ and $K_h(X_i - x) > 0$, while the other has $Y_i < \tilde{\mu}(x)$ and again $K_h(X_i - x) > 0$. For compactly supported kernels there are at least $O(nh)$ points with $K_h(X_i - x) > 0$ and for other kernels there may be n such points. The convex hull condition is very quickly satisfied in this case.

Easy calculations show that $V_n = \text{Var}(Z_{in}) = O(h^{-1})$ as $n \to \infty$ and $h = h(n) \to 0$. In this case V_n is its own smallest and largest eigenvalue. The ratio of largest to smallest eigenvalues is thus constant with n and so does not raise difficulties. Next, we require a limit of zero for

$$\frac{E((Z_{in} - E(Z_{in}))^4)}{nV_n^2}. \tag{5.5}$$

Mild moment assumptions give a rate of $O(h^{-3})$ for the numerator in (5.5) and then the ratio itself is $O((nh)^{-1})$. We already needed $nh \to \infty$ to control the variance of the estimate. Thus the triangular array ELT applies to kernel smoothing under weak conditions.

The fact that $E(\sum_{i=1}^n Z_{in}) = 0$ not at $\mu(x)$, but at $\tilde{\mu}(x)$, is more problematic. There are several approaches to dealing with this problem, none of them completely satisfactory.

The first approach is based on undersmoothing. The value of h is taken small enough that the bias in $\hat{\mu}$ is negligible compared to its standard deviation. Then the error $\mu - \hat{\mu}$ is primarily due to sampling fluctuations and not bias, making empirical likelihood inferences on $\tilde{\mu}$ relevant to μ. The disadvantages of this approach are that the undersmoothed estimate $\hat{\mu}$ is less accurate than it would be with the usual choice of h, and that the curve $\hat{\mu}$ that results is more wiggly than the usual one. In applications where we are most concerned about $\mu(x_0)$ for one or a small number of specific x_0 values, this roughness is less important. Undersmoothing is well suited to problems where getting a reliable confidence region for $\mu(x)$ is more important than getting the best possible point estimate of $\mu(x)$.

The second approach is to accept the empirical likelihood inferences as confidence statements about $\tilde{\mu}$, a smoothed version of μ. The disadvantage here is that $\tilde{\mu}$ is not very interpretable and the error $\tilde{\mu} - \mu$ may be as large as, or larger than, the diameter of the confidence region. As n and h change, $\tilde{\mu}$ changes too, and so it is not even a feature of the joint (X, Y) distribution. There are some settings in which this approach is adequate. Sometimes the curve $\hat{\mu}$ is used for qualitative interpretation, and not strictly as an estimate of μ. In settings like this, one might prefer an oversmoothed estimate $\hat{\mu}$, using a larger value of h than the usual one. Empirical likelihood confidence regions for $\tilde{\mu}$ can be used to assess how much sampling fluctuation might contribute to features in $\hat{\mu}$.

A third approach is to compute an estimate of the bias $\tilde{\mu}(x) - \mu(x)$, and subtract this from the estimate $\hat{\mu}(x)$ and from the boundary of the confidence set. The bias can be estimated by using a kernel of higher order. The disadvantage of this approach is that the point estimate of $\mu(x)$ is essentially produced by a higher order kernel, while the confidence set around it is constructed for the original kernel.

5.4 Blood pressure trajectories

The complete blood pressure data, after removing 63 incomplete cases, has the age and blood pressures (systolic and diastolic) of 7532 men and 2934 women. As people age, their blood pressures tend to increase, though the two blood pressures increase in different ways, and the pattern is different for men and women. Figure 5.2 shows the trajectories taken by the average blood pressure measurements for both men and women. A Gaussian kernel with a bandwidth of $h = 5$ years was used for both trajectories. Here we have age $X \in \mathbb{R}$ and blood pressure $Y \in \mathbb{R}^2$, so the conditional means μ_M and μ_F, for men and women, respectively, are space curves. Figure 5.3 replots the same kernel smooths in perspective to show them as space curves.

The changes in mean blood pressure with age tend to be small compared to the fluctuations between people at a single age. A relatively small shift in the blood pressure of an entire population can, however, be a very significant public health issue.

Both average blood pressures increase for men, until some point around 50 to 55 years of age. Then the systolic blood pressure keeps on increasing, while the diastolic blood pressure starts to decrease.

Women's average blood pressures also increase with age, until about age 55. Then the diastolic blood pressure stays roughly constant while the systolic increases then decreases. The final decrease is for the highest ages, and is estimated with a smaller sample.

The women's blood pressure curve starts lower than the men's. Their diastolic blood pressure increases more slowly, especially during their 30's and 40's and

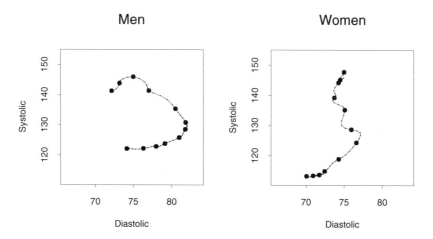

Figure 5.2 *Shown are the trajectories taken by the kernel smoothed systolic and diastolic blood pressures of men and women in New Zealand. There is a point for each integer age from 25 to 80 inclusive. The large points are for ages that are multiples of 5 years. In both cases an age of 25 years corresponds to the lowest systolic blood pressures.*

does not rise to the levels reached by men's average diastolic blood pressure. The rapid decrease in average diastolic blood pressure for men brings that average towards the women's average by about age 70.

Some of the reasons for these patterns are well understood. For example, estrogen protects younger women from hypertension. Other reasons are more complicated. The men's decrease in diastolic blood pressure could be due to increased mortality among those with higher blood pressure, or to increased use of medication to reduce blood pressure. Notice that men's systolic blood pressure does not show the same pattern. Chapter 5.7 revisits the pattern in men's diastolic blood pressure.

The bandwidth for Figures 5.2 and 5.3 is 5 years. Figure 5.4 shows pointwise empirical likelihood confidence regions for μ_M and μ_F both at age 40. The ellipses are narrower for men, because there are more men in the data set. In both cases a positive correlation between systolic and diastolic blood pressure is evident.

5.5 Conditional quantiles

Average blood pressure is perhaps less interesting than extreme blood pressure. The α-quantile of $Y \in \mathbb{R}$ conditional on $X = x_0 \in \mathbb{R}^p$ may be estimated by the

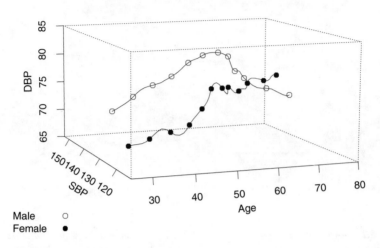

Figure 5.3 *The blood pressure trajectories from Figure 5.2 are replotted in a perspective plot to depict them as space curves. The circles show the men's trajectory, the disks show the women's trajectory.*

solution $Q^\alpha(x_0)$ of the estimating equation

$$0 = \frac{1}{n}\sum_{i=1}^{n} K_h(X_i - x)(1_{Y_i \leq Q^\alpha(x)} - \alpha). \tag{5.6}$$

Table 5.1 on page 120 shows estimated conditional 90th percentiles of women's systolic blood pressure at ages from 25 to 80 by steps of 5 years. A Gaussian kernel with a bandwidth of 5 years was used to smooth the data. From Figure 5.2 it appears that women's average systolic blood pressure starts to move sharply upwards at around age 40 to 45. The estimates in Table 5.1 show that large increases in the 90th percentile may start earlier, between the ages of 35 and 40. Figure 5.5 shows the empirical likelihood function for $Q^{0.90}(x_0)$ at ages x_0 from 30 to 80 by steps of 10 years.

5.6 Simultaneous inference

For some purposes, we seek a confidence region for the whole function $\mu(x)$ over x in a domain $\mathcal{D} \subset \mathbb{R}^p$ of interest. Examples for $\mathcal{D}$ include finite sets of $k \geq 1$ points, intervals when $p = 1$, and hyper-rectangles, spheres, or balls when $p > 1$. In principle the choice for $\mathcal{D}$ is very open, but in practice it is necessary to be able

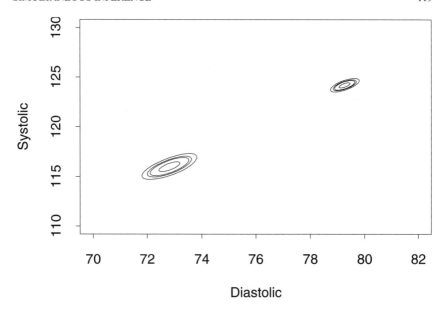

Figure 5.4 *Systolic blood pressure is plotted versus diastolic. The ellipses shown give confidence regions for the mean blood pressure of 40-year-old men and 40-year-old women. The ellipses for men are at higher blood pressures than those for women. The contours correspond to confidence levels of 50, 90, 95, and 99 percent, based on a $\chi^2_{(2)}$ calibration.*

to optimize over $\mathcal{D}$. The confidence region is the set of functions μ given by

$$\left\{ \mu : \mathcal{D} \longrightarrow \mathbb{R}^q \mid \sup_{x \in \mathcal{D}} -2\log\left(\mathcal{R}_x(\mu(x))\right) \leq C \right\},$$

where $\mathcal{R}_x$ is given by (5.4). For $p = 1$ and $\mathcal{D}$ an interval, this is a confidence band or confidence tube depending on whether $q = 1$ or $q > 1$. For $p = 2$ and $q = 1$ and $\mathcal{D}$ a rectangle, this is a confidence sandwich. Clearly the threshold C should be larger than the $1 - \alpha$ point of a $\chi^2_{(q)}$ distribution.

There is some extreme value theory for choosing C, but in practice, it is probably better to use bootstrap calibration. Suppose that $p = 1$, $\mathcal{D} = [a, b]$ is an interval, and h is given. Let $\hat{\mu}(x)$ be the kernel estimate (5.2). Let (X_i^b, Y_i^b) be independent samples from the EDF $\hat{F}$ of (X, Y) pairs, for $i = 1, \ldots, n$, and $b = 1, \ldots, B$. Let

$$C^b = \sup_{x \in \mathcal{D}} -2\log\left(\mathcal{R}_x^b(\hat{\mu}(x))\right),$$

Age	Estimate	Lower	Upper
25	126.20	125.0	127.9
30	127.80	126.0	130.0
35	130.50	130.0	134.0
40	135.55	132.5	137.0
45	139.55	137.0	140.0
50	145.50	142.5	147.5
55	151.50	149.0	155.0
60	157.45	155.0	160.9
65	164.45	160.0	169.0
70	169.70	165.0	172.0
75	171.45	169.5	176.0
80	173.50	170.0	181.0

Table 5.1 *Shown are kernel-based estimates of the 90th percentile of women's systolic blood pressure at ages separated by 5-year intervals. The lower and upper 95% confidence bounds for these quantiles are based on the empirical likelihood with a $\chi^2_{(1)}$ calibration.*

where $\mathcal{R}^b_x(\mu)$ is

$$\max\left\{\prod_{i=1}^n nw_i \mid \sum_{i=1}^n w_i K_h\left(X_i^b - x\right)\left(Y_i^b - \mu\right) = 0, w_i \geq 0, \sum_{i=1}^n w_i = 1\right\},$$

and define the order statistics $C^{(1)} \leq C^{(2)} \leq \cdots \leq C^{(B)}$. Then the order statistic $C^{((1-\alpha)B)}$ provides bootstrap calibration at the level $1 - \alpha$.

For the New Zealand men's blood pressure data, 1000 bootstrap values of C^b were computed using for $\mathcal{D}$ a grid of ages ranging from 20 to 80 inclusive by steps of 5 years. The distribution of $-2\log(\mathcal{R}^b_x(\hat{\mu}(x)))$ fits the $\chi^2_{(2)}$ very closely at each x from 20 to 80, as might be expected for such a large sample. For simultaneous coverage over a set of ages, we are interested in the distribution of C^b, a maximum of correlated random variables, each with nearly the $\chi^2_{(2)}$ distribution. In this instance, the 95th percentile of C^b was 10.41. Using this threshold, we can produce a confidence tube for the mean blood pressure trajectories, with simultaneous coverage over $x_0 \in \mathcal{D}$ of 95%. That tube is displayed in Figures 5.6 and 5.7.

AN ADDITIVE MODEL

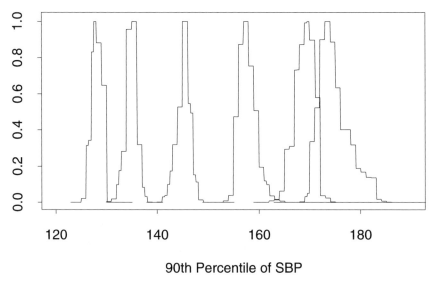

Figure 5.5 *The horizontal axis represents the 90th percentile of women's systolic blood pressure. The vertical axis is the empirical likelihood ratio. The curves, from left to right, are for ages 30 through 80 by steps of 10 years. A Gaussian kernel with bandwidth of 5 years was used. The sample is very sparse near 80 years of age, and this accounts for the greater uncertainty regarding the blood pressure there.*

5.7 An additive model

One particularly interesting feature in the blood pressure data is the eventual decline in men's diastolic blood pressure at greater ages. This decline starts to set in at around the same age where the blood pressures in Figure 5.1 stop showing rounding to multiples of 10. The data come from two studies done on different populations and with different measurement methods, as described in Chapter 5.8. The populations have different age ranges, but there is some overlap. A graphical exploration of the age range where the studies overlap indicates that the male diastolic blood pressures at a given age tend to run higher in the Fletcher study, which had younger subjects. Thus some or all of the decline in blood pressure could be an artifact of age differences between the men in the two studies.

It would be interesting to know whether the decline is solely attributable to the study difference or not. To handle this, we formulate a model in which the mean diastolic blood pressure has the form $\mu(x, z) = s(x) + z\beta$ where x is age, $z = 1$ for the Fletcher-Challenge study, $z = 0$ for the Auckland Heart & Health study, β is a scalar coefficient, and s is a smooth function. Then if $s(x)$ shows the eventual decline with age, we can be more certain that it is real.

A convenient way to encode the smooth function $s(x)$ is through a cubic spline.

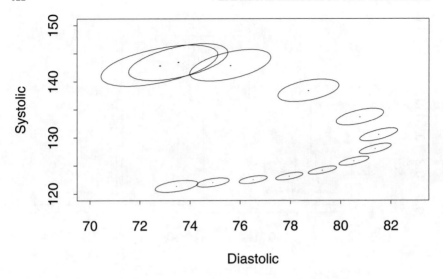

Figure 5.6 *Systolic blood pressure is plotted versus diastolic. The ellipses shown give simultaneous 95% confidence regions for the kernel smoothed mean blood pressure of men ranging in age from 20 to 80 years at intervals of 5 years. Increasing age corresponds to a counter-clockwise movement from the lower left. At extreme ages there are fewer data points, and consequently larger confidence regions.*

This is a function that is piecewise cubic between points called knots, and has two continuous derivatives at the knots. For this example, knots were placed at ages 30, 40, 50, 60, and 70. The truncated power basis for $s(x)$ takes the form

$$s(x) = \beta_0 + \beta_1 x + \beta_2 x^2 + \beta_3 x^3 + \sum_{j=1}^{5} \beta_{3+j} [x - 10(2+j)]_+^3, \qquad (5.7)$$

where $[x - t]_+ = x - t$ if $x \geq t$ and is 0 if $x \leq t$. This basis can be very badly conditioned numerically, and so another basis for the same family of curves was constructed using the S-PLUS function bs(). This B-spline basis is much more stable numerically, but the individual functions in it are not as interpretable as those in (5.7).

Consider the additive model

$$E(Y_i | X_i = x_i, Z_i = z_i) = \beta_0 + \sum_{j=1}^{8} \beta_j \phi_j(x_i) + \beta_9 z_i,$$

where ϕ_j are the spline basis functions, X_i is age, Y_i is the diastolic blood pressure, and Z_i is the study indicator variable described above. The least squares estimate of the study effect is $\hat{\beta}_9 = 6.35$ and the least squares estimate of the age

AN ADDITIVE MODEL

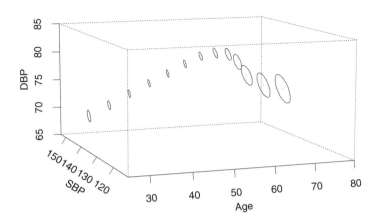

Figure 5.7 *The simultaneous confidence ellipses from Figure 5.6 are plotted in the same coordinate space as used in Figure 5.3 to display mean blood pressure trajectories. They outline the shape of a 95% confidence tube for the smoothed path of mean blood pressure with age.*

effect

$$\gamma \equiv s(75) - s(50) = \sum_{j=1}^{8} \beta_j(\phi_j(75) - \phi_j(50))$$

is

$$\hat{\gamma} = \sum_{j=1}^{8} \hat{\beta}_j(\phi_j(75) - \phi_j(50)) = 3.52.$$

In Figure 5.2, DBP appears to decrease by about 10 mmHg from age 50 to age 75. The data for 75-year-olds are almost entirely from the Auckland Heart & Health study, while most of the data for 50-year-olds are from the Fletcher-Challenge study. Therefore at least a large part of the 10-mmHg decline can be attributed to the difference between the two studies. We expect $\hat{\gamma} + \hat{\beta}_9$ should be close to 10 but it need not match exactly because kernel and spline smoothers are slightly different, and because the kernel smooth at age 50 mixes data from both studies.

The empirical log likelihood for γ has $-2\log(\mathcal{R}(0)) = 15.48$, providing very strong evidence that the age effect γ, comparing 75- and 50-year-old men, is not 0. A 95% confidence interval for the age effect γ, using a $\chi^2_{(1)}$ calibration, extends from 1.77 to 5.26.

5.8 Bibliographic notes

The blood pressure data, kindly supplied by Thomas Yee of the University of Auckland, had two sources. One source was a study on the Fletcher-Challenge company for which most subjects were between 20 and 60 years of age. The other source was the Auckland Heart & Health study (Jackson et al. 1995), which sampled people from the electoral roles. Their ages were evenly distributed between 35 and 85 years. In the Fletcher-Challenge study, blood pressure was measured with a standard Hg sphygmomanometer and recorded by a nurse. In the Auckland Heart & Health study, blood pressure was measured with a Hawsley random sphygmomanometer for which the nurse was not aware of the blood pressure value.

The smoothing presented here used a bandwidth of 5 years in all the blood pressure examples. For such a large data set, this is likely to be oversmoothing. It might be better to use a small bandwidth, or better yet, a local linear smooth as described in Chapter 5.9. The analysis of women's systolic blood pressure did not include an indicator variable for the studies. Such an indicator might improve the analysis, though there only seemed to be a small study effect for women's systolic blood pressure.

The Nadaraya-Watson estimator was proposed, independently, by Nadaraya (1965) and Watson (1964). Sufficient conditions for the bias and variance expansions of kernel estimates may be found in Härdle (1990). Basic references on smoothing include Hastie & Tibshirani (1990) and Fan & Gijbels (1996). Specialized accounts of bandwidth estimation appear in Härdle & Marron (1985) and Härdle, Hall & Marron (1988).

Stone (1977) describes conditional M-estimators defined by local reweighting. Owen (1987) describes these local weights as distributions on the space of X values and gives conditions for consistency and asymptotic normality of estimates in terms of convergence of these distributions to δ_x. The monograph Loader (1999) on local likelihood, considers local versions of statistical methods in depth.

Hall & Owen (1993) consider empirical likelihood confidence bands for kernel density estimates. They obtain an asymptotic calibration for the critical likelihood based on extreme value theory. They also propose bootstrap calibration for this problem. Relatively minor changes are required to translate empirical likelihood results from densities to regressions, and vice versa. Hall & Owen (1993) display confidence bands for the probability density function applied to the Old Faithful geyser data.

Chen (1996) studies coverage levels of undersmoothed kernel density estimates. In his aerial transect sampling problems the population density of blue-fin tuna depends on a density function at the origin, corresponding to a zero distance between the school of fish and the spotting plane. He shows that empirical likelihood is effective at forming a confidence interval for the desired quantity $f(0)$, and that Bartlett correction is possible, though even more undersmoothing is required. Zhang (1998) shows that there is no asymptotic benefit from global side constraints in kernel density estimation. Chen (1997) imposes side constraints (zero mean and skewness) on kernel density estimates of tuna densities estimated from aerial surveys. He shows that there is no asymptotic benefit to imposing those side constraints, but finds an improvement in finite samples.

Fan & Gijbels (1996) is a comprehensive reference on local polynomial smoothing. Chen & Qin (2000) consider empirical likelihood confidence intervals for local linear kernel smoothing. For an undersmoothed estimator, they get a $\chi^2_{(1)}$ limit with a coverage error of $O(nh^5 + h^2 + (nh)^{-1})$. They remark that the use of empirical likelihood produced the same rate of convergence for coverage error at the endpoints as in the middle of the predictor range. An alternative approach (Chen & Qin 2001) using first and second moments has coverage error $O(nh^5 + h^2 + (nh)^{-1})$ in the interior and $O(nh^5 + h + (nh)^{-1})$ near the boundary.

The approach taken here to smoothing started by considering problems in which the curve could be analyzed pointwise, producing at $x_0 \in \mathbb{R}^p$ a confidence region for $\mu(x_0) \in \mathbb{R}^q$. If we were interested in the joint behavior of $\mu(x_1), \ldots, \mu(x_k)$ at k points, then we could maximize the empirical likelihood while forcing all of the reweighted values to be 0 simultaneously. We would expect an asymptotic χ^2 distribution with kq degrees of freedom, and we would expect, at least for modest k, that this test should have better power than one based on the supremum of k pointwise empirical likelihoods. But suppose we want to test a hypothesis such as $\mu(x) = 0$ for all $x \in \mathbb{R}^p$. We cannot expect to reweight the data and get $\hat\mu(x) = 0$ at infinitely many points x. The approach in Chapter 5.6 is based on the supremum over x of $R_x(0)$. It is reasonable to expect that a better method exists. Sieve techniques letting $k \to \infty$ with n, as described in Chapter 9.10, may help.

5.9 Exercises

Exercise 5.1 Let $\tilde\mu(x)$ satisfy $E(K_h(X - x)(Y - \tilde\mu(x))) = 0$. Suppose that $K_h(z) = K(z/h)/h$ where K is a symmetric probability density. Give an informal argument that the bias $\tilde\mu(x) - \mu(x)$ is

$$\frac{h^2}{2f(x)} [\mu''(x)f(x) + 2\mu'(x)f'(x) + \mu''(x)f(x)] \int z^2 K(z)dz + O(h^4)$$

as $h \to 0$. Here X has probability density function f and the expected value of Y given $X = x$ is $\mu(x)$.

Exercise 5.2 Local linear and local polynomial regression are effective ways to smooth data, because they adapt to the local spacing of the X_i. For local linear regression, let $\theta_0 = \theta_0(x)$ and $\theta_1 = \theta_1(x)$ minimize

$$\sum_{i=1}^n K_h\left(\frac{X_i - x_0}{h}\right)\left(Y_i - \theta_0 - \theta_1(X_i - x_0)\right)^2,$$

where $X_i, Y_i, x_0 \in \mathbb{R}$. The smooth value at x is $\theta_0(x)$. Write estimating equations for θ_0 and θ_1. Extend the local linear estimating equations to local polynomial estimating equations. Extend the local linear estimating equations to $X_i \in \mathbb{R}^p$ and $Y_i \in \mathbb{R}^q$.

Exercise 5.3 Formulate estimating equations for an α quantile of Y that is locally linear in X. Here $Y \in \mathbb{R}$ and $X \in \mathbb{R}^p$.

CHAPTER 6

Biased and incomplete samples

The previous chapters considered empirical likelihood based on observations from the distribution (or distributions) of interest. This chapter considers empirical likelihood inference in some nonstandard sampling settings. In biased sampling, the data are sampled from a distribution different from the one we want to study. In censoring, some of the observations are not completely observed, but are known only to belong to a set. The prototypical example is the time until an event. For an event that has not happened by time t, the value is known only to be in (t, ∞). Truncation is a more severe distortion than censoring. Where censoring replaces a data value by a subset, truncation deletes that value from the sample if it would have been in a certain range. Truncation is an extreme form of biased sampling where certain data values are unobservable.

These incomplete sampling ideas are closely related. They have also been widely studied in varied settings. A lot is known about NPMLE's for incomplete sampling, while there is a comparatively small body of knowledge about the corresponding likelihood ratios.

6.1 Biased sampling

It is common in applied statistics for data to be sampled from a distribution other than the one for which inferences are to be drawn. Sometimes this is an undesirable feature, as with measuring equipment for which the chance of recording a value depends on what that value is. Other times it is an intentional device to gain more informative data, as in retrospective sampling of people with rare diseases, or importance sampling in simulations. Finally, there are settings like the sampling of families by independent sampling of children. There, averages over the sampled families are biased towards larger families, while averages over the sampled children are not biased.

A concrete and common example is length biased sampling. Some methods of sampling cotton fibers, sample them with probability proportional to their length. If one samples people waiting in a hospital room at a random time, those with longer waits, and presumably less serious ailments, are more likely to be in the sample. If one samples entries in an Internet log file, the longer sessions are over-represented.

Suppose that a random variable Y has distribution F_0, but that we obtain a length biased sample. Let X be one of our observations. Then X has distribution

G_0 with CDF

$$G_0((-\infty, x]) = \frac{\int_0^x y\,dF_0(y)}{\int_0^\infty y\,dF_0(y)}.$$

This is a proper distribution if $E(Y)$ is positive and finite.

More generally, suppose that $Y \in \mathbb{R}^d$ has distribution F_0 and that $X \in \mathbb{R}^d$ has distribution G_0, where for $A \subseteq \mathbb{R}^d$

$$G_0(A) = \frac{\int_A u(y)\,dF_0(y)}{\int u(y)\,dF_0(y)},$$

for a biasing function $u(y) \geq 0$, with $0 < \int u(y)\,dF_0(y) < \infty$. When $0 \leq u(y) \leq 1$, biased sampling has an acceptance sampling interpretation. The value Y is first sampled from F_0, and then with probability $u(Y)$ it is accepted, while with probability $1 - u(Y)$ this value of Y is rejected. Sampling continues until the first time a Y is accepted. That first accepted Y is the observed value of X. If $u(y) \leq B$ for some $B > 0$ then u-biased sampling gives the same data distribution as v-biased sampling with $v(y) = u(y)/B$, and so the acceptance sampling interpretation carries over to any bounded biasing function u.

The nonparametric likelihood for F is

$$L(F) = \prod_{i=1}^n \frac{F(\{X_i\})u(X_i)}{\int u(x)\,dF(x)}.$$

Suppose that $u(x) > 0$ for all x, and let $u_i = u(X_i)$. Then the NPMLE is easily shown to be

$$\widehat{F} = C\sum_{i=1}^n \frac{\delta_{X_i}}{u_i}, \quad C^{-1} = \sum_{i=1}^n \frac{1}{u_i}. \tag{6.1}$$

The NPMLE weights each observation in inverse proportion to its sampling probability. In the acceptance sampling setting this has the natural interpretation that every accepted value X represents $1/u(X)$ sampled values of which on average one was accepted. This downweighting is familiar in survey sampling, where it includes stratified sampling and the more general Horvitz-Thompson estimator. It is also well known in Monte Carlo simulation, where the method of importance sampling samples from a distribution other than the nominal one. When $u(x) = c > 0$ for all x, then the NPMLE $\widehat{F}$ reduces to the usual NPMLE F_n.

For length biased sampling the NPMLE of the mean of F_0 is

$$\frac{\sum_{i=1}^n u_i^{-1} X_i}{\sum_{i=1}^n u_i^{-1}} = \left(\frac{1}{n}\sum_{i=1}^n X_i^{-1}\right)^{-1}.$$

This is the harmonic mean of the sample.

If $u(x) = 0$ is possible, then the NPMLE is not unique. Any mixture distribution $\alpha H + (1-\alpha)\widehat{F}$, where H puts all its probability on $\{x \mid u(x) = 0\}$, is also an NPMLE, for $0 \leq \alpha < 1$. If there is such a thing as a cotton fiber of zero length, then F_0 could put probability $\alpha \in [0,1)$ on such fibers and the distribution of the data would be the same for any value of α. The mean fiber length would be affected by α, and any value between 0 and the harmonic mean would be an NPMLE for $\int x dF(x)$. A pragmatic approach is to fix $\alpha = 0$ and consider any inferences to be on F restricted to the set $\{X \mid u(X) > 0\}$.

By considering the estimating equation

$$0 = \int m(x,\theta) dF(x) = \int \frac{m(x,\theta)}{u(x)} dG(x),$$

we find that we can work directly with biased data, simply by replacing the estimating function $m(x,\theta)$ by $\widetilde{m}(x,\theta) \equiv m(x,\theta)/u(x)$. In particular, the NPMLE is the solution $\hat{\theta}$ to

$$\sum_{i=1}^{n} \widetilde{m}(X_i, \theta) = 0,$$

and the profile empirical likelihood ratio function for θ is

$$\mathcal{R}(\theta) = \max \left\{ \prod_{i=1}^{n} nw_i \mid \sum_{i=1}^{n} w_i \widetilde{m}(X_i, \theta) = 0, \sum_{i=1}^{n} w_i = 1, w_i \geq 0 \right\}.$$

Tests and confidence regions for θ depend on the distribution of $\widetilde{m}(X,\theta)$ under the sample distribution G_0.

Figure 6.1 displays the widths of 46 shrubs, as reported in Muttlak & McDonald (1990). These shrubs were obtained by transect sampling. Any shrub intersecting a line on the ground was sampled. The probability of a shrub entering the sample is thus proportional to its width. The top histogram shows the observed widths. The bottom histogram shows the data weighted inversely to its sampling probability. The height of each bar is proportional to the sum of $1/X_j$, summed over X_j in the corresponding interval.

The mean μ and variance σ^2, of the shrub widths are defined by

$$0 = \int [x - \mu] \, dF(x), \quad \text{and}$$

$$0 = \int \left[(x-\mu)^2 - \theta \right] dF(x),$$

and so, accounting for the bias, the NPMLE's are defined through

$$0 = \sum_{i=1}^{n} X_i^{-1}(X_i - \mu), \quad \text{and}$$

$$0 = \sum_{i=1}^{n} X_i^{-1} \left[(X_i - \mu)^2 - \sigma^2 \right].$$

Figure 6.2 shows the profile empirical likelihood ratio function for the mean shrub width μ. Figure 6.3 shows the profile empirical likelihood ratio function

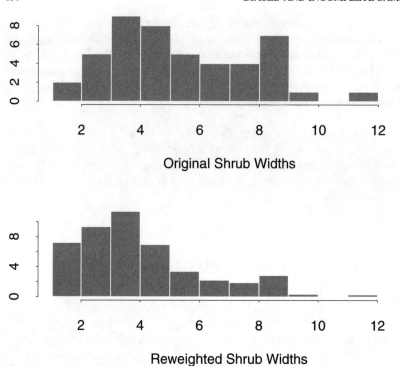

Figure 6.1 *The top histogram shows the widths of 46 shrubs found by transect sampling. The bottom histogram has the same total area, and the same bins, but each shrub is weighted inversely to its width to correct for sampling bias.*

for the standard deviation σ of shrub width. Taking account of the sampling bias makes a big difference in the inferences, reducing the mean shrub width μ and the standard deviation σ.

6.2 Multiple biased samples

Now suppose that s samples are available, $X_{ij} \in \mathbb{R}^d$, for $i = 1, \ldots, s$ and $j = 1, \ldots, n_i$. All the observations are independent, but there are s different biases: $X_{ij} \sim G_{i0}$, where

$$G_{i0}(A) = \frac{\int_A u_i(x) dF_0(x)}{\int u_i(x) dF_0(x)}. \tag{6.2}$$

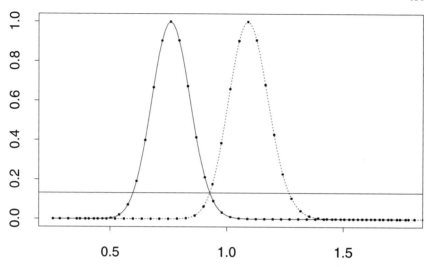

Figure 6.2 *The solid curve shows the empirical likelihood ratio for the mean shrub width, after accounting for length biased sampling. The dotted curve does not account for length biased sampling. The horizontal reference line designates a 95% confidence level using an $F_{1,45}$ criterion.*

The functions $u_i(x)$ are real valued and nonnegative with

$$0 < \nu_i(F_0) \equiv \int u_i(x) dF_0(x) < \infty. \qquad (6.3)$$

Data of this kind could arise in s clinical trials with different enrollment criteria for subjects, or with measurements on the same underlying phenomenon from s different devices. They also arise in choice-based sampling in marketing. When studying the brand preferences of consumers, sample i might correspond to consumers whose brand preferences are known to belong in the i'th in a list of s subsets of brands.

The NPMLE of F_0 is useful for the data fusion problem of combining these differently biased observations. We will assume that the domain of X is given by $\mathcal{X} = \{x \mid \sum_{i=1}^{s} u_i(x) > 0\}$. There can be no data points sampled from outside of this domain. Our inferences are implicitly on F_0' where $F_0'(A) = F_0(\mathcal{X} \cap A)/F_0(\mathcal{X})$.

Let $Z_1, \ldots, Z_h$ be the distinct observations among the sample X_{ij} values, and let $n_{ik} = \#\{X_{ij} = Z_k \mid 1 \leq j \leq n_i\}$. Let F be a distribution putting probability

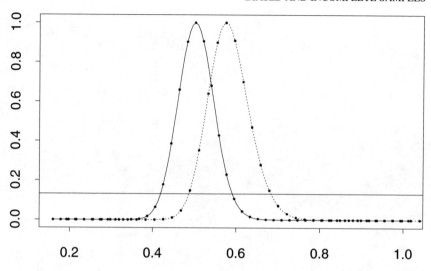

Figure 6.3 *The solid curve shows the empirical likelihood ratio for the standard deviation of shrub width, after accounting for length biased sampling. The dotted curve does not account for length biased sampling. The horizontal reference line is for 95% confidence using an $F_{1,45}$ criterion.*

$p_k \geq 0$ on Z_k, with $\sum_{k=1}^{h} p_k = 1$. Then the likelihood for F is

$$\prod_{k=1}^{h} \prod_{i=1}^{s} \left(\frac{u_i(Z_k) p_k}{\sum_{l=1}^{h} p_l u_i(Z_l)} \right)^{n_{ik}}.$$

As in the single sample unbiased case of Chapter 2.3, it is possible to ignore ties and work with observation specific weights $w_{ij} \geq 0$ on X_{ij}. The weights generate p_k if

$$p_k = \sum_{i=1}^{s} \sum_{j=1}^{n_i} w_{ij} 1_{X_{ij}=Z_k}.$$

The likelihood in terms of the weights w_{ij} is

$$\prod_{i=1}^{s} \prod_{j=1}^{n_i} \frac{u_i(X_{ij}) w_{ij}}{\sum_{l=1}^{s} \sum_{t=1}^{n_l} w_{lt} u_i(X_{lt})} = \prod_{i=1}^{s} \prod_{j=1}^{n_i} \frac{u_i(X_{ij}) w_{ij}}{\sum_{k=1}^{h} p_k u_i(Z_k)}. \quad (6.4)$$

Notice that the denominator is unaffected by how the probability p_k is allocated among weights w_{ij} for $X_{ij} = Z_k$. It follows that for fixed p_k, the maximizing weights are $w_{ij} = p_k/m_k$ where m_k is the number of observations in the combined samples for which $X_{ij} = Z_k$. Interestingly, the weight w_{ij} does not depend on which sample contributed the value X_{ij}. For any p_k the likelihood in terms of $w_{ij} = p_k/m_k$ is a constant multiple of the likelihood in terms of p_k, and this constant cancels when forming nonparametric likelihood ratios. The same argument goes through if $\sum_{k=1}^{h} p_k < 1$, except that then $\nu_i(F)$ cannot be written as a weighted sum of sample values.

The factors $u_i(X_{ij})$ in the numerator of (6.4) do not depend on w_{ij} and so they may be ignored. The log likelihood may be taken to be

$$\sum_{i=1}^{s}\sum_{j=1}^{n_i}\log(w_{ij}) - \sum_{k=1}^{s} n_k \log\left(\sum_{i=1}^{s}\sum_{j=1}^{n_i} w_{ij} u_i(X_{ij})\right) \qquad (6.5)$$

To find the NPMLE we must maximize the expression in (6.5) over $w_{ij} \geq 0$ subject to $\sum_i \sum_j w_{ij} = 1$.

For $s = 1$, the sampling probability of an observation is known, apart from a constant factor, and in Chapter 6.1 we saw that the NPMLE weights the data in inverse proportion to that probability. For $s \geq 2$, matters are more complicated. There is not always an NPMLE, and when an NPMLE exists it is not always unique. But under mild conditions a unique NPMLE exists:

Theorem 6.1 *A unique NPMLE exists if and only if for every proper subset $B \subset \{1,\ldots,s\}$*

$$\left(\bigcup_{i \in B}\{X_{i1},\ldots,X_{in_i}\}\right) \cap \left(\bigcup_{i \notin B}\{X \mid u_i(X) > 0\}\right) \neq \emptyset.$$

Proof. Vardi (1985). □

In words: there will not be a unique NPMLE if the s sample data sets can be partitioned into two subsets, where no observation in the first subset could possibly have been observed in the second subset. This does not mean that each pair of sampling distributions has to overlap. A pair that does not overlap might be bridged by a third sample. If even one of the samples, say sample i, has $u_i(X) > 0$ for all X, then a unique NPMLE will exist.

To illustrate why the NPMLE might not be unique, take $s = 2$, let $\mathcal{X}_i = \{x | u_i(x) > 0\}$ for $i = 1, 2$ and suppose that $\mathcal{X}_1 \cap \mathcal{X}_2 = \emptyset$. If the distribution of X_{ij} from G_{i0} is consistent with acceptance sampling from F_0 with acceptance probability $u_i(x)$ then it is equally consistent with an acceptance probability of $u_i(x)/100$. Such samples, even as $n_i \to \infty$, cannot help us determine the relative weight $F(\mathcal{X}_i)/(F(\mathcal{X}_1) + F(\mathcal{X}_2))$ that belongs on the domain of sample i. If however $\mathcal{X}_1 \cap \mathcal{X}_2 = Z$ and $\Pr(X_{ij} \in Z) > 0$ for both $i = 1, 2$, then we may be able to estimate $F(Z)/F(\mathcal{X}_i)$ from the X_{ij} and put these together to estimate $F(\mathcal{X}_i)/(F(\mathcal{X}_1) + F(\mathcal{X}_2))$. It is possible to have a degenerate NPMLE with $F(\mathcal{X}_i) = 1$ if sample i has no observations in Z.

For $s > 2$, if two samples intersect, then we can estimate the relative probabilities of their domains $\mathcal{X}_i$. If we can estimate the ratios $F(\mathcal{X}_i)/F(\mathcal{X}_{i'})$ and $F(\mathcal{X}_{i'})/F(\mathcal{X}_{i''})$ then we can estimate $F(\mathcal{X}_i)/F(\mathcal{X}_{i''})$. We can estimate the relative probabilities of other domains if there is a chain of ratios connecting them.

Theorem 6.2 *Suppose that F_0 is a distribution $\mathbb{R}^d$, that for $i = 1, \ldots, s$, the functions u_i satisfy (6.3), and that distributions G_{i0} are defined by (6.2). Let $\mathcal{G}$ be a graph on s vertices with an edge connecting vertices i and i' if and only if $\int u_i(x)u_{i'}(x)dF_0 > 0$. Then F_0 is uniquely determined as a function of G_{i0} for $i = 1, \ldots, s$, if and only if the graph $\mathcal{G}$ is connected. In that case the probability that a unique NPMLE exists tends to 1 as $\min_i n_i \to \infty$.*

Proof. Gill, Vardi & Wellner (1988). □

To maximize the nonparametric likelihood, suppose for a moment that we know the values $\nu_i = \int u_i(X)dF_0(X)$. Let $N = \sum_{i=1}^s n_i$, $\nu = (\nu_1, \ldots, \nu_s)$, and $U_{ij} = (u_1(X_{ij}), \ldots, u_s(X_{ij}))$. Then let $\ell(\nu)$ maximize $\sum_i \sum_j \log(w_{ij}) - \sum_i n_i \log(\nu_i)$ subject to

$$\sum_i \sum_j w_{ij} = 1, \quad \text{and}$$

$$\sum_i \sum_j w_{ij} U_{ij} = \nu.$$

The computation of $\ell(\nu)$ reduces to empirical likelihood maximization for a mean as described in Chapter 3.14, and the solution has

$$w_{ij}(\nu) = \frac{1}{N} \frac{1}{1 + \delta'(U_{ij} - \nu)}$$

where the Lagrange multiplier $\delta = \delta(\nu)$ satisfies

$$\sum_i \sum_j \frac{U_{ij} - \nu}{1 + \delta'(U_{ij} - \nu)} = 0.$$

To find the NPMLE we maximize

$$\ell(\nu) = \sum_i \sum_j \log(w_{ij}(\nu)) - \sum_i n_i \log(\nu_i)$$

over $\nu = (\nu_1, \ldots, \nu_s)$.

To incorporate estimating equations, we must now impose the additional constraint

$$\sum_{i=1}^s \sum_{j=1}^{n_i} w_{ij} m(X_i, \theta) = 0.$$

Given ν, the solution is

$$w_{ij}(\nu, \theta) = \frac{1}{N} \frac{1}{1 + \lambda' m(X_{ij}, \theta) + \delta'(U_{ij} - \nu)}$$

where $\lambda(\nu, \theta)$ and $\delta(\nu, \theta)$ satisfy

$$\sum_i \sum_j \frac{U_{ij} - \nu}{1 + \lambda' m(X_{ij}, \theta) + \delta'(U_{ij} - \nu)} = 0$$

$$\sum_i \sum_j \frac{m(X_{ij}, \theta)}{1 + \lambda' m(X_{ij}, \theta) + \delta'(U_{ij} - \nu)} = 0.$$

The profile empirical likelihood ratio for θ is then the ratio

$$\mathcal{R}(\theta) = \frac{\max_\nu \prod_i \prod_j w_{ij}(\nu, \theta)}{\max_\nu \prod_i \prod_j w_{ij}(\nu)}.$$

Under mild conditions, the asymptotic distribution of $-2\log(\mathcal{R}(\theta_0))$ is $\chi^2_{(p)}$ where p is the dimension of θ. See the references in Chapter 6.9.

6.3 Truncation and censoring

Truncation is an extreme version of biased sampling, where the bias function is

$$u(x) = \begin{cases} 1 & \text{if } x \in T \\ 0 & \text{if } x \notin T, \end{cases} \quad (6.6)$$

for some set T. Consider historical data on the heights of men drafted into an army, where a minimum height restriction H was in effect. For conclusions on the heights of draft-aged men, such data represent a sample truncated to $T = [H, \infty)$.

Censoring is a milder form of information loss than truncation. An observation is censored to the set C if instead of observing X we only observe the fact that $X \in C$. A censored point is known to have existed, whereas a truncated point produces no observation. In the example above, if we knew the number of draft-aged men rejected because of the height restriction, then their heights would be censored to the set $C = (0, H)$.

Suppose θ is a quantity that depends in part on how a truncated random variable X is distributed over values $X \notin T$. Then some assumptions are necessary to get an estimate of θ. The truncated data can never contain any $x \notin T$, but perhaps there is a way to extrapolate from $x \in T$ to $x \notin T$. In practice our extrapolation might introduce a systematic error that we can neither check nor correct. Yet it may be better to patch in a possibly flawed extrapolation than to ignore the truncation completely.

One way to extrapolate is to fit a parametric model for X with density or mass function $f(x; \theta)$. Our sample is from $X \sim f(x; \theta)$ conditional on $X \in T$. For such a model, the likelihood is

$$L_{\text{TRUN}}(\theta; X_1, \ldots, X_n) = \prod_{i=1}^n \frac{f(X_i; \theta)}{\int_T dF(x; \theta)},$$

with estimating equations

$$\sum_{i=1}^{n} \left(\frac{\frac{\partial}{\partial \theta} f(X_i; \theta)}{f(X_i; \theta)} - \frac{\frac{\partial}{\partial \theta} \int_T dF(x; \theta)}{\int_T dF(x; \theta)} \right) = 0. \qquad (6.7)$$

Both parametric and empirical likelihood inferences can be based on (6.7). Both inference methods give the same maximum likelihood estimate $\hat{\theta}$. Whether $\hat{\theta}$ is estimating the desired quantity can depend on how accurate the parametric model is. Empirical likelihood confidence regions will have the right asymptotic coverage for the quantity being estimated by $\hat{\theta}$ under very weak conditions, whereas parametric likelihood regions will have coverage levels sensitive to the parametric model used.

Now suppose that we have two parametric models $f_j(x; \theta_j)$ for $j = 1, 2$ with $\theta_j \in \mathbb{R}^{p_j}$, and that the quantity we are interested in can be written as $\tau_j(\theta_j)$ under model j. For example f_1 might be a normal distribution and f_2 might be a gamma distribution. The values $\tau_j(\hat{\theta}_j)$ will not in general agree with each other. The nature of the discrepancy can be investigated using an empirical likelihood confidence region for $(\tau_1(\theta_1), \tau_2(\theta_2))$ or for $\tau_1(\theta_1) - \tau_2(\theta_2)$. This will not indicate which, if either, of the parametric models provides a reliable extrapolation. But it does allow us to judge whether the extrapolated answer is sensitive to the extrapolation formula, without knowing which, if either, parametric model is right.

In applications with censored and truncated data, the nature of the censoring and truncation rules may vary from observation to observation. The general case has X_i truncated to T_i then censored to C_i, taking $C_i = \{X_i\}$ for uncensored data, and T_i equal to the domain of X_i, usually a subset of $\mathbb{R}^d$, for untruncated data. The sets T_i may be random. Apart from trivial exceptions, the set C_i has to be random, because it depends on X_i. We consider coarsening at random (CAR), in which T_i has been partitioned at random and independently of Y_i into a number of sets. The set that happens to contain X_i is observed as C_i.

Some of the most widely studied types of censoring are listed below. Of these, Examples 6.1 and 6.2 will be considered at greater length. We will find that a form of conditional likelihood is most suitable for them.

Example 6.1 (Right censoring) Here, the distribution of the real-valued random variables X_i is of direct interest. For each X_i there is a $Y_i \in \mathbb{R}$. This Y_i may be random. If $X_i \le Y_i$ we observe X_i, otherwise X_i is censored to (Y_i, ∞). We say that X_i is right censored by Y_i. For example, X_i could be survival time after an operation, with Y_i the time from the operation to the end of the study.

Example 6.2 (Left truncation) The pair (X_i, Y_i) is observed if and only if $X_i \ge Y_i$. The Y_i may be random. We say that X_i is left truncated by Y_i. In astronomy, the brightness X of an object may be left truncated by some function $Y = h(Z)$ of its distance Z from Earth.

TRUNCATION AND CENSORING

Example 6.3 (Left truncation and right censoring) The random variable X_i is right censored by Y_i, and left truncated by $Z_i < Y_i$. If $Z_i \leq X_i < Y_i$ then X_i is observed directly. If $Z_i < Y_i \leq X_i$ then X_i is censored to (Y_i, ∞) and if $Z_i > X_i$ then none of X_i, Y_i or Z_i are observed. For example, X_i could be the survival time after a transfusion, Y_i a corresponding censoring time, and Z_i the time between the transfusion and the beginning of a study of transfused patients.

Example 6.4 (Double censoring) The random variable $X_i \in \mathbb{R}$ is observed if $Z_i \leq X_i \leq Y_i$, is right censored to (Y_i, ∞) if $X_i > Y_i$, and is left censored to $(-\infty, Z_i)$ if $X_i < Z_i$. Here $Z_i \leq Y_i$ and either or both may be random.

Example 6.5 (Interval censoring) The random variable X_i is censored to the set $(Z_{i,k}, Z_{i,k+1}]$ for $Z_{i,1} < Z_{i,2} < Z_{i,K_i}$. For example, $Z_{i,k}$ could be times at which patients are studied or equipment is inspected, and X_i the time of some change in status. Interval-censored data are also known as current status data. The usual likelihood for data from a continuous parametric distribution is motivated by arguing that each component of each observation was interval censored to a small interval.

Suppose that $X_1, \ldots, X_n$ are sampled from a common distribution F, and are conditionally independent given right censoring times $Y_1, \ldots, Y_n$. Let $Z = \min(X, Y)$ and let $\delta = 1_{X \leq Y}$ indicate an uncensored failure. By convention, X is not considered censored when $X = Y$. Similarly, if several observations are tied at the same value of Z, the censoring times are deemed to follow the failure times by an infinitesimal amount.

Let $\mathcal{X} = (X_1, \ldots, X_n)$ and $\mathcal{Y} = (Y_1, \ldots, Y_n)$. The likelihood for F and G from right-censored data is the product of a marginal and conditional likelihood

$$L(F, G; \mathcal{X}, \mathcal{Y}) = L(F, G; \mathcal{Y}) \times L(F, G; \mathcal{X} \mid \mathcal{Y}) \tag{6.8}$$

where $L(F, G; \mathcal{Y}) = G(Y_1, \ldots, Y_n)$ and

$$L(F, G; \mathcal{X} \mid \mathcal{Y}) = \prod_{i:\delta_i=1} F(\{X_i\}) \prod_{i:\delta_i=0} F((Y_i, \infty))$$

$$= \prod_{i=1}^{n} F(\{Z_i\})^{\delta_i} F((Z_i, \infty))^{1-\delta_i}. \tag{6.9}$$

Any factor of 0^0 in (6.9) is understood to be 1. These likelihoods are nonparametric, but are easily modified if F or G are known to belong to parametric families.

It is usual to base inferences for F on the conditional likelihood (6.9). That conditional likelihood does not depend on G, and it can be computed from the Z_i and δ_i without knowing the Y_i from uncensored X_i. Because the marginal likelihood of the Y_i does not depend on F, using the conditional likelihood does not lead to a loss of information on F. In the absence of a known functional relationship between F and G, the conditional likelihood (6.9) gives the same likelihood ratio function for F as the full likelihood (6.8). We did not need to assume that the Y_i were independent in order to settle on the conditional likelihood. Of course strong dependence in Y_i could make the conditional likelihood very uninformative.

For left-truncated data, we never observe $X < Y$, but if Y is independent of X then we may reasonably hope to learn the distribution of X, or at least that part of it larger than the left end point of the Y distribution. A conditional likelihood approach is also applicable to left-truncated data, though the derivation is more complicated.

Suppose that X and Y are independent from distributions F and G, respectively, and that independent (X, Y) pairs are truncated to the set $\{(x, y) \mid x \geq y\}$. The likelihood is then

$$L(F, G \; ; \; \mathcal{X}, \mathcal{Y}) = \alpha^{-n} \prod_{i=1}^{n} 1_{X_i \geq Y_i} F(\{X_i\}) G(\{Y_i\}) \qquad (6.10)$$

where

$$\alpha = \iint_{x \geq y} dF(x) dG(y) = \int G((-\infty, u]) dF(u) = \int F([u, \infty)) dG(u)$$

is the probability that $X \geq Y$.

The likelihood (6.10) may be factored into a product of marginal and conditional likelihoods by either of

$$L(F, G; \mathcal{X}, \mathcal{Y}) = L(F, G; \mathcal{X}) \times L(F, G; \mathcal{Y} \mid \mathcal{X})$$
$$= L(F, G; \mathcal{Y}) \times L(F, G; \mathcal{X} \mid \mathcal{Y})$$

where

$$L(F, G; \mathcal{X}) = \alpha^{-n} \prod_{i=1}^{n} G((-\infty, X_i]) F(\{X_i\})$$

$$L(F, G; \mathcal{Y} \mid \mathcal{X}) = \prod_{i=1}^{n} 1_{X_i \geq Y_i} \frac{G(\{Y_i\})}{G((-\infty, X_i])}$$

$$L(F, G; \mathcal{Y}) = \alpha^{-n} \prod_{i=1}^{n} F([Y_i, \infty)) G(\{Y_i\})$$

$$L(F, G; \mathcal{X} \mid \mathcal{Y}) = \prod_{i=1}^{n} 1_{X_i \geq Y_i} \frac{F(\{X_i\})}{F([Y_i, \infty))}.$$

Suppose that interest centers on F. The conditional likelihood based on X_i given Y_i depends on F but not on G, and hence is available for inference. Unlike the case of right censoring, the marginal distribution of Y_i depends on F, so that even without a known link between F and G, there may be an information loss from using the conditional likelihood. The conditional distribution of Y_i given X_i does not involve F, suggesting that the marginal distribution of X_i has all the information on F, but the marginal distribution of X_i involves G.

If G is known, then the full likelihood (6.10) for F may be written

$$L(F) = \prod_{i=1}^{n} \frac{1_{X_i \geq Y_i} F(\{X_i\}) G(\{Y_i\})}{\int G((-\infty, u]) dF(u)}$$

$$\propto \prod_{i=1}^{n} \frac{F(\{X_i\})}{\int G((-\infty, u]) dF(u)}. \qquad (6.11)$$

The likelihood (6.11) is proportional to the likelihood of biased sampling with $u(x) = G((-\infty, u])$ and so the NPMLE for F with G known is given by (6.1) with $u_i = G((-\infty, X_i])$. The following theorem lends support to the use of conditional likelihood for F when the distribution G is not known.

Theorem 6.3 *Suppose that X_i is observed with independent left truncation by Y_i. In the joint NPMLE $(\hat{F}, \hat{G})$ of (F, G), the distribution $\hat{F}$ is the maximizer of the conditional likelihood $L(F, G; \mathcal{X} \mid \mathcal{Y})$ above.*

Proof. Wang (1987) and Keiding & Gill (1990). □

6.4 NPMLE's for censored and truncated data

For $X_i \in \mathcal{X} \subseteq \mathbb{R}^d$ independently sampled from F and censored, by coarsening at random, to C_i, including $C_i = \{X_i\}$ for uncensored data, the conditional (on C_i) likelihood may be written

$$L_c(F) = \prod_{i=1}^{n} \int_{C_i} dF(x) = \prod_{i=1}^{n} F(C_i), \qquad (6.12)$$

and for truncated and censored data the conditional likelihood is

$$L_c(F) = \prod_{i=1}^{n} \frac{\int_{C_i} dF(x)}{\int_{T_i} dF(x)} = \prod_{i=1}^{n} \frac{F(C_i)}{F(T_i)}. \qquad (6.13)$$

The censored data likelihood (6.12) does not always have a unique maximum. There are 2^n disjoint sets of the form

$$E_j = \bigcap_{i=1}^{n} D_{ij}$$

where each D_{ij} is either C_i or $C_i^c = \mathcal{X} - C_i$. The union of these E_j is $\mathcal{X}$. Letting $w_j = F(E_j)$, $M = 2^n$, and $H_{ij} = 1_{E_j \subset C_i}$ we can write the conditional likelihood from (6.12) as

$$L_c(F) = \prod_{i=1}^{n} \left(\sum_{j=1}^{M} H_{ij} w_j \right). \qquad (6.14)$$

Thus F is determined only up to the values of w_j.

Theorem 6.4 *There is a unique set of weights $w_j \geq 0$ with $\sum_{j=1}^{M} w_j = 1$ that maximize (6.14).*

Proof. Let $w = (w_1, \ldots, w_M)'$ belong to the closed, bounded, and convex set $S = \{w \mid w_j \geq 0, \sum_{j=1}^{M} w_j = 1\}$. After identifying $F(E_j)$ with w_j, the function $L_c(w)$ is a continuous function on the compact set S, so it attains a maximum there. This maximum is nonzero, and so $\ell(w) = \log(L_c)$ also attains a finite maximum ℓ_m on S. It remains to show that this maximum is unique.

Suppose to the contrary that $\ell(u) = \ell(v) = \ell_m$ and that $u_{j'} > v_{j'}$ for some j'. If there is no i' with $H_{i'j'} = 1$, then taking $w_{j'} = 0$ and $w_j = u_j/(1 - u_{j'})$ for $j \neq j'$ we find $\ell(w) > \ell(u)$ contradicting the maximality of $\ell(u)$. So let i' satisfy $H_{i'j'} = 1$ and put $w = (u+v)/2$. Convexity of S implies that $w \in S$. Now $\ell(w) - \ell_m = \ell(w) - (\ell(u) + \ell(v))/2$ is a sum of n nonnegative terms, one for each i. The term for i' is strictly positive, contradicting the maximality of ℓ_m. □

A censored-data NPMLE is not necessarily a good estimator of F. See Chapter 6.9 for an example with bivariate censoring and for a remedy.

It is not practical to keep track of 2^n probability weights. For uncensored data, at most n of the E_j are nonempty. As the next two theorems show, a great simplification occurs when X_i are real values and C_i are all intervals, with or without truncation.

Theorem 6.5 *Let $C_i = [L_i, R_i]$, and let $E_j = [p_j, q_j]$ for $j = 1, \ldots, m$ be the set of intervals with endpoints taken from $U = \cup_{i=1}^{n} \{L_i, R_i\}$ and that contain no interior points from U. Then there are uniquely determined probabilities $w_j \geq 0$ on E_j with $\sum_{j=1}^{m} w_j = 1$ such that F maximizes L if and only if $F(E_j) = w_j$.*

Proof. Peto (1973). □

Theorem 6.6 *Let $X_i \in \mathbb{R}$ be truncated to $T_i \subseteq \mathbb{R}$ and then censored to C_i, a finite union of disjoint closed intervals $[L_{ik}, R_{ik}]$, $k = 1, \ldots, K_i$. Let $E_j = [p_j, q_j]$ for $j = 1, \ldots, m$ be the set of intervals with endpoints taken from $U = \cup_{i=1}^{n} \cup_{k=1}^{K_i} \{L_{ik}, R_{ik}\}$ and that contain no interior points from U. Let $D = \cup_{j=1}^{m} E_j$. Then:*

1. *Any NPMLE F has $F(D) = 1$, unless $T_i \cap D = C_i \cap D$ for all i.*
2. *The likelihood depends on F only through $w_j = F(E_j)$, $j = 1, \ldots, m$.*
3. *The likelihood is*
$$L_c(F) = \prod_{i=1}^{n} \frac{\sum_{j=1}^{m} H_{ij} w_j}{\sum_{j=1}^{m} K_{ij} w_j}$$
   where $H_{ij} = 1_{E_j \subseteq C_i}$ and $K_{ij} = 1_{E_j \subseteq T_i} \geq H_{ij}$.
4. *There are unique NPMLE weights w_j, unless*

   (a) *$H_{ij} = H_{ij'}$, for some $1 \leq j < j' \leq m$, and all $i = 1, \ldots, n$, or,*
   (b) *There is a subset R with $C_i \cap D \subset R$ or $C_i \cap D \subset R^c$ for all i.*

Proof. The results above are collected from Turnbull (1976). □

If 4(a) above happens, the non-uniqueness is that the sum $w_j + w_{j'}$ is determined but not w_j itself. Condition 4(b) is like the graph condition in Theorem 6.2. When it happens, $F(R)$ and $F(R^c)$ cannot be determined.

Another simplification can be achieved if we restrict attention to those distributions with a given set of support points. In the sieved NPMLE, we take a list of points $x_1, \ldots, x_{n'} \in \mathcal{X}$, including at least one element from each C_i. If there are indices $j \neq j'$ with $x_{j'} \in C_i$ whenever $x_j \in C_i$ then remove any one such x_j from the list. We repeat this removal process until no more points can be removed, and relabel the remaining points as $x_1, \ldots, x_m$. The sieved likelihood, for censored but not truncated data, is $L_{\text{SIEVE}}(F) = \prod_i F(C_i) = \prod_i \sum_{j=1}^m 1_{x_j \in C_i} F(\{x_j\})$.

Theorem 6.7 *Given points $x_1, \ldots, x_m$, there is a unique sieved-NPMLE maximizing $L_{\text{SIEVE}}(F)$.*

Proof. van der Laan (1995, Chapter 3.3). □

6.5 Product-limit estimators

For real-valued X subject to right censoring or left truncation, there is an explicit closed form for the NPMLE. These NPMLE's are more conveniently derived through the hazard function, defined below, than through the cumulative distribution function.

The survival function is $S(t) = F([t, \infty)) = 1 - F((-\infty, t))$. It is widely used in medical applications, where it is natural to consider the fraction of subjects surviving past time t. For continuous distribution functions with density f, the hazard function is defined as

$$\lambda(t) = \lim_{\varepsilon \to 0^+} \frac{1}{\varepsilon} \Pr(X \leq t + \varepsilon \mid X \geq t) = \frac{f(t)}{S(t+)} = \frac{f(t)}{S(t)}.$$

The product $\lambda(t)dt$ gives the probability of failure before time $t + dt$, conditional on surviving at least to time t. For continuously distributed data $\lambda(t) = -d\log(S(t))/dt$, and so $S(t) = \exp(-\int_0^t \lambda(u)du)$.

For discrete distributions with $F(\{t_j\}) > 0$, for a finite (or countably infinite) number of t_j, the hazard function is defined as

$$\lambda_j = \Pr(X = t_j \mid X \geq t_j) = \frac{F(\{t_j\})}{S(t_j)}.$$

For discrete distributions $S(t) = \prod_{t_j \leq t}(1 - \lambda_j)$, $F(\{t_i\}) = \lambda_j \prod_{t_j < t_i}(1 - \lambda_j)$, and $F((-\infty, t]) = 1 - \prod_{t_j \leq t}(1 - \lambda_j)$.

The cumulative hazard is defined as

$$\Lambda(t) = \int_0^t \frac{dF(u)}{F((-\infty, u))},$$

which simplifies to $\int_0^t \lambda(u)du$ for continuous distributions, to $\sum_{t_j \leq t} \lambda_t$ for discrete distributions, and to a sum of discrete and continuous hazards for distributions with discrete and continuous parts.

For right-censored data, the NPMLE $\hat{F}$ must have positive probability at each observed failure time. Let the observed failure times be $t_1 < t_2 < \cdots < t_k$, suppose that $t_1 > 0 \equiv t_0$, and define $t_{k+1} = \infty$. Let $d_j \geq 1$ be the number of failures at t_j and suppose that m_j observations were censored in the interval $[t_i, t_{i+1})$. The NPMLE puts 0 probability inside the interval (t_i, t_{i+1}) for $i < k$, because moving such probability to t_{i+1} would increase at least one factor in L_c and would not decrease any of them. The number

$$r_j = (d_i + m_i) + \ldots + (d_k + m_k)$$

denotes the number of subjects at risk of failure just prior to t_j.

Let $\lambda_j = \Pr(X = t_j \mid X \geq t_j)$ denote the hazard probabilities of the distribution F. The conditional likelihood given by (6.9) or (6.12) may be written

$$L_c(F) = \prod_{j=1}^{k} \lambda_j^{d_j}(1 - \lambda_j)^{r_j - d_j},$$

and so the NPMLE has $\hat{\lambda}_j = d_j/r_j$. The CDF of the NPMLE may be written

$$F((-\infty, t]) = 1 - \prod_{j \mid t_j \leq t} \frac{r_j - d_j}{r_j} \tag{6.15}$$

This is the celebrated Kaplan-Meier product-limit estimator.

If the largest observed failure time is greater than the largest observed censoring time, then the NPMLE is unique. Otherwise $F((-\infty, t_k]) < 1$ and any distribution that satisfies (6.15) for $t \leq t_k$ is also an NPMLE. A common convention to force uniqueness is to place probability $1 - F((-\infty, t_k])$ on the largest observed censoring time when that time is larger than t_k.

There is also a product-limit estimator for left truncation of X by an independent Y. Let F and G be distributions of X and Y, respectively. Let a_G and b_G be the smallest and largest observable values of Y. Formally $a_G = \inf\{y \mid G((y,\infty)) < 1\}$, and $b_G = \sup\{y \mid G((y,\infty)) > 0\}$. The values a_F and b_F are defined similarly. If $a_G > a_F$, then the lower end of the F distribution cannot be observed. We assume that either $a_G \leq a_F$, or that we are satisfied with inferences on $\Pr(X \leq t)/\Pr(X \geq a_G)$. We also assume that $b_G \leq b_F$. If $b_G > b_F$, then the upper end of the G distribution cannot be observed.

The assumption of independence between X and Y is often reasonable, but is not to be made lightly. In astronomy this assumption follows from a simplifying idea that, at very large scales, space is the same everywhere and in every direction (the cosmological principle). This independence is thought to be nearly, though perhaps not exactly, correct. We suppose that G is unknown, and so we use the conditional likelihood based on the distribution of X given Y.

It is convenient to work in terms of ordered observations $X_{(1)} \leq X_{(2)} \leq \cdots \leq$

$X_{(n)}$. Let $Y_{(i)}$ denote their concomitants, that is $(X_{(i)}, Y_{(i)})$ for $i = 1, \ldots, n$ are the points (X_i, Y_i), $i = 1, \ldots, n$, after sorting on X_i. Of course, the $Y_{(i)}$ are not necessarily in increasing order. To simplify the derivation, we suppose that there are no i and j for which $X_i = Y_j$. Candidates for the NPMLE $\hat{F}$ put nonnegative probability on every observed value $X_{(i)}$ and put no probability anywhere else. For such distributions, $F((Y_{(i)}, \infty)) = F((Z_i, \infty))$, where $Z_i = \max\{X_{(j)} \mid X_{(j)} < Y_{(i)}\}$. The conditional likelihood is

$$L_c(F) = \prod_{i=1}^{n} \frac{F(\{X_{(i)}\})}{F((Z_i, \infty))} = \prod_{i=1}^{n} \frac{F(\{X_{(i)}\})}{F((X_{(i)}, \infty))^{K_i}},$$

where

$$K_i = \#\{j \mid Z_j = X_{(i)}\}$$
$$= \#\{j \mid X_{(i)} < Y_{(j)} \leq X_{(i+1)}\},$$

with $X_{(n+1)} = \infty$.

Writing in terms of $\lambda_i = F(\{X_{(i)}\})/F((X_{(i)}, \infty))$,

$$L_c(F) = \prod_{i=1}^{n} \frac{\lambda_i \prod_{j=1}^{i-1}(1-\lambda_j)}{\left[\prod_{j=1}^{i}(1-\lambda_j)\right]^{K_i}}$$
$$= \prod_{i=1}^{n} \lambda_i (1-\lambda_i)^{n-i-\sum_{j=i}^{n} K_j}.$$

This is maximized by values

$$\hat{\lambda}_i = \frac{1}{n - i + 1 - \sum_{j=i}^{n} K_j}$$
$$= \frac{1}{\#\{j \mid X_{(j)} \geq X_{(i)}\} - \#\{j \mid Y_{(j)} > X_{(i)}\}}$$
$$= \frac{1}{\#\{j \mid Y_{(j)} < X_{(i)} \leq X_{(j)}\}},$$

so that

$$\hat{F}((-\infty, t]) = 1 - \prod_{i=1}^{n}\left(1 - \frac{1_{X_i \leq t}}{\sum_{l=1}^{n} 1_{Y_l < X_i \leq X_l}}\right). \tag{6.16}$$

Equation (6.16) is known as the Lynden-Bell estimator. The Lynden-Bell estimator can be degenerate: $\hat{F}((-\infty, X_{(i)}]) = 1$ for some $i < n$ is possible.

6.6 EL for right censoring

Table 6.1 presents the AML data. Most of the values indicate the time until relapse of a patient whose leukemia has gone into remission. Those values with a + sign designate right-censored times.

Maintained	9	13	13+	18	23	28+
	31	34	45+	48	161+	
Non-Maintained	5	5	8	8	12	16+
	23	27	30	33	43	45

Table 6.1 *Shown are the number of weeks until relapse for patients whose acute myelogenous leukemia (AML) has gone into remission. One group of patients received maintenance chemotherapy, the other did not. Source: Embury et al. (1977).*

The standard 95% confidence interval for $S(t)$ is $\hat{S}(t) \pm 1.96(\widehat{\mathrm{Var}}(\hat{S}(t)))^{1/2}$, using Greenwood's formula

$$\widehat{\mathrm{Var}}(\hat{S}(t)) = \hat{S}(t)^2 \sum_{j \mid t_j \leq t} \frac{d_j}{r_j(r_j - d_j)}. \qquad (6.17)$$

These intervals are based on a central limit theorem. They do not respect range restrictions, in that they can extend outside of the interval $[0, 1]$. When $S(t)$ takes an extreme value like 0.99 or 0.01, then symmetric intervals do not seem as natural as intervals that extend a greater distance towards $1/2$ than away from $1/2$. These standard intervals also tend to have poor coverage accuracy for moderate n.

Let t be a fixed time point for which $S(t)$ is of interest. Define the profile empirical likelihood function

$$\mathcal{R}(s,t) = \max \left\{ \prod_{j=1}^{k} \frac{\lambda_j^{d_j}(1-\lambda_j)^{r_j-d_j}}{\hat{\lambda}_j^{d_j}(1-\hat{\lambda}_j)^{r_j-d_j}} \;\bigg|\; 0 \leq \lambda_j \leq 1, \prod_{t_j \leq t}(1-\lambda_j) = s \right\},$$

for $S(t)$. A Lagrange multiplier argument shows that the maximizing λ_j satisfy

$$\lambda_j = \frac{d_j}{r_j + \gamma 1_{t_j \leq t}}, \qquad (6.18)$$

for a multiplier γ satisfying

$$\sum_{j \mid t_j \leq t} \log\left(\frac{r_j - d_j - \gamma}{r_j + \gamma}\right) - \log(s) = 0.$$

Theorem 6.8 shows that the empirical likelihood ratio may be used to construct pointwise confidence intervals for $S(t)$.

Theorem 6.8 *For $i = 1, \ldots, n$, let $X_i, Y_i \in \mathbb{R}$ be independent random variables with $X_i \sim F$ and $Y_i \sim G$. Let $(Z_i, \delta_i) = (\min(X_i, Y_i), 1_{X_i \leq Y_i})$, $i = 1, \ldots, n$*

be observed. Assume that $G((-\infty, t)) < 1$ and $0 < S(t) < 1$. Then

$$-2 \log \mathcal{R}(S(t), t) \to \chi^2_{(1)},$$

in distribution as $n \to \infty$.

Proof. Thomas & Grunkemeier (1975), Li (1995b), and Murphy (1995). □

The top plot in Figure 6.4 shows the empirical likelihood function for $S(20)$, the probability that remission lasts for at least 20 weeks, in each of the two groups in the AML data. The likelihood curves overlap considerably. The survival difference is apparently not very large and the sample size is also small.

The middle plot in Figure 6.4 shows the empirical likelihood ratio curve for the difference $\Delta = S_M(20) - S_N(20)$ in survival probabilities between the maintained (subscript M) and non-maintained (subscript N) groups. The probability of going 20 weeks or more in remission could reasonably be larger for either group. This likelihood ratio is defined as

$$\mathcal{R}(\Delta) = \frac{\max_{\theta_1 - \theta_2 = \Delta} \mathcal{L}(\theta_1, \theta_2)}{\max_{\theta_1, \theta_2} \mathcal{L}(\theta_1, \theta_2)}$$

where

$$\mathcal{L}(\theta_1, \theta_2) = \max \left\{ \prod_{i=1}^n (nw_i)^{\delta_i} \left(\sum_{j|t_j > t_i} nw_j \right)^{1-\delta_i} \mid w_i \geq 0, \sum_{i=1}^n w_i = 1, \right.$$

$$\left. \sum_{i=1}^n w_i M_i (1_{Y_i \geq 20} - \theta_1) = \sum_{i=1}^n w_i (1 - M_i)(1_{Y_i \geq 20} - \theta_2) = 0, \right\}$$

where M_i is 1 for the maintained group and 0 for the non-maintained group.

The bottom plot in Figure 6.4 shows the empirical likelihood curve for the difference in medians between the two groups. This plot was computed by first computing

$$\mathcal{L}(\theta_1, \theta_2) = \max \left\{ \prod_{i=1}^n (nw_i)^{\delta_i} \left(\sum_{j|t_j > t_i} nw_j \right)^{1-\delta_i} \mid w_i \geq 0, \sum_{i=1}^n w_i = 1, \right.$$

$$\left. \sum_{i=1}^n w_i M_i (1_{Y_i \leq \theta_1} - 1/2) = \sum_{i=1}^n w_i (1 - M_i)(1_{Y_i \leq \theta_2} - 1/2) = 0, \right\}$$

on a fine grid of (θ_1, θ_2) values, then taking

$$\mathcal{R}(\Delta) = \frac{\max_{\theta_1 - \theta_2 = \Delta} \mathcal{L}(\theta_1, \theta_2)}{\max_{\theta_1, \theta_2} \mathcal{L}(\theta_1, \theta_2)},$$

as before. To maximize over θ_1 with $\theta_1 - \theta_2$ fixed at Δ is to profile out a variable that does not enter the estimating equations in a smooth way. There are few results of this kind, but the present case is covered by Theorem 10.1. The median duration of remission does not differ significantly between these two groups.

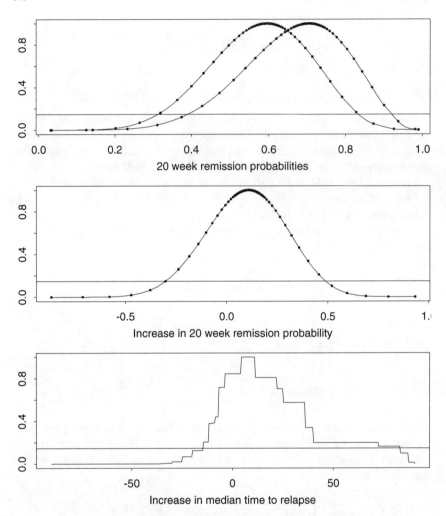

Figure 6.4 *The top figure shows empirical likelihood curves for the probability that remission lasts at least 20 weeks. There is one curve for each of the treatment groups. The curve for the maintained group peaks at 0.7045, lying slightly to the right of the curve for the non-maintained group which peaks at 0.5956. The middle figure shows the empirical likelihood ratio function for the difference between the probabilities of remission lasting at least 20 weeks. This curve peaks at 0.1088. The bottom curve is for the difference in median times to relapse between the two groups. Each plot has a horizontal reference line at the approximate 95% confidence level using a $\chi^2_{(1)}$ criterion.*

6.7 Proportional hazards

The most convenient way to incorporate predictor variables into survival time likelihoods is through the hazard function. Suppose that $X \in \mathbb{R}$ is a survival time, $Y \in \mathbb{R}$ is a corresponding censoring time, and $U \in \mathbb{R}^d$ is a vector of predictors. Let $Z = \min(X, Y)$ and $\delta = 1_{X \le Y}$ as before. Cox's proportional hazards model has survival functions

$$S(X_i \mid U_i = u_i) = S_0(X_i)^{\exp(u_i'\beta)},$$

for a baseline survival function S_0, and a vector of parameters β. If S_0 is a continuous survival function, then this model has hazard function

$$\lambda(X_i \mid U_i = u_i) = \lambda_0(X_i) \exp(u_i'\beta).$$

The exponential model for covariates keeps the hazard function nonnegative. It also means that the effect of changing U_i is to make a proportional increase or decrease in the hazard rate, for all times. The baseline survival distribution corresponds to a random variable X with covariate vector $U = 0$. To make the baseline correspond to a default value U_0, it is only necessary to replace each U_i by $U_i - U_0$.

The data can be organized through two sequences of events unfolding in time. The first sequence specifies the time of the next failure to occur, and the second sequence specifies which of the subjects currently under study is the one to fail at that time. With S_0 completely unknown, it is reasonable that only the second sequence contains information on β. Suppose that there are no ties among the Z_i. For $j = 1, \ldots, k = \sum_i \delta_i$, let the item labeled (j) be the one with the j'th largest of the observed failure times $\{Z_i \mid \delta_i = 1\}$. Let $R_j = \{i \mid X_{(j-1)} < Z_i \le X_{(j)}\}$ be the set of individuals at risk of failure, just prior to time $X_{(j)}$, taking $X_{(0)} = 0$. The partial likelihood is

$$L_P(\beta) = \prod_{j=1}^{k} \frac{\exp(u_{(j)}'\beta)}{\sum_{i \in R_j} \exp(u_i'\beta)},$$

after canceling $\lambda_0(X_{(j)})\Delta t$ from the numerator and denominator in each of the k factors. This partial likelihood can be extended, with some difficulty, to take account of ties in Z_i, for observations not necessarily having having tied U_i and δ_i.

The partial likelihood behaves like an ordinary parametric likelihood. Maximizing it provides consistent asymptotically normal estimates of β, under mild assumptions, and the profile likelihood formed by maximizing over S_0 can be used to construct confidence regions for β.

6.8 Further empirical likelihood ratio results

Asymptotic χ^2 distributions have been obtained for numerous truncation and censoring settings. Some theorems are quoted here. Some more are described in Chapter 6.9.

For left-truncated data of Example 6.2 let $L_c(F)$ be the conditional likelihood $L(F, G; \mathcal{Y} \mid \mathcal{X})$ given on page 138. Define

$$\mathcal{R}_t(p) = \frac{\max\{L_c(F) \mid F((-\infty, t]) = p\}}{\max_F\{L_c(F)\}}.$$

Theorem 6.9 *Let $X \sim F_0$ and $Y \sim G_0$ be independent from continuous distributions. Assume that $\inf\{x \mid F_0(x) > 0\} > \inf\{x \mid G_0(x) > 0\}$ and that $F_0(t) = p_0 > 0$. Then $-2\log \mathcal{R}_t(p_0) \to \chi^2_{(1)}$ as $n \to \infty$.*

Proof. Li (1995a, Theorem 1). □

Suppose that $(X_i, Y_i, Z_i) \in \mathbb{R}^3$ are independent and identically distributed, with X doubly censored, by Y_i on the right and by X_i on the left, as in Example 6.4. Let F_X, F_Y, and F_Z be the distributions of X, Y, and Z, respectively. If $Z < X \leq Y$, let $U = X$ and $\delta = 0$, if $X > Y$, let $U = Y$ and $\delta = 1$, and if $X \leq Z$, let $U = Z$ and $\delta = -1$.

Then the conditional likelihood function for F_X is

$$L_c(F) = \prod_{i=1}^n F(\{U_i\})^{\delta_i=0} F((U_i, \infty))^{\delta_i=1} F([0, U_i))^{\delta_i=-1},$$

using x^A as a shorthand for x^{1_A}. Let $\theta = \int q(x) dF_X(x)$, for a known function q, and define

$$\mathcal{R}(\theta) = \frac{\max\{L_c(F) \mid \int q(x) dF(x) = \theta\}}{\max_F\{L_c(F)\}}.$$

Theorem 6.10 *Let $\mathcal{R}$ and θ be as described above. Suppose that F_X, F_Y, and F_Z are continuous distributions, that $F_X([A, B]) = 1$, for some $0 \leq A < B < \infty$, that $\Pr(Z < u \leq Y) > 0$ for all $u \in [A, B]$, that $F_Z([0, B]) = 1$, and that $F_Y([0, A)) = 0$. Suppose that q is a left continuous function of bounded variation on $[A, B]$ with $\int q^2(x) dF_X(x) - (\int q(x) dF_X(x))^2 > 0$. Then $-2\log \mathcal{R}(\theta_0) \to \chi^2_{(1)}$.*

Proof. Murphy & van der Vaart (1997, Theorem 2.1). □

Now consider current status data as in Example 6.5, with only one observation time Z, but with a Cox model covariate $U \in \mathbb{R}$. The event time is X, and we observe Z and $\delta = 1_{X_i \leq Z_i}$. Suppose that conditionally on U, the hazard function of X is $\lambda(t)\exp(\theta U)$ for a parameter $\theta \in \mathbb{R}$. The conditional likelihood in terms of the cumulative hazard of X is

$$L(\Lambda, \theta) = \prod_{i=1}^n \Big(1 - \exp(-e^{\theta U_i}\Lambda(Z_i))\Big)^{\delta_i} \Big(\exp(-e^{\theta U_i}\Lambda(Z_i))\Big)^{1-\delta_i}.$$

Now define

$$\mathcal{R}(\theta) = \frac{\max_\Lambda\{L(\Lambda, \theta)\}}{\max_{\Lambda, \eta}\{L(\Lambda, \eta)\}},$$

where Λ is maximized over the space of nondecreasing functions continuous from the right with limits from the left and taking values in $[0, C]$ for a known bound C, and θ is maximized over a parameter set Θ.

Theorem 6.11 *For the current status setting above let the observation time Z have a continuous positive density function on $[A, B]$ for some $0 < A < B < \infty$, where the true cumulative hazard Λ_0 of X satisfies $\Lambda_0(A-) > 0$ and $\Lambda_0(B) < C$. Let Λ_0 be differentiable on $[A, B]$ with a derivative everywhere above some $\epsilon > 0$, let U be bounded with $E(\mathrm{Var}(U \mid Z)) > 0$, and assume that the true value θ_0 is interior to Θ. Finally, assume that $\phi(Z)$ has a bounded derivative on $[A, B]$ where*

$$\phi(Z) = \frac{E(U\psi(U, X, \delta)^2 \mid Z)}{E(\psi(U, X, \delta)^2 \mid Z)}, \quad \text{and}$$

$$\psi(U, X, \delta) = e^{\theta_0 U}\left[\delta \frac{\exp(-\exp(\theta_0 U))\Lambda(X)}{1 - \exp(-\exp(\theta_0 U))\Lambda(X)} - (1 - \delta)\right].$$

Then $-2\log \mathcal{R}(\theta_0) \to \chi^2_{(1)}$ as $n \to \infty$.

Proof. Murphy & van der Vaart (1997, Theorem 2.2). □

A form of cumulative hazard estimating equation can be used to define parameters in survival analysis. There, the parameter θ solves $\int q(x, \theta) d\Lambda(x) = C$ for some constant C.

Theorem 6.12 *For right-censored data as described in the conditions of Theorem 6.8, suppose that $q(x)$ is a left continuous function with*

$$0 < \int \frac{|q(x, \theta)|^m}{F([x, \infty))G([x, \infty))} d\Lambda(x) < \infty, \quad m = 1, 2.$$

Then $-2\log \mathcal{R}(\theta_0) \to \infty$ where $\mathcal{R}(\theta)$ is an empirical likelihood ratio defined through the hazard function at observed failure times, and maximized subject to $\int g(x, \theta) d\Lambda(x) = C$.

Proof. Pan & Zhou (2000, Theorem 5). □

6.9 Bibliographic notes

Biased sampling

Bratley, Fox & Schrage (1987) is a standard reference on Monte Carlo that includes importance sampling. Cochran (1977) and Lohr (1998) are standard references on finite population sampling. Cox (1967) proposed the harmonic mean for length biased data. The shrub data are from Muttlak & McDonald (1990). Jones (1991) uses them to investigate kernel density estimation from length biased samples.

The problem of constructing an NPMLE from one length biased and one unbiased sample was considered by Vardi (1982). Cosslett (1981) proposes an NPMLE for choice-based sample data. Vardi (1985) and Gill et al. (1988) consider multiple samples each with their own biasing conditions. Vardi (1985) gives an algorithm for computing the NPMLE, and shows that the NPMLE $\hat{F}$ is a sufficient statistic for the unknown F_0.

Empirical likelihood for a combination with one biased and one unbiased sample was considered by Qin (1993). Qin proves an ELT for the mean with one biased and one unbiased sample and states that the result holds more generally.

Qin & Zhang (1997) study a problem in which the bias functions in multiple biased sampling contain parameters. In group i the data are a biased sample defined through $u_i(x, \eta_i)$, incorporating unknown parameter vectors η_i. They were motivated by case-control studies, where samples are taken of people with and without a rare condition. Empirical likelihood methods can be used to draw inferences on the η_i, as well as to test the goodness of fit of the parametric specification of the bias function. Fokianos, Peng & Qin (1999) use this idea to test the goodness of a logistic link function. Qin (1999) considers three samples, $X_i \sim F$, $Y_i \sim G$, and $Z_i \sim \lambda F + (1 - \lambda)G$ for an unknown mixture proportion λ. The distributions F and G have densities f and g that are assumed to satisfy $\log(g(x)/f(x)) = \beta_0 + x\beta_1$. Qin (1999) finds asymptotic chisquared distributions for likelihood ratio tests of λ and β_1.

Qin (1998) considers empirical likelihood inference in upgraded mixture models. This setting combines a "good sample" of directly observed data $Z \sim F$, with a "bad sample" of data from a density $p(x) = \int p(x|z)dF$ for a conditional density $p(x|z)$. These models originate with Hasminskii & Ibragimov (1993). The name is from van der Vaart & Wellner (1992), who develop a discrete consistent estimator of F. An example from Vardi & Zhang (1992) has $X = ZU$ for $U \sim U(0, 1)$ independently of Z. The KDD CUP 2000 data mining competition (Kohavi, Brodley, Frasca, Mason & Zheng 2001) featured Internet log entries, including some complete sessions of length Z and some clipped sessions that only included the first $\lceil UZ \rceil$ entries for a uniform U. The winning entry of Rafal Kustra, Jorge Picazo, and Bogdan Popescu showed that clipping produced an artifact wherein the rate at which visitors left a site appeared to increase with the duration of their session, although the true departure rate declined with duration. Qin (1998) shows how to use empirical likelihood inferences on upgraded mixture models, and how to incorporate parameterized data distortions $p(x \mid z; \theta)$.

Censoring and truncation

Peto (1973) considers NPMLE's for general patterns of interval censored real values, and proves Theorem 6.5. The proof of Theorem 6.4 is adapted from the argument in Peto (1973).

An example of a bad NPMLE arises for bivariate failure times $X_i = (X_{i1}, X_{i2})$ where X_{ij} is subject to right censoring by Y_{ij}. If X_{i1} is censored but X_{i2} is not, then $C_i = \{(x_1, X_{i2}) | Y_{i1} < x_1 < \infty\}$ is a ray of infinite length in the plane. If X_i have a continuous distribution F then there will never be an $X_{i'} \in C_i$ for $i' \neq i$. As Tsai, Leurgans & Crowley (1986) note, such additional points are necessary to properly distribute probability within C_i, and without them the NPMLE is not even consistent. van der Laan (1996) proposes a way to fix this problem in which a ray like C_i is replaced by a thin strip, with a width that decreases to 0 as $n \to \infty$.

Turnbull (1976) is a definitive reference on NPMLE's for combinations of censored and truncated real-valued data. Turnbull (1976)'s description of censoring is a form of coarsening at random. Heitjan & Rubin (1991) define and illustrate coarsening at random and show that under coarsening at random, the conditional likelihood is proportional to the full likelihood. Coarsening at random includes information loss due to missing data components or rounding.

Turnbull (1976) provides a self-consistency algorithm for finding the NPMLE. This algorithm is an example of the EM algorithm (see Dempster, Laird & Rubin (1977) and Baum (1972)). Efron (1967) used self-consistency to derive the Kaplan-Meier estimator. There is as yet no ELT for the general setting Turnbull considers.

The survival, hazard, and cumulative hazard functions are defined in Fleming & Harrington (1991), as well as in Kalbfleisch & Prentice (1980), who use a survivor function $\Pr(X \geq t)$ instead of the survival function $\Pr(X > t)$.

Kaplan-Meier estimator

Kaplan & Meier (1958) introduced the product-limit estimator for right-censored survival times, using an NPMLE argument. Similar estimators had previously been used by actuaries. The variance estimate of the Kaplan-Meier estimate is from Greenwood (1926). The NPMLE derivation is based on Kalbfleisch & Prentice (1980), who also present a derivation of Greenwood's formula.

Thomas & Grunkemeier (1975) gave a heuristic proof of Theorem 6.8. This was later made rigorous by Li (1995b) and by Murphy (1995). Murphy (1995) proves an ELT for inferences on the cumulative hazard function. Equation (6.18) was obtained by Thomas & Grunkemeier (1975) and independently by Cox & Oakes (1984, Chapter 4.3) who use it to derive Greenwood's formula (6.17) from the curvature of the censored data empirical log likelihood.

Adimari (1997) considers empirical likelihood inferences for the mean of a distribution under independent right censoring. He finds an asymptotic chisquared distribution for $2n \sum_{i=1}^{n} \widetilde{p}_i \log(1 + \lambda'(T_i - \mu))$ where $\widetilde{p}_i$ is the Kaplan-Meier probability of the observed failure time T_i and λ satisfies $\sum_{i=1}^{n} \widetilde{p}_i (T_i - \mu)/(1 + \lambda'(T_i - \mu)) = 0$.

Pan & Zhou (2000) prove Theorem 6.12. They also establish a chisquared calibration for parameters $\int q_n(x) d\Lambda(x)$ where q_n is a data-dependent function. Such parameters often arise where a data-based estimate of one quantity is plugged into an equation for another.

The AML data come from Embury et al. (1977). They are reproduced in Miller, Gong & Munoz (1981).

Lynden-Bell and astronomy

Efron & Petrosian (1994) explore some data where objects that are either too bright or too dim are truncated. The dim objects are not visible, while the bright ones are possibly not the sort of object of interest. They introduce a nonparametric estimate of the distribution function of brightness for such doubly truncated data and provide a bootstrap-based test of the cosmological principle.

Keiding & Gill (1990) and Woodroofe (1985) provide a detailed analysis of left-truncated sampling. The NPMLE in this case was found by Lynden-Bell (1971) and is known as the Lynden-Bell estimator. Lynden-Bell considered the more general setting in which an (X,Y) pair was observed with probability $u(X,Y)$, allowing a model in which the probability of observing an object of given brightness decreases smoothly from 1 to 0 as its distance from Earth increases. Lynden-Bell (1971) gives a small data set of 40 3CR quasars. The NPMLE is degenerate, putting positive weight on only three of the quasars. Woodroofe (1985) describes conditions leading to this degeneracy and conditions in which the probability of a degenerate NPMLE vanishes as $n \to \infty$. Lynden-Bell (1971) also implements a fix in which the histogram of an intermediate quantity is replaced by the nearest unimodal one.

Wang (1987) proves Theorem 6.3. Keiding & Gill (1990) provide another proof and add the caveat that the maximizer $\widetilde{F}$ of the conditional likelihood is not a component of the joint NPMLE $(\hat{F}, \hat{G})$ in cases where $\widetilde{F}$ is degenerate. Li (1995a) proves Theorem 6.9. Li, Qin & Tiwari (1997) consider the case where there is a known parametric family of distributions for G, but not for F. They use the marginal distribution of the X_i because in this setting it can have more information than the conditional distribution of the X_i given the Y_i. They also show how to get empirical likelihood ratio confidence regions for the probability α that an observation is not truncated.

Other

The proportional hazards model in Chapter 6.7 was proposed by Cox (1972). The partial likelihood argument for it is due to Cox (1975). Bailey (1984) considered the joint likelihood for β and S_0 taking jumps at observed failure times. He showed that the estimate of β obtained by maximizing the likelihood over β and S_0 is asymptotically equivalent to the one obtained by maximizing the partial likelihood. The resulting estimate of the cumulative hazard is equivalent to the one in Tsiatis (1981). Confidence regions for β or for the cumulative hazard (at finitely many points) can be obtained from the curvature of the log likelihood. Bailey (1984) remarks that the presence of a large number of nuisance parameters does not lead to difficulty. Murphy & van der Vaart (2000) consider the problem of infinite dimensional nuisance parameters more generally.

EXERCISES 153

Murphy & van der Vaart (1997) prove Theorems 6.10 and 6.11. They also establish χ^2 limits for some frailty models incorporating random effects into the proportional hazards framework.

In this chapter, survival analysis was viewed as analysis of life times that might be missing or partially observed. The modern treatment of survival analysis treats each subject's data as a counting process observed over a time window. The number of deaths for an individual is a counting process that starts at 0 and may increase to 1 in the time window of observation. A second counting process takes the value 1 if the individual is at risk of failure and 0 otherwise, whether the reason be failure or censoring. For a more comprehensive treatment of survival analysis, using the theory of counting processes, see Fleming & Harrington (1991) and Andersen, Borgan, Gill & Keiding (1993), with a very accessible applied presentation in Therneau & Grambsch (2000). Counting process models extend naturally to handle competing risks from different causes of death, events such as infections which can recur for individuals, and transitions between states such as cancer and remission.

The dual likelihood of Mykland (1995) is an extension of empirical likelihood to martingales. Dual likelihood inferences should cover many or most of the counting process examples, though this is outside the scope of the present text.

6.10 Exercises

Exercise 6.1 Suppose that $X_1, \ldots, X_n$ are IID with the exponential probability density function $f(x; \theta) = \theta \exp(-\theta x) 1_{x>0}$. Thus θ is the failure rate per unit time and $1/\theta = E(X)$. Suppose that $Y_1, \ldots, Y_n$ are censoring times independent of $X_1, \ldots, X_n$, and let the observations be $Z_i = \min(X_i, Y_i)$ and $\delta_i = 1_{X_i < Y_i}$. Write an expression for the parametric conditional likelihood of $X_1, \ldots, X_n$ given $Y_1, \ldots, Y_n$, in terms of Z_i and δ_i. Find the conditional MLE $\hat{\theta}$. Is this quantity interpretable, in the case where X_i are not exponentially distributed?

Exercise 6.2 Suppose that F puts weight $w_i \geq 0$ on x_i and that G puts weight $v_j \geq 0$ on y_j, where $\sum_{i=1}^{n} w_i = \sum_{j=1}^{n} v_j = 1$. Let $u_{ij} = 1_{x_i > y_j}$, and define $\alpha = \Pr(X > Y) = \sum_{i=1}^{n} \sum_{j=1}^{n} w_i v_j u_{ij}$. Show by Lagrange multipliers that

$$w_i = \left(\sum_{j=1}^{n} \frac{u_{ij}}{\sum_{k=1}^{n} w_k u_{kj}} \right)^{-1}$$

for the NPMLE in the Lynden-Bell setup.

CHAPTER 7

Bands for distributions

This chapter considers confidence bands for a distribution function and some related functions. Chapter 5.8 describes bands for kernel density estimates. For $X \in \mathbb{R}$, the cumulative distribution function is

$$\mathsf{F}(x) = F((-\infty, x]) = \Pr(X \le x)$$

taken as a function of x.

A confidence band for $\mathsf{F}(x)$ is a pair of functions $L(x)$ and $H(x)$ for which

$$\Pr(L(x) \le \mathsf{F}(x) \le H(x),\ \forall x \in \mathbb{R}) = 1 - \alpha \qquad (7.1)$$

under independent sampling of $X_i \sim F$. The randomness in (7.1) arises from the fact that L and U depend on $X_1, \ldots, X_n$, although this is suppressed from the notation. Some exact confidence bands are available, others are asymptotic.

If the inequalities in (7.1) were imposed only at B points x, the result could be described as a B-dimensional hyper-rectangular confidence region. Bands are essentially infinite dimensional hyper-rectangles. As such, they are do not necessarily correspond to tests with the greatest power. Ellipsoids or other shapes are often better. Bands have the advantage that they can be easily plotted.

Bands are also of interest for some related functions. The quantile function $Q(u)$ is defined through

$$Q(u) = \mathsf{F}^{-1}(u) \equiv \inf\{x \mid u \le \mathsf{F}(x)\}, \quad 0 < u < 1. \qquad (7.2)$$

The definition (7.2) makes Q unique even when $\mathsf{F}(x) = \mathsf{F}(x') = u$ for $x \ne x'$.

For independent real-valued data $X_1, \ldots, X_n \sim F$ and $Y_1, \ldots, Y_m \sim G$, the QQ plot is formed by plotting an estimate of $QQ(x) = \mathsf{G}^{-1}(\mathsf{F}(x))$. If the sample QQ plot lies far from the 45° line $QQ(x) = x$, then the distributions F and G differ.

For three or more samples from distributions $F_1, \ldots, F_k$, we can select one of the distributions, say F_1, as a baseline, and define a $k - 1$ dimensional quantile-quantile function by $(\mathsf{F}_2^{-1}(\mathsf{F}_1(x)), \ldots, \mathsf{F}_k^{-1}(\mathsf{F}_1(x)))$, over x.

The survival function is $S(t) = F((t, \infty)) = 1 - F((-\infty, t])$. It is widely used in medical applications, as is the cumulative hazard function

$$\Lambda(t) = \int_0^t \frac{dF(u)}{F((-\infty, u))}.$$

These are discussed in Chapter 6.5.

7.1 The ECDF

The empirical CDF is the value $\hat{F}(x) = \#\{X_i \le x\}/n$, taken as a function of x. The 95% Kolmogorov-Smirnov bands for F are of the form $\hat{F}(x) \pm D_n^{0.95}$, where $D_n^{1-\alpha}$ is defined in terms of the random variable

$$D_n \equiv \sup_{-\infty < x < \infty} \left| \hat{F}(x) - F(x) \right|, \qquad (7.3)$$

by $\Pr(D_n \le D_n^{1-\alpha}) = 1 - \alpha$.

Such bands can have exact coverage levels for finite n, because the distribution of D_n for $X_i \sim F$ is the same for any continuous distribution F. If F is not continuous, then Kolmogorov-Smirnov bands have greater than the nominal coverage level. To see why the distribution of D_n does not depend on F, write the order statistics of the sample as $X_{(1)} \le X_{(2)} \le \ldots \le X_{(n)}$, and introduce random variables $U_i = F(X_i)$. The U_i are independent observations from the $U(0,1)$ distribution, and have order statistics $U_{(i)} = F(X_{(i)})$. For continuous F the supremum in (7.3) occurs either immediately to the left or right of an observation $X_{(i)}$, so

$$D_n = \max_{1 \le i \le n} \max \left(\left| \frac{i-1}{n} - F(X_{(i)}) \right|, \left| \frac{i}{n} - F(X_{(i)}) \right| \right)$$

$$= \max_{1 \le i \le n} \max \left(\left| \frac{i-1}{n} - U_{(i)} \right|, \left| \frac{i}{n} - U_{(i)} \right| \right).$$

For any continuous F, D_n can be expressed in terms of the order statistics of a uniform sample, and so $D_n^{1-\alpha}$ can be calculated for one distribution, such as $F = U(0,1)$, and then applied to any continuous distribution. The hypothesis that X_i have CDF F is rejected at level α when F is not contained within the bands at all t.

Where the upper band goes above 1 it is replaced by 1, and similarly the lower band is replaced by 0 where it goes below 0. The Kolmogorov-Smirnov bands are widely used, but they are not particularly sensitive in the tails. To address this problem, weighted Kolmogorov-Smirnov bands, of the form

$$D_{n\psi} = \sup_{-\infty < x < \infty} \psi(F(x)) \left| \hat{F}(x) - F(x) \right|,$$

have been proposed. For example, the choice

$$\psi(z) = (z(1-z))^{-1/2} \qquad (7.4)$$

weights each point x in inverse proportion to the standard deviation of $\hat{F}(x)$, and so puts more weight on the tail regions.

The random variable $n\hat{F}(x)$ has the binomial distribution with parameters n and $p = F(x)$. Kolmogorov-Smirnov bands are based on the most extreme discrepancy between the observed and expected binomial random variables. The weighted version with weights (7.4) takes account of the unequal variances of

EXACT CALIBRATION OF ECDF BANDS

those binomial random variables. Empirical likelihood bands may be constructed using the most extreme binomial likelihood at any x.

Empirical likelihood for $F(x)$ at a single point x was presented in Chapter 3.6. For $0 < p < 1$, and $-\infty < q < \infty$, define

$$\mathcal{R}(p,q) = \max\left\{\prod_{i=1}^{n} nw_i \mid \sum_{i=0}^{n+1} w_i Z_i(p,q) = 0, w_i \geq 0, \sum_{i=0}^{n+1} w_i = 1\right\},$$

with $Z_i(p,q) = 1_{X_i \leq q} - p$, taking $X_0 = -\infty$ and $X_{n+1} = \infty$, so that $Z_0 = 1 - p$ and $Z_{n+1} = -p$. An asymptotic confidence interval for $F(x)$ is $\{p \mid -2\log \mathcal{R}(p,x) \leq \chi^2_{(1)}\}$.

To get a confidence band for F, we consider the distribution of the most extreme pointwise likelihood, via

$$E_n = \sup_{-\infty < x < \infty} -\log \mathcal{R}(\mathsf{F}(x), x).$$

Let $c_n^{1-\alpha}$ satisfy $\Pr(E_n \leq c_n^{1-\alpha}) = 1 - \alpha$. Then the band $(L(x), H(x))$ with

$$L(x) = \min\left\{p \mid -\log \mathcal{R}(p,x) \leq c_n^{1-\alpha}\right\}$$
$$H(x) = \max\left\{p \mid -\log \mathcal{R}(p,x) \leq c_n^{1-\alpha}\right\}$$

is a $100(1-\alpha)\%$ confidence band for $F(x)$. First we consider constructing L and H given $c_n^{1-\alpha}$, then we consider how to find $c_n^{1-\alpha}$.

7.2 Exact calibration of ECDF bands

It is computationally easy to obtain an exact calibration for empirical likelihood bands. The reason is that for any set of numbers $a_1, \ldots, a_n$ and $b_1, \ldots, b_n$, there is a recursive algorithm to compute

$$\Pr\left(a_i \leq U_{(i)} \leq b_i, \quad i = 1, \ldots, n\right).$$

See the discussion of Noé's recursion in Chapter 7.4. Noé's recursion also applies to weighted Kolmogorov-Smirnov confidence bands.

From equation (3.15) in Chapter 3.6,

$$-\frac{1}{n}\log \mathcal{R}(p,x) = \hat{p}\log(\hat{p}/p) + (1-\hat{p})\log((1-\hat{p})/(1-p)), \tag{7.5}$$

where $\hat{p} = \hat{p}(x) = \#\{X_i \leq x\}/n = F_n((-\infty, x])$, and $p = \mathsf{F}(x)$. For fixed $\hat{p}$, equation (7.5) is a convex function of p with a minimum of 0 at $p = \hat{p}$. Thus $L(x)$ and $H(x)$ can be easily found by safeguarded searches, like those described in Chapter 2.9, starting in the intervals $(0, \hat{p})$ and $(\hat{p}, 1)$, respectively. Convexity in p of (7.5) implies that $-\log \mathcal{R}(p,x) \leq c_n^{1-\alpha}$ if and only if $L(x) \leq p \leq H(x)$. The bands $L(x)$ and $H(x)$ are piecewise constant functions, taking jumps at the n observed values $X_{(i)}$. Therefore, it is only necessary to compute them at $n+1$ different points. Let L_i and H_i be the values of $L(x)$ and $H(x)$, respectively, on the open interval $(X_{(i)}, X_{(i+1)})$, for $i = 0, \ldots, n$, with $X_{(0)} = -\infty$ and $X_{(n+1)} = \infty$.

Having found either the L_i or the H_i, the other ones can be found by symmetry through

$$L_i = 1 - H_{n-i}.$$

Note that $L(X_{(i)}) = \min(L_{i-1}, L_i) = L_{i-1}$ and $H(X_{(i)}) = \max(H_{i-1}, H_i)$, for $1 \leq i \leq n$. Therefore, $H(x)$ is continuous from the right and $L(x)$ is continuous from the left.

To calibrate the curves we need to find $c_n^{1-\alpha}$. The extreme value of E_n must take place at or just to the left of an order statistic $X_{(i)}$. Thus

$$E_n = \max_{1 \leq i \leq n} \max \left(-\log \mathcal{R}\left(\mathsf{F}(X_{(i)}), X_{(i)}-\right), -\log \mathcal{R}\left(\mathsf{F}(X_{(i)}), X_{(i)}\right) \right).$$

Suppose that F is continuous. Then $\mathcal{R}(p, q)$ with $X_i \sim F$ is the same as $\mathcal{R}(p, \mathsf{F}(q))$ on data $U_i = \mathsf{F}(X_i)$. Thus we may write

$$E_n = \max_{1 \leq i \leq n} \max \left(-\log \mathcal{R}\left(U_{(i)}, \frac{i}{n}-\right), -\log \mathcal{R}\left(U_{(i)}, \frac{i}{n}\right) \right)$$

$$= \max_{1 \leq i \leq n} \max \left(-\log \mathcal{R}\left(U_{(i)}, \frac{i-1}{n}\right), -\log \mathcal{R}\left(U_{(i)}, \frac{i}{n}\right) \right).$$

Now $E_n \leq c_n^{1-\alpha}$ is equivalent to

$$a_i \equiv L_{i-1} \leq U_{(i)} \leq H_{(i)} \equiv b_i, \quad i = 1, \ldots, n.$$

It follows that Noé's algorithm can be employed to find the coverage probability for any value of $c_n^{1-\alpha}$. A one-dimensional numerical search can then be employed to find the value of $c_n^{1-\alpha}$.

Critical values $c_n^{1-\alpha}$ can be precomputed and tabulated. It may be more convenient to store them as a function of n. The function values in Table 7.1 give very accurate coverage for the standard coverage levels 0.95 and 0.99, for sample sizes up to 1000.

7.3 Asymptotics of bands

The confidence bands of the previous section were constructed without employing any asymptotics. This was made possible by Noé's recursion. These bands have good power properties. Suppose that X_i have a continuous distribution F. Then the empirical likelihood confidence band of level $1-\alpha$ has better asymptotic power for rejecting an alternative $\tilde{F} \neq F$ than a weighted Kolmogorov-Smirnov band of level $1 - \alpha$. This holds simultaneously for all weighted Kolmogorov-Smirnov bands and all alternatives $\tilde{F} \neq F$. Such universal optimality is surprising because $\tilde{F}$ might only differ from F in a narrow interval, and a weighted Kolmogorov-Smirnov band might be constructed to be particularly sensitive to departures from F in just that one interval. See Chapter 7.4. The power consid-

ASYMPTOTICS OF BANDS

Coverage 95% to 95.01%
  Sample size $n = 1$:
    2.9957
  Sample sizes $1 < n \leq 100$:
    $3.0123 + 0.4835 \log(n) - 0.00957 \log(n)^2 - 0.001488 \log(n)^3$
  Sample sizes $100 < n \leq 1000$:
    $3.0806 + 0.4894 \log(n) - 0.02086 \log(n)^2$

Coverage 99% to 99.01%
  Sample size $n = 1$:
    4.60517
  Sample sizes $1 < n \leq 100$:
    $4.626 + 0.541 \log(n) - 0.0242 \log(n)^2$
  Sample sizes $100 < n \leq 1000$:
    $4.71 + 0.512 \log(n) - 0.219 \log(n)^2$

Table 7.1 *Shown are approximate critical values $c_n^{1-\alpha}$, for empirical likelihood confidence bands for the CDF from Owen (1995). The nominal coverage level is $1 - \alpha$, either 0.95 or 0.99. The actual coverage level is between the nominal level, and the nominal plus 0.0001. The sample sizes are from $n = 1$ to $n = 1000$.*

ered is of large deviations type. Further large deviations results are described in Chapter 13.5.

The empirical likelihood confidence bands are based on the distribution of the most extreme of $2n$ binomial p-values, arising from an upper and a lower bound at each of n points. These p-values are strongly correlated with each other because they are based on the same data. It is interesting to compare the critical value of the likelihood used in setting bands with the finite degrees of freedom case. Figure 7.1 plots $c_n^{0.95}$ versus n for $1 \leq n \leq 1000$. The effective degrees of freedom corresponding to c_n are defined to be d such that $\Pr(\chi^2_{(d)} \leq 2c_n) = 0.95$. The factor of 2 enters because in parametric settings the test statistic is minus twice a log likelihood where $c_n^{0.95}$ was developed for a negative log likelihood. Chisquareds on fractional degrees of freedom are Gamma distributions.

For $n = 1$, the effective degrees of freedom are $d = 2$. The effective degrees of freedom increase very slowly with n, to $d = 3$ at $n = 7$, to $d = 4$ at $n = 62$, and to $d = 5$ at some $n > 1000$. The effective degrees of freedom would be slightly different at a confidence level other than 0.95. The effective degrees of freedom are very nearly linear in c_n.

The case $n = 1$ is interesting. It involves just one quantile. As $n \to \infty$ for one quantile a $\chi^2_{(1)}$ limit is appropriate. The effect of $n = 1$ instead of $n = \infty$ is to change the degrees of freedom from 1 to 2.

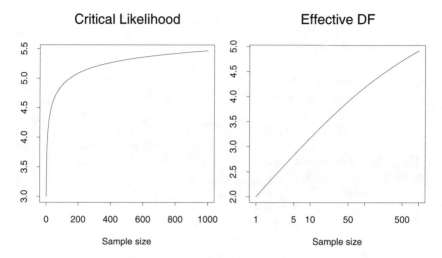

Figure 7.1 *The left plot shows the critical likelihood threshold for exact 95% empirical likelihood confidence bands for the distribution function. The sample sizes range from 1 to n. A critical likelihood of c corresponds to an effective degrees of freedom of d where* $\Pr(\chi^2_{(d)} \leq 2c) = 0.95$. *The right plot shows effective degrees of freedom versus sample size. The two quantities have nearly the same dependence on sample size. This is nearly linear on a log scale as shown in the right plot.*

7.4 Bibliographic notes

Exact confidence bands for the CDF based on empirical likelihood were published by Owen (1995). Hollander, McKeague & Yang (1997) find asymptotic confidence bands for the survival function, $1 - F$, from right-censored data.

The weights (7.4) were proposed by Anderson & Darling (1952). The better known Anderson-Darling statistic is based on an integral over x, not an extreme as presented here. It corresponds to an infinite dimensional ellipsoidal region instead of an infinite dimensional hyper-rectangle.

The recursive algorithm for finding the probability that the ECDF from a $U(0,1)$ sample stays within a given band is due to Noé (1972). It takes $O(n)$ space, and appears to be numerically stable for $n \leq 1000$. Noé's algorithm is given in Shorack & Wellner (1986). The fact that the bands described here give a test with better asymptotic power than any weighted Kolmogorov-Smirnov test at any alternative to $U(0,1)$ was proved by Berk & Jones (1979) using the notion of relative optimality discussed in Berk & Jones (1978).

Qin & Lawless (1994) show that the error in estimating a distribution function is smaller if side information is used. Zhang (1996a) and Zhang (1999) describe confidence bands for the distribution function, given some side information expressed through estimating equations.

BIBLIOGRAPHIC NOTES

Switzer (1976) computes a confidence band for the QQ function by inverting Smirnov's two sample rank test. Confidence bands for the quantile function are given by Zhang (1997), by resampling from the NPMLE. Li, Hollander, McKeague & Yang (1996) present confidence bands for the quantile function from censored data. Einmahl & McKeague (1999) create empirical likelihood-based confidence tubes for QQ plot relating samples from two or more populations.

CHAPTER 8

Dependent data

Empirical likelihood was motivated by independent identically distributed data. As Theorem 4.1 shows, the requirement for identically distributed data can be relaxed. When the observations are dependent, then this usually has to be accounted for in constructing confidence regions and tests. The ways of handling dependent data with empirical likelihood parallel the methods from parametric likelihood and the bootstrap. Failure to account for dependence among the data can destroy the coverage properties of confidence regions.

In order to construct nonparametric confidence regions for dependent data, we must assume something about the nature of the dependence. For some time series problems, we assume that the dependent data are driven by an unobserved set of independent random variables. For some other time series, we assume that there is possibly very strong dependence between relatively few pairs of observations. By contrast, some finite population sampling settings have very weak dependence between many or even all pairs of observations.

8.1 Time series

Chapter 8.10 gives some background references on time series. Here we provide some definitions. A time series is a sequence of observations $Y_i \in \mathbb{R}^d$, $i = 1, \ldots, T$, where Y_{i+1} is observed one time unit after Y_i. The time unit could be a fixed amount of real time, such as a day or year, or it could simply indicate the order in which values were observed.

Models with independent Y_i are seldom appropriate for time series data. There is generally some dependence among series values to account for. Time series are usually modeled as realizations of stochastic processes in which $(Y_1, \ldots, Y_T)$ is drawn from a joint distribution on $\mathbb{R}^{dT}$.

For some joint distributions of $Y_1, \ldots, Y_T$, there is clearly no way to learn about the underlying stochastic process. As an extreme example, suppose that $Y_i = Z + e_i$ for all $i \geq 1$, where e_i are independent of each other and of the random variable Z. Such data are less informative about the mean of Y_i than is one single data value from the distribution of Z. Another hard case has $Y_i = \mu_i + e_i$ where μ_i is an arbitrary unknown sequence of values in $\mathbb{R}^d$, and e_i are independent with mean zero. Some assumptions are needed in order that the amount of information in the Y_i about the underlying process should increase with T.

A widely used assumption is that Y_i for $i = 1, \ldots, T$ are T consecutive observations from an infinite series $\ldots, Y_{-1}, Y_0, Y_1, \ldots$ having a stationary distribution. This means that the joint distribution of any finite set of observations is unaffected by a time shift of k units. Thus $(Y_r, Y_s, \ldots, Y_t)$ has the same distribution as $(Y_{r+k}, Y_{s+k}, \ldots, Y_{t+k})$, for any k and any $r, s, \ldots, t$. A weaker assumption has some low order stationary moments. For example, in a real-valued time series there may be some set of $a, b, \ldots, c$ values for which $E(Y_r^a \times Y_s^b \times \cdots \times Y_t^c)$ is unaffected by a time shift of k units.

A stationarity assumption addresses the problem in the second hard case. A stationary series would have to have a common value $\mu_i = \mu$. The first hard case, $Y_i = Z + e_i$, is stationary if the e_i are IID, so another condition is needed to rule this case out.

Under a mixing condition, the dependence between observations before and including time t and observations from $t + k$ onward becomes negligible as $k \to \infty$. For a rigorous description of mixing, see the references in Chapter 8.10. Let A be a random variable that takes the value 0 or 1 depending on what the series does at times up to time t. Let B be a 0 or 1 random variable depending on what the series does from time $t + k$ on. It is natural to write $\Pr(A)$ for $E(A)$, identifying A with the event that $A = 1$. If the future is independent of the past then $\Pr(A \cap B) = \Pr(A) \Pr(B)$. Now measure the dependence through

$$\alpha(k) = \sup_{t} \sup_{A,B} |\Pr(A \cap B) - \Pr(A)\Pr(B)|. \tag{8.1}$$

The Y_i are α-mixing if $\alpha(k) \to 0$ as $k \to \infty$. If Y_i is stationary, then it is not necessary to maximize over t in (8.1).

A mixing condition rules out the first hard case. The series $Z + e_i$ is not α-mixing outside of trivial cases. Theorems that use mixing usually also stipulate that $\alpha(k)$ goes to zero sufficiently fast as $k \to \infty$.

The discussion above emphasizes the time domain approach to time series. In the frequency domain approach, we study how much of the variance in the series comes from oscillations at different frequencies. Suppose that $Y_t \in \mathbb{R}$ is a stationary time series, with mean $E(Y_t) = \mu$ and autocovariances $\gamma_k = E((Y_t - \mu)(Y_{t+k} - \mu))$. The spectral density function of Y_t is defined as

$$f(\omega) = \frac{1}{\pi}\left[\gamma_0 + 2\sum_{k=1}^{\infty} \gamma_k \cos(\omega k)\right], \quad 0 \le \omega \le \pi,$$

when this exists. The variance of Y_t is $\gamma_0 = \int_0^\pi f(\omega) d\omega$ and the interpretation of $f(\omega)$ is that frequencies in the interval $[\omega_1, \omega_2]$ contribute $\int_{\omega_1}^{\omega_2} f(\omega) d\omega$ of this variance.

The periodogram is a sample version of the spectral density function

$$I(\omega_j) = \frac{1}{\pi}\left[\hat{\gamma}_0 + 2\sum_{k=1}^{\infty} \hat{\gamma}_k \cos(\omega k)\right],$$

using estimates $\hat{\gamma}_k = T^{-1} \sum_{i=1}^{T-k}(Y_i - \overline{Y})(Y_{i+k} - \overline{Y})$. Under mild conditions,

REDUCING TO INDEPENDENCE

$E(I(\omega)) \to f(\omega)$ as $T \to \infty$, but because $\text{Var}(I(\omega_j))$ does not converge to 0 as $T \to \infty$, the periodogram is usually smoothed somehow, when an estimate of $f(\omega)$ is required.

8.2 Reducing to independence

Parametric likelihood methods often tackle dependent data by expressing the observations as functions of some other variables assumed to be statistically independent. One approach to empirical likelihood is to use the estimating equations from those models.

The autoregressive model is widely used in parametric modeling of time series data. As the name describes, the data series is generated by a regression on its own past. Suppose for example that $e_i \sim N(0, \sigma^2)$ are independent, that

$$Y_i = \beta_0 + \beta_1 Y_{i-1} + e_i, \quad i = 1, \ldots, T, \tag{8.2}$$

and that the series started off with an unobserved normally distributed Y_0 independent of the e_i. If $|\beta_1| < 1$, the distribution of Y_i tends to an equilibrium distribution $N(\mu, \sigma_y^2)$ as $i \to \infty$, where $\mu = \beta_0/(1 - \beta_1)$ and $\sigma_y^2 = \sigma^2/(1 - \beta_1^2)$. If $Y_0 \sim N(\mu, \sigma_y^2)$, then the Y_i all have the same distribution.

We will suppose that $|\beta_1| < 1$ and then reparameterize equation (8.2) as

$$Y_i - \mu = \beta_1(Y_{i-1} - \mu) + e_i, \quad i = 1, \ldots, T. \tag{8.3}$$

The parametric likelihood for the autoregressive model (8.3) is

$$L = \prod_{i=1}^{T} f(Y_i \mid Y_1, \ldots, Y_{i-1}; \mu, \beta_1, \sigma)$$

$$= \frac{e^{-\frac{1}{2\sigma_y^2}(Y_1 - \mu)^2}}{\sqrt{2\pi}\sigma_y} \prod_{i=2}^{T} \frac{e^{-\frac{1}{2\sigma^2}((Y_i - \mu) - \beta_1(Y_{i-1} - \mu))^2}}{\sqrt{2\pi}\sigma}. \tag{8.4}$$

The special treatment of Y_1 in (8.4) is awkward. The conditional (on Y_1) likelihood

$$L_c = \prod_{i=2}^{T} f(Y_i \mid Y_1, \ldots, Y_{i-1}; \mu, \beta_1, \sigma)$$

$$= \prod_{i=2}^{T} \frac{1}{\sqrt{2\pi}\sigma} \exp\left(-\frac{1}{2\sigma^2}\left((Y_i - \mu) - \beta_1(Y_{i-1} - \mu)\right)^2\right) \tag{8.5}$$

treats the data more symmetrically. Using L_c instead of L sacrifices some of the information available from Y_1. This information loss is small, especially when $|\beta_1|$ is close to 1. Furthermore, there is the possibility that the series has not yet reached the equilibrium distribution. Then L is not the likelihood but L_c is still the conditional likelihood given Y_1.

We can use the conditional likelihood to generate estimating equations. For

$i \geq 2$, let $e_i = e_i(\mu, \beta_1) = Y_i - \mu - \beta_1(Y_{i-1} - \mu)$, $\theta = (\mu, \beta_1, \sigma)'$ and

$$Z_i = Z_i(\theta) = (e_i, (Y_i - \mu)e_i, e_i^2 - \sigma^2)'.$$

Then the estimating equations are $\sum_{i=2}^{T} Z_i = 0$.

The empirical likelihood approach is then based on

$$\mathcal{R}(\theta) = \sup\left\{\prod_{i=1}^{n} w_i \mid w_i > 0, \sum_{i=1}^{n} w_i = 1, \sum_{i=1}^{n} w_i Z_{1+i} = 0\right\},$$

where $n = T - 1$. If the e_i are independent $N(0, \sigma^2)$ then the limiting distribution of $-2\log \mathcal{R}$ with r parameters constrained is $\chi^2_{(r)}$. The empirical likelihood inferences go through under some weaker conditions, using results for the dual likelihood described below. The e_i do not have to be normal, nor identically distributed. They do have to be nearly independent, so that $(1/n)\sum_i Z_i Z_i'$ estimates the variance matrix of $(1/\sqrt{n})\sum_i Z_i$.

It is convenient that inferences may be based on the limiting distribution of $\log \mathcal{R}(\theta)$, though it is is troubling that in time series models $\mathcal{R}$ is no longer a likelihood ratio. If θ_0 is the true value of the parameter then $Z_i(\theta_0)$ are independent, but for $\theta \neq \theta_0$ $Z_i(\theta)$ are not independent, and so it is hard to consider $\prod_i w_i$ to be the probability of the observations. The dual likelihood is one way to explain why $\mathcal{R}$ has likelihood asymptotics. Write

$$\mathcal{D}_\theta(\lambda) = \prod_{i=1}^{n}(1 + \lambda' Z_i(\theta))^{-1}.$$

For the correct value of θ, the Z_i are independent and the test for $\theta = \theta_0$ using $\mathcal{R}$ is the same as the test for $\lambda = 0$ using $\mathcal{D}_\theta$.

The autoregressive model (8.3) is known as the AR(1) model because it uses a regression on one past data point. In an AR(k) model we write

$$Y_i - \mu = \sum_{j=1}^{k} \beta_j(Y_{i-j} - \mu) + e_i. \tag{8.6}$$

The estimating equations for the AR(k) are a natural extension of those for AR(1). If an AR(k) series has uncorrelated e_i with mean 0 and constant variance, then it will approach an equilibrium distribution, under conditions on β_j. Let $u_1, \ldots, u_k$ be the solutions to

$$1 - \sum_{j=1}^{k} \beta_j u^j = 0. \tag{8.7}$$

The u_i are complex numbers, not necessarily all distinct. The AR(k) series approaches an equilibrium if and only if all u_j lie outside the unit circle in the complex plane.

We can form estimating equations for very general regressions, linear or nonlinear, relating Y_i to past values Y_{i-j} as well as past and present values of covariates

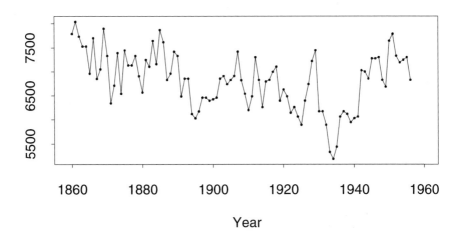

Figure 8.1 *Flow of the St. Lawrence river in cubic meters per second, at Ogdensburg, New York. The data are annual values from 1860 to 1957. Source: Yevjevich (1963).*

X_i, An asymptotic χ^2 distribution for empirical likelihood inferences holds very generally for non-explosive series using dual likelihood.

For a model including predictors at lags of up to k time steps, we obtain $n = T - k$ estimating function values to reweight. In software implementations, it can be a nuisance to have the number of estimating function values differ from the number of data values, or differ from model to model. A simple remedy is to define $Z_i = 0$ for $1 \leq i \leq k$. It is easy to show that maximizing $\prod_{i=1}^{T} \log(nw_i)$ subject to $\sum_{i=1}^{T} w_i Z_i = 0$ and $\sum_{i=1}^{T} w_i = 1$ places weight $1/T$ on any Z_i that equals zero, and that the empirical likelihood ratio based on $Z_{k+1}, \ldots, Z_T$ does not change when $Z_1 = \cdots = Z_k = 0$ are adjoined to the sample.

Figure 8.1 shows annual flow of water in the St. Lawrence river. These values are clearly not independent. The correlation between one year's flow and the next is about 0.71.

We consider an AR(3) model for this data set. Let $Y_1, \ldots, Y_{97}$ be the raw values, and $X_i = Y_i - \mu$ be centered values. We use estimating equations

$$Z_i = \begin{cases} (X_i, 0, 0, 0, 0)', & 1 \leq i \leq 3 \\ (X_i, e_i X_{i-1}, e_i X_{i-2}, e_i X_{i-3}, e_i^2 - \exp(2\tau))', & 4 \leq i \leq 97 \end{cases} \quad (8.8)$$

where $e_i = X_i - \sum_{j=1}^{3} \beta_j X_{i-j}$. These describe an autoregression of Y_i on its past 3 lags, with an error standard deviation of $\exp(\tau)$. Instead of taking Z_1 through Z_4 equal to 0, the first component was modified slightly. It is customary in autore-

j	$\hat{\beta}_j$	$-2\log \mathcal{R}_j(0)$
1	0.627	30.16
2	−0.093	0.48
3	0.214	4.05

Table 8.1 *An AR(3) model was fit to the St. Lawrence River flow data. Shown are the coefficient estimates $\hat{\beta}_j$, and the empirical likelihood values for testing that $\beta_j = 0$.*

gressive modeling to estimate μ by $\overline{Y} = (1/T)\sum_{i=1}^{T} Y_i$. Neither the conditional nor the unconditional likelihood leads to $\hat{\mu} = \overline{Y}$, but equations (8.8) do.

The mean flow is estimated to be $\hat{\mu} = 6818.6$ cubic meters per second. The standard deviation of e_i is estimated to be $\exp(\hat{\tau}) = \exp(6.006) = 405.9$ cubic meters per second. This describes the uncertainty in a linear prediction of one year's river flow, based on the previous three years' data. Table 8.1 presents the estimated autoregressive coefficients as well as the empirical likelihood test statistics for each coefficient to be zero. The lag 1 coefficient β_1 is clearly nonzero, β_2 could reasonably be zero. A $\chi^2_{(1)}$ test rejects $\beta_3 = 0$ at just below the 5% level as does an F test.

It is interesting to consider whether $\beta_2 = \beta_3 = 0$ is tenable. Imposing both constraints can only reduce the empirical likelihood compared to the test of $\beta_3 = 0$ alone. This lower likelihood must, however, be compared to a distribution appropriate to a two-dimensional hypothesis. As is well known, omnibus tests that can detect multiple kinds of departure from a hypothesis often do so with reduced power compared to more specific tests. In this instance, a test of the hypothesis $\beta_2 = \beta_3 = 0$ has p-value somewhat above 5%, while a test of $\beta_3 = 0$ has a p-value below 5%.

Not being able to reject $\beta_2 = \beta_3 = 0$ is not the same as proving that they are zero. We retain these coefficients in the model, judging that there is more to lose in dropping them should they matter than in retaining them if they do not. People can reasonably differ in these judgments. The model then gives a 95% confidence interval for τ of $(5.871, 6.134)$ using a $\chi^2_{(1)}$ calibration. Exponentiating, we get a confidence interval of $(354.6, 461.3)$ for σ.

8.3 Blockwise empirical likelihood

We do not always know a model in which the data are generated from a series of independent observations. A weaker assumption is that the data have a stationary distribution, or stationary moments, as described in Chapter 8.1. Stationarity alone cannot support a good asymptotic theory. An additional condition, such as one on the α-mixing coefficients described in Chapter 8.1, is required.

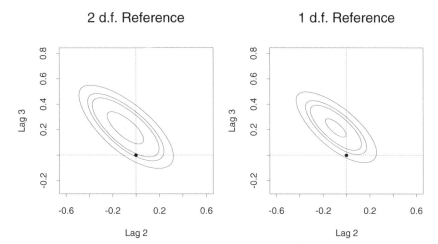

Figure 8.2 *Both figures show empirical likelihood contours for two of the autoregressive parameters for the St. Lawrence river flows. The figure on the left uses two degrees of freedom, appropriate for two parameters. The contour levels correspond to 50, 90, 95, and 99 percent confidence. The solid circle shows that the origin has a statistical significance level between 5 and 10 percent. Here the hypothesis $\beta_2 = \beta_3 = 0$ is not rejected at the 5% level, but the hypothesis $\beta_3 = 0$ is rejected. The figure on the right illustrates how this can happen, by redrawing the contours using the $\chi^2_{(1)}$ calibration. The hypothesis $\beta_3 = 0$ was rejected because a horizontal line segment through $\beta_3 = 0$ lies outside the 95% confidence contour based on 1 degree of freedom.*

A bootstrap method for handling stationary mixing time series is to resample the data in blocks of length $M > 1$. By concatenating randomly sampled blocks of consecutive data points, the resampled series can capture some of the structure from the original series. For k small compared to M, resampled observations k apart are likely to be genuine observation pairs separated by k units in the original data. When k is large compared to M then resampled observation pairs k units apart are essentially independent in the resampled series, matching the weak dependence in the original series.

Suppose now that θ is a parameter of the joint distribution of $r \ge 1$ consecutive observations $Y_{t-r+1}, \ldots, Y_t$, defined by

$$E(m(X_t, \theta)) = 0$$

where $X_t = (Y'_{t-r+1}, \ldots, Y'_t)'$ bundles r consecutive observations from the original series. Thus if $Y_t \in \mathbb{R}^d$, then $X_t \in \mathbb{R}^{dr}$.

The blocking idea can also be used in empirical likelihood. Starting with the series X_t, form blocks of length M with starting points separated by L time units.

That is,
$$B_i = (X_{(i-1)L+1}, \ldots, X_{(i-1)L+M}), \quad i = 1, \ldots, n$$
where
$$n = \left\lfloor \frac{T-M}{L} + 1 \right\rfloor.$$

Here $\lfloor z \rfloor$ denotes the largest integer that is less than or equal to z. Values L between 1 and M inclusive are reasonable. With $L = M$, the blocks do not overlap. Taking $L > M$ would leave some X_t values unused.

Now define the blockwise estimating function
$$b(B_i, \theta) = \frac{1}{M} \sum_{j=1}^{M} m(X_{(i-1)L+j}, \theta).$$

Of course, if $E(m(X_t, \theta)) = 0$ then $E(b(B_i, \theta)) = 0$ too. If $L = M \to \infty$, as $T \to \infty$, then with some assumptions, the dependencies among $b(B_i, \theta)$ become negligible. Now blockwise empirical likelihood inferences for θ are based on the empirical likelihood ratio
$$\mathcal{R}(\theta) = \sup \left\{ \prod_{i=1}^{n} w_i \mid w_i > 0, \sum_{i=1}^{n} w_i = 1, \sum_{i=1}^{n} w_i b(B_i, \theta) = 0 \right\}.$$

For $L \doteq \alpha M$ with $\alpha < 1$, the dependencies do not become negligible, because there is a fixed fraction of overlap between consecutive blocks.

Theorem 8.1 *Under conditions given in Kitamura (1997), including $M \to \infty$ and $MT^{-1/2} \to 0$*
$$-2 \left(\frac{T}{nM} \right) \log \mathcal{R}(\theta_0) \to \chi^2_{(q)}$$
as $T \to \infty$, where q is the dimension of θ.

Proof. Kitamura (1997). □

The factor $T/(nM)$ accounts for the overlap in the blocks. It would have to be there even if the time series were IID. If T is a multiple of the block size M and if $L = M$, so the blocks do not overlap, then $T/(nM) = 1$. For $L = \alpha M$, with $\alpha < 1$, the blocks overlap, and to compensate
$$\frac{T}{nM} = \frac{T}{\left\lfloor \frac{T-M}{\alpha M} + 1 \right\rfloor M} \doteq \alpha.$$

Thus when the blocks overlap, the asymptotic distribution of $-2 \log \mathcal{R}(\theta_0)$ is approximately $\alpha^{-1} \chi^2_{(q)}$ ranging from $\chi^2_{(q)}$ to $M \chi^2_{(q)}$ as L ranges from M to 1.

Figure 8.3 shows the 5405 years of bristlecone pine tree ring widths from Campito Mountain in California. The last year in the data set is 1969. The units are 0.01 millimeters. The series values range from 0 to 99. There are several interesting features in this data, one of which is that downward spikes tend to be

Campito tree ring data

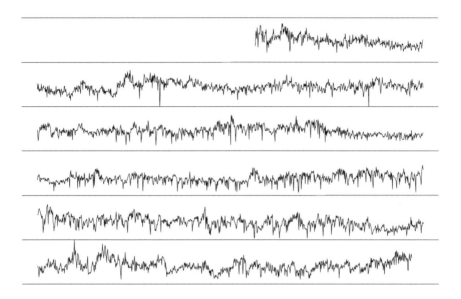

Figure 8.3 *Shown are 5405 years of bristlecone pine tree ring data from Campito Mountain, California. Time increases from top to bottom of the figure, going from left to right within each of 6 ranges. The bottom range is for the years 1001 through 1969, where the series ends. Moving up one range corresponds to going back 1000 years. The data values are between 0 and 99. Within a range the data are plotted between lower and upper reference lines corresponding to values 0 and 100. The data are in units of 0.01 mm. The data are from Fritts et al. (1971) and are available on Statlib.*

larger than upward ones. There were 39 years in which the tree ring width was more than 0.2mm larger than the average of the previous 10 years but 145 years in which the width was more than 0.2mm smaller than the average of the previous 10 years. We could not capture such asymmetry in an AR model with normally distributed errors.

The natural estimate of the probability of such a downward spike is $145/(5405-10) = 0.027$, because there are 145 successes in 5395 trials. A binomial confidence interval for this probability would not be appropriate because it would ignore the dependence in the data.

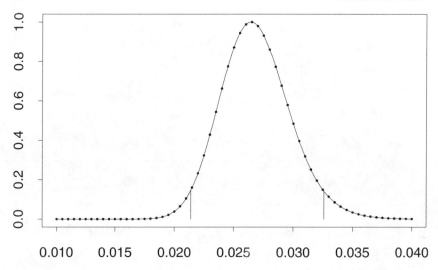

Figure 8.4 *The horizontal axis shows the probability that the Campito tree ring width decreases by more than 0.2mm from its average over the previous 10 years. The vertical axis shows the block empirical likelihood ratio, including an adjustment for overlapping blocks. A 95% confidence interval extends from 0.0214 to 0.0326, as indicated by two vertical segments.*

Define derived time series

$$W_i = Y_i - \frac{1}{10}\sum_{j=1}^{10} Y_{i-j}, \quad \text{and}$$

$$Z_i = \begin{cases} 1, & W_i < -20 \\ 0, & W_i \geq -20. \end{cases}$$

If Y_i is a stationary series then so are W_i and Z_i. We are interested in inferences on $E(Z_i)$. We take a block size of $M = 50$ years, and starting points separated by $L = 10$ years, and construct B_i as above.

For these data, the W and Z series have length 5395 so that is the appropriate value of T. The value of n is $\lfloor (5395 - 50)/10 + 1 \rfloor = 535$. Figure 8.4 shows the empirical likelihood curve for these blocked data. The empirical log likelihood is multiplied by $T/(nM) = 5395/(535 \times 50) = 0.2017$ to adjust for block overlap. This is equivalent to multiplying the $\chi^2_{(1)}$ threshold value by $1/0.2017 = 4.96$. The 95% confidence interval for the probability of a large downward spike ranges from 2.14% to 3.26%.

Choosing M and L can be difficult. Here are some guidelines, with the caveat that blocked empirical likelihood is a new method. Suppose at first that we take

$L = M$ and look for the right size of non-overlapping block. The asymptotic theory suggests that M should tend to infinity in order to control the dependence among blocks. If $T/M = c$, then for very large T the coverage of empirical likelihood should be like that for the mean of c independent blocked random vectors B_i. Sending c to infinity is necessary in order for the asymptotic coverage to set in. If we felt that ordinary empirical likelihood coverage properties were satisfactory for sample sizes c and above, then we might take M as the nearest integer to T/c. The value c that we would use would depend on the dimension of B_i.

Suppose that $m(X_i, \theta)$ were really independent and normally distributed, and that we have grouped them. Then we have lost some efficiency. Instead of T observations with mean 0 and variance V_m say, we have only T/M observations $b(B_i, \theta)$ having mean 0 and variance V_m/M. Our confidence regions for blocked data will not be as good as for the unblocked data, but it will primarily be the difference between using T/M degrees of freedom instead of T degrees of freedom. The sample size reduction by M is largely compensated by a variance reduction of M. Similar comments apply if the data are independent, but not normally distributed. This efficiency loss is explored in Exercise 8.6.

In time series examples the data are not usually independent. The errors in treating small blocks as independent, when they are not, can be very large. Thus it seems that caution would dictate larger values of M.

The value of L would seem to be less crucial. L can range from 1 to M. Smaller values of L are usually more statistically efficient, though diminishing returns seem likely. The number of blocks grows as L decreases, increasing the computational effort.

8.4 Spectral method

The moving average model of order 1, denoted by MA(1) has

$$Y_i - \mu = e_i + \alpha_1 e_{i-1}$$

where $|\alpha_1| < 1$, and e_i are independent identically distributed random variables with mean 0 and variance σ^2. By substituting for e_{i-1} we find that the MA(1) model is an AR model of infinite order with $\beta_k = \alpha_1^k$ for $k \geq 1$. We could base our inferences for $\theta = (\mu, \sigma, \alpha_1)'$ on the expected values of $Y_i - \mu$, $(Y_i - \mu)(Y_{i-1} - \mu)$ and $(Y_i - \mu)^2$, written in terms of θ. But such inferences are not efficient. They do not capture the information about α_1 in higher order lags than the first.

The moving average model of order ℓ, MA(ℓ) has $Y_i - \mu = e_i + \sum_{j=1}^{\ell} \alpha_j e_{i-j}$, and the ARMA($k, \ell$) model has $Y_i - \mu$ described as an AR(k) model with MA(ℓ) errors. Estimating equations for MA and ARMA models are more complicated than those for AR models. See Chapter 8.10.

An alternative to reweighting the estimating equations of an ARMA model is to proceed through the spectrum, as outlined here. The spectrum may be written in terms of the parameters of the ARMA model, although μ does not enter. The periodogram of a time series provides a noisy estimate of the spectrum. We can take the estimate at $\lfloor (T-1)/2 \rfloor$ different frequencies. Asymptotically, these are independent exponential random variables with means equal to the true spectrum at the corresponding frequencies. The terms in the exponential log likelihood can be treated as a log likelihood and reweighted with a χ^2 limit. See Chapter 8.10 for references.

8.5 Finite populations

In many applications the sample is taken from a finite population. Theory and methods for sampling finite populations have historically had their impetus in survey sampling. The same problems now arise in some data mining applications. Computers that analyze a population of records now have to keep up with other computers that generate the data, making it attractive to work with a sample.

Suppose that the sample has n observations made on a population of N individuals. The statistical problem may be to estimate something about the whole finite population from the sample. In other settings, we seek inferences on an infinite superpopulation from which the N finite sample observations were drawn before we sampled n of them.

Several features make sampling finite populations different from the usual statistical problems. First, for inferences on the population, we get the answer without error if $n = N$. A related feature is that the problems of estimating a population total, or even of estimating N when it is unknown, may arise for finite populations, but not for infinite ones. Next, in finite populations it is especially common for there to be some variables with known population means or totals. These may be variables that were measured in a census, or they may be quantities that are constantly updated as records are added to a database. By taking account of the known population values, we can get sharper estimates for things that we do not know. Finally, there are a variety of strategies that can be employed in sampling to get better answers at lower cost. In stratification, we take separate samples within subpopulations perhaps overweighting an important rare group, such as records for fraudulent credit card transactions. In cluster sampling, we partition the population into groups of contiguous individuals, and take a sample of the groups.

In a simple random sample all $N!/[n!(N-n)!]$ ways of selecting n of N observations are equally probable. When $n \ll N$, then the fact that the population is finite may often be ignored. But if n/N is not negligible, the finiteness of the population introduces a dependence described below that should not be ignored.

The customary notation for a simple random sample is that the population values are $Y_i \in \mathbb{R}^q$, measured on individuals $i = 1, \ldots, N$. The simple random sample is then denoted by y_i, vectors measured on sampled individuals $i = 1, \ldots, n$. There is understood to be no connection between the sample and population indices. In particular, y_1 is not necessarily Y_1.

FINITE POPULATIONS

The population mean of Y is

$$\overline{Y} = \frac{1}{N} \sum_{i=1}^{N} Y_i,$$

which is estimated by

$$\overline{y} = \frac{1}{n} \sum_{i=1}^{n} y_i = \frac{1}{n} \sum_{i=1}^{N} Y_i Z_i, \qquad (8.9)$$

where Z_i is one if population element i is in the sample and zero otherwise. The population variance of Y is defined to be

$$S_{YY} = \frac{1}{N-1} \sum_{i=1}^{N} (Y_i - \overline{Y})(Y_i - \overline{Y})',$$

which we suppose is invertible. By elementary calculations with equation (8.9), in which Z_i are random and Y_i are fixed, we find that

$$\mathrm{Var}(\overline{y}) = \frac{1}{n}\left(1 - \frac{n}{N}\right) S_{YY}. \qquad (8.10)$$

The finite population correction factor $1 - n/N$ arises from negative correlations among the Z_i. For two population members, if one is sampled then it is less likely that the other one is sampled. These negative correlations are usually very small, but there are $O(N^2)$ of them, and together they cause an important variance reduction when n/N is not small.

The standard approach to inference for simple random samples is to obtain an unbiased estimate

$$s_{yy} = \frac{1}{n-1} \sum_{i=1}^{n} (y_i - \overline{y})(y_i - \overline{y})'.$$

of S_{YY} and plug it into (8.10), getting

$$\widehat{\mathrm{Var}}(\overline{y}) = \frac{1}{n}\left(1 - \frac{n}{N}\right) s_{yy}. \qquad (8.11)$$

Then, under a central limit theorem for $\overline{y}$, we have $(\overline{y} - \overline{Y})'(\widehat{\mathrm{Var}}(\overline{y}))^{-1}(\overline{y} - \overline{Y})$ is asymptotically $\chi^2_{(q)}$. When $Y_i \in \mathbb{R}$, the standard 95% confidence intervals for $\overline{Y}$ are $\overline{y} \pm 1.96(\widehat{\mathrm{Var}}(\overline{y}))^{1/2}$.

Central limit theorems for finite sampling are a bit more subtle than those for infinite populations. As we let $n \to \infty$, we must also have $N \to \infty$ to keep $n \le N$. Indeed, we assume that $N - n \to \infty$, for otherwise $\overline{y}$ is determined by the average of a small number of excluded points. Finally there has to be a condition on the sequence of finite populations so that in the limit each $Z_i Y_i$ is asymptotically negligible compared to $\sum_{i=1}^{n} y_i$. A commonly used condition is that $(1/N) \sum_{i=1}^{N} |Y_i|^3 \le B$ for all N.

8.6 MELE's using side information

In finite population settings, the effective use of side information (called auxiliary information in this context) is a very important issue. Maximum empirical likelihood estimates for finite populations have been more thoroughly studied than empirical likelihood ratios. MELE's allow us to incorporate side information while obeying range restrictions.

Suppose now the population consists of vectors $U_i \in \mathbb{R}^u$ of underlying quantities with $Y_i = \mathbb{Y}(U_i) \in \mathbb{R}^p$ for some function $\mathbb{Y}$. We take a simple random sample of values u_i, observing $y_i = \mathbb{Y}(u_i)$. Similarly, let $X_i = \mathbb{X}(U_i) \in \mathbb{R}^q$ and $x_i = \mathbb{X}(u_i)$. We suppose that the population mean $\overline{X}$ is known to us. Introducing the function $\mathbb{X}$ allows us to encode a known mean or quantile of a component of U_i through $\mathbb{X}(U_i) = U_{ij}$ or $\mathbb{X}(U_i) = 1_{U_{ij} < Q} - \alpha$. If one component U_{ij} represents a categorical variable taking a finite number c of values, and we know the population proportions in the c categories, we can encode this knowledge through $c - 1$ components of $\mathbb{X}(U_i)$ taking values 0 or 1. In general, $\mathbb{X}$ encodes a finite number q of quantities whose population means are known.

The population variance of X, denoted S_{XX}, is defined analogously with S_{YY}. To avoid inessential complications, assume that S_{XX} has full rank. Define

$$S_{XY} = \frac{1}{N-1} \sum_{i=1}^{N} (X_i - \overline{X})(Y_i - \overline{Y})'.$$

Suppose at first that $p = 1$. For any vector β of q components

$$\widehat{\overline{Y}}_\beta = \frac{1}{n} \sum_{i=1}^{n} y_i - (x_i - \overline{X})' \beta$$

is an unbiased estimate of $\overline{Y}$. The usual estimate (8.9) which ignores the x_i corresponds to $\beta = 0$. The variance of $\widehat{\overline{Y}}_\beta$ is at a minimum for $\beta_{\text{LS}} = S_{XX}^{-1} S_{XY}$, which is usually unknown. The regression estimator of $\overline{Y}$ is $\widehat{\overline{Y}}_{\text{REG}} = \widehat{\overline{Y}}_{\hat\beta}$ where

$$\hat\beta = \left(\sum_{i=1}^{n} (x_i - \overline{x})(x_i - \overline{x})' \right)^{-1} \sum_{i=1}^{n} (x_i - \overline{x})(y_i - \overline{y}).$$

We may write the regression estimator as a weighted combination of data values

$$\widehat{\overline{Y}}_{\text{REG}} = \frac{1}{n} \sum_{i=1}^{n} \left(1 - \frac{n}{n-1} (x_i - \overline{X})' s_{xx}^{-1} (\overline{x} - \overline{X}) \right) y_i, \qquad (8.12)$$

where s_{xx} is the sample version of S_{XX}. For large n we expect the estimator $\hat\beta$ to be close to β_{LS} and then $\widehat{\overline{Y}}_{\hat\beta}$ has variance near the optimal value.

If $p > 1$, then we may still estimate $\overline{Y}$ by the weighted combination (8.12). The optimality results for a univariate $\overline{Y}$ apply to any linear combination of components of the multivariate $\overline{Y}$.

SAMPLING DESIGNS

A practical concern with equation (8.12) is that some of the y_i can receive negative weights. The result is that range restrictions are not obeyed. Estimated variances can be negative, and estimated cumulative distribution functions can be decreasing over some intervals.

An empirical likelihood approach enforces nonnegative weights. We find w_i satisfying

$$\sum_{i=1}^{n} w_i(x_i - \overline{X}) = 0, \quad \text{and} \quad \sum_{i=1}^{n} w_i = 1 \tag{8.13}$$

and maximizing $\sum_{i=1}^{n} \log(nw_i)$ subject to (8.13). Then we estimate $\overline{Y}$ by the maximum empirical likelihood estimator

$$\widehat{\overline{Y}}_{\text{MELE}} = \sum_{i=1}^{n} w_i Y_i. \tag{8.14}$$

For small n or large q it may happen that $\overline{X}$ is not in the convex hull of $x_1, \ldots, x_n$. Then the weights required for $\widehat{\overline{Y}}_{\text{MELE}}$ do not exist, and so neither does the MELE. In that case we may have to use regression, and accept some negative weights, or decide not to impose some or all of the known population means.

The MELE has similar asymptotic properties to the regression estimator, but it respects range restrictions. Below is a theorem for the case $p = q = 1$.

Theorem 8.2 *Suppose that n, N, and $N - n$ increase to ∞ such that*

$$\frac{1}{N} \sum_{i=1}^{N} |Y_i|^3 < B, \quad \text{and} \quad \frac{1}{N} \sum_{i=1}^{N} |X_i|^3 < B$$

for some $B < \infty$. Then

$$\sqrt{n} \frac{\widehat{\overline{Y}}_{\text{MELE}} - \overline{Y}}{\sigma_{Y|X}} \to N(0, 1)$$

where

$$\sigma_{Y|X}^2 = \left(1 - \frac{n}{N}\right) \left(S_{YY} - \frac{S_{XY}^2}{S_{XX}}\right).$$

Proof. Chen & Qin (1993). □

The asymptotic variance of $\widehat{\overline{Y}}_{\text{MELE}}$ in Theorem 8.2 is the same as that of $\widehat{\overline{Y}}_{\text{REG}}$. Variance estimates for $\widehat{\overline{Y}}_{\text{REG}}$ can be used for $\widehat{\overline{Y}}_{\text{MELE}}$.

8.7 Sampling designs

A simple random sample is not always the most efficient way to gather data. In stratified sampling one divides the population into a finite number H of strata.

Let the population have elements $Y_{hi} = \mathbb{Y}(U_{hi})$ corresponding to individuals $i = 1, \ldots, N_h$ in strata $h = 1, \ldots, H$. Stratified sampling takes a simple random sample of n_h observations from stratum h, yielding observations $y_{hi} = \mathbb{Y}(u_{hi})$. Each stratum is sampled independently.

In cluster sampling, the population is partitioned into groups as for stratified sampling. The groups comprise individuals that are conveniently sampled together, such as inhabitants of a block, or files created in a time window. Instead of taking a simple random sample from within each group, a simple random sample of the groups is taken. From each sampled group, we might take all individuals, or possibly a simple random sample, or even a cluster sample.

Clustering and stratification can be combined in complicated ways, as for example, clusters of clusters within strata, with auxiliary variables having known population and/or stratum means. The Horvitz-Thompson and generalized regression estimators provide a unified approach to estimating $\overline{Y}$. For each unit in the population let Π_i be the probability that it is included in the sample. Similarly, let π_i be these probabilities for the n observations actually included in the sample. The Horvitz-Thompson estimator of $\overline{Y}$ is

$$\widehat{\overline{Y}}_{\text{HT}} = \frac{1}{N} \sum_{i=1}^{n} \frac{y_i}{\pi_i} = \frac{1}{N} \sum_{i=1}^{N} \frac{Y_i Z_i}{\Pi_i}, \tag{8.15}$$

where Z_i is again an indicator variable for inclusion in the sample. The estimator (8.15) does not require that we know all the Π_i, only those for the individuals actually sampled. It weights items inversely to their inclusion probability, just as was done in Chapter 6.1.

Given known values for $\overline{X}$, the generalized regression estimator is

$$\widehat{\overline{Y}}_{\text{GREG}} = \widehat{\overline{Y}}_{\text{HT}} - (\widehat{\overline{X}}_{\text{HT}} - \overline{X})' \hat{\beta}_{\text{HT}} \tag{8.16}$$

where, for scalar Y_i, $\hat{\beta}_{\text{HT}}$ minimizes the weighted sum of squares

$$\sum_{i=1}^{n} \pi_i^{-1} \left((y_i - \widehat{\overline{Y}}_{\text{HT}}) - (x_i - \widehat{\overline{X}}_{\text{HT}})' \beta \right)^2.$$

Known stratum sizes can be incorporated into X as known population means of stratum indicator variables.

The estimator (8.16) may be written as a weighted sum of y_i values, and the same weights are employed for multivariate Y_i. The generalized regression estimator does not necessarily respect range restrictions. Once again, a solution is available using an MELE, subject to a convex hull condition. Non-existence of the MELE does provide a diagnostic that the GREG estimator is using some negative weights, constituting a form of extrapolation.

To construct an MELE that takes account of π_i, we maximize

$$L(w) = \sum_{i \in s} d_i \log w_i \tag{8.17}$$

EMPIRICAL LIKELIHOOD RATIOS FOR FINITE POPULATIONS 179

where $d_i = 1/\pi_i$ is called a design weight, subject to constraints $\sum_{i \in s} w_i = 1$ and $\sum_{i \in s} w_i(X_i - \overline{X}) = 0$. Here s is the sample, and writing the summation limits as $i \in s$ reminds us that the sample size may be random. The motivation for L is that it is an unbiased estimate of the log likelihood $\sum_{i=1}^{N} \log W_i$ that we would use for inferences on a superpopulation, if we had observed the entire finite population.

Using some foresight, we construct the Lagrangian

$$G = \sum_{i \in s} d_i \log(w_i) - D\lambda' \sum_{i \in s} w_i(X_i - \bar{X}) + \gamma \Big(1 - \sum_{i \in s} w_i\Big),$$

where $D = \sum_{i \in s} d_i$. Setting $\sum_{i \in s} w_i \partial G / \partial w_i = 0$ gives $\gamma = D$, and $\partial G / \partial w_i = 0$ gives

$$w_i = \frac{d_i}{D} \frac{1}{1 + \lambda'(X_i - \overline{X})}$$

where λ satisfies

$$0 = \sum_{i \in s} \frac{d_i(X_i - \overline{X})}{1 + \lambda'(X_i - \overline{X})}.$$

The MELE

$$\widehat{\overline{Y}}_{\text{MELE}} = \sum_{i \in s} w_i Y_i,$$

respects range restrictions and is close to the generalized regression estimator:

Theorem 8.3 *If as N and n increase to ∞, $\max_{i \in s} \|X_i - \overline{X}\| = o_p(n^{-1/2})$,*

$$\Big(\sum_{i \in s} d_i(X_i - \overline{X})(X_i - \overline{X})'\Big)^{-1} \Big(\sum_{i \in s} d_i(X_i - \overline{X})\Big) = O_p(n^{-1/2}),$$

and $\max_{1 \le i \le N} \|Y_i\|$ is bounded, then $\widehat{\overline{Y}}_{\text{MELE}} = \widehat{\overline{Y}}_{\text{GREG}} + o_p(n^{-1/2})$.

Proof. Chen & Sitter (1999) show that the MELE and GREG weights differ by $o_p(n^{-1/2})$ and then the bound on Y_i completes the proof. □

Theorem 8.3 applies to various forms of cluster sampling. For probability sampling within L strata, we introduce $L(w) = \sum_{h=1}^{L} \sum_{i \in s_h} d_{hi} \log(w_{hi})$. See Chapter 8.10.

8.8 Empirical likelihood ratios for finite populations

Now we consider empirical likelihood ratios. For simple random sampling, let

$$\mathcal{R}(\mu) = \max \Big\{ \prod_{i=1}^{n} n w_i \,\Big|\, w_i \ge 0, \sum_{i=1}^{n} w_i = 1, \sum_{i=1}^{n} w_i y_i = \mu \Big\}.$$

From Theorem 8.2, it is reasonable to expect that $-2(1 - n/N)^{-1} \log \mathcal{R}(\overline{Y})$ will have an asymptotic $\chi^2_{(p)}$ distribution. A more general result, including stratified sampling is cited in Chapter 8.10. The factor of $1-n/N$ plays a similar role to the factor used in blockwise empirical likelihood in Chapter 8.3. It is not surprising that a correction must be employed, because $\mathcal{R}$ was derived for $N = \infty$ and is not a finite population likelihood ratio.

One way to build the finite population assumption into the likelihood is to suppose that there are K distinct values of Y_i in the population. Let them be $\mathcal{Y}_i$, for $i = 1, \ldots, K$ and suppose that $\mathcal{Y}_i$ appears N_i times in the population and n_i times in a simple random sample. If the $\mathcal{Y}_i$ are known then the population is described by the parameter $(N_1, \ldots, N_K)$, which has a hypergeometric likelihood

$$L(N_1, \ldots, N_K) = \binom{N}{n}^{-1} \prod_{i=1}^{K} \binom{N_i}{n_i}. \tag{8.18}$$

This likelihood is difficult to work with, because it is only defined over integer N_i. However, in the limit with $N_i/N \to w_i$ and $n/N \to 0$, a likelihood proportional to $\prod_{i=1}^{n} w_i$ emerges.

8.9 Other dependent data

This section describes some other settings with dependent data, where an empirical likelihood analysis might add value.

Longitudinal data arise as repeated measures, usually over time, on a series of subjects. Such data are commonly found in biomedical applications. They can be arranged into multiple time series, one per subject. The statistical issues may be to describe the typical time trend of a subject, the subject-to-subject variation in the trends, or the effects of covariates such as treatments.

Random fields are generalizations of time series to higher dimensional index spaces. The observations may be taken on a grid in $\mathbb{R}^g$, or at scattered sites, or continuously, or on some hybrid such as along line transects, as was done for the shrub width data in Chapter 6.1. Spatial point processes are scattered observations, such as the locations of trees or galaxies, where the random locations themselves are under study.

8.10 Bibliographic notes

Time series

Box, Jenkins & Reinsel (1994), Cryer (1986), and Anderson (1994) provide background material on time series. Politis, Romano & Wolf (1999) provide an appendix with results on mixing on which equation (8.1) is based. There are many different nomenclatures for describing spectral densities and their estimates. The account in Chapter 8.1 follows Chatfield (1989).

The idea of describing a dependent series through a series of independent "shocks" is a powerful one that Box et al. (1994) attribute to Yule (1927). This approach underlies most parametric work on time series in the time domain. Efron

& Tibshirani (1986) propose a bootstrap based on resampling residuals from an autoregression.

The dual likelihood, due to Mykland (1995), takes the Lagrange multiplier λ to be the parameter. For each fixed parameter value θ the test of θ corresponds to a test of $\lambda = 0$ for the $Z_i = m(X_i, \theta)$. For IID data the dual and empirical likelihoods coincide. The dual likelihood applies also to time series and survival analysis settings with martingale estimating equations. Very generally the dual likelihood statistic is close to a quadratic statistic (like the Euclidean likelihood), and has a χ^2 limit. Mykland (1995) also presents a notion of Bartlett correctability for this martingale setting.

Hipel & McLeod (1994) analyze the St. Lawrence river flow data. They identified an AR(3) model for it, and recommend constraining the lag 2 coefficient to be zero. The data are from Yevjevich (1963). They are repeated in Hipel & McLeod (1994) and are also available from Statlib.

Chuang & Chan (2001) study unstable autoregressions in which (8.7) has at least one root on the unit circle, but no roots inside the unit circle. They show that both the empirical log likelihood ratio statistic and the usual quadratic test statistic have the same (non χ^2) limiting distribution.

Politis et al. (1999) trace the development of blockwise approaches for the bootstrap. Carlstein (1986) proposed non-overlapping blocks for variance estimation. Künsch (1989) and Liu & Singh (1992) developed versions for confidence regions. Politis & Romano (1994) propose a method of sampling with random block lengths, so that the resampled series are stationary, conditionally on the observed one. Politis et al. (1999) remark that more overlap among blocks (smaller L) gives more efficiency.

Blockwise empirical likelihood was developed by Kitamura (1997), for estimating equations and for smooth functions of means. Kitamura (1997) also extends the results from Qin & Lawless (1994) to stationary time series, and establishes Bartlett correctability for some time series versions of empirical likelihood. The Bartlett correction supposes that M is of exact order $T^{1/3}$. Then Bartlett correction improves the order of coverage error from $T^{-2/3}$ to $T^{-5/6}$. The blocking used in Chapter 8.3 takes simple averages over blocks. Kitamura (1997) raises the possibility of taking weighted averages within blocks and relates this idea to kernel methods of smoothing the spectrum.

Kitamura (1999) applies a pre-whitening filter to the time series before applying blockwise empirical likelihood. The filter subtracts a linear combination of past series values Y_{t-k} from Y_k. The linear combination is chosen to make the filtered series more nearly, even if not exactly, uncorrelated, allowing a smaller block size.

Hipel & McLeod (1994) give an estimate of the spectrum for the Campito tree ring data. Those data are available on Statlib, with an attribution to Fritts et al. (1971).

Some properties of a time series, such as its spectrum, are not functions of a finite dimensional margin, but depend instead on the whole infinite dimensional joint distribution of the data. For these Kitamura (1997) describes an approach based on blocks of blocks, paralleling the blocks of blocks bootstrap of Politis & Romano (1992).

Estimation in MA and ARMA models is described in Box et al. (1994) and Hipel & McLeod (1994). Maximum likelihood algorithms usually require back forecasting of error terms from before the start of the data and the formation of a sum of squares of estimated errors. As a result, the estimating equations being solved are not explicit.

Forming a likelihood from the distribution of the periodogram is known as Whittle's method after Whittle (1953). Bootstrap-resampled periodograms have been used by Ramos (1989) to generate new estimators by Rao-Blackwellization. Franke & Härdle (1992), Janas (1994), and Dahlhaus & Janas (1996) propose inferences based on resampled periodograms.

The spectral approach to empirical likelihood is due to Monti (1997), who also proposes a Bartlett correction. Monti (1997) presents a confidence region for the parameters of an ARMA(1, 1) model fit to a series of 197 chemical process concentration readings from Box et al. (1994). The region is asymmetric, extending farther toward the origin where the series would be independent than away from it where the series would be explosive. The parametric region is elliptical.

Monti (1997) simulates some MA(1) processes with parameter values in $(0, 1)$. Two error distributions are considered: $N(0, 1)$ and $\chi^2_{(5)} - 5$. For simulations with Gaussian errors, methods based on the Gaussian likelihood did best, but for non-Gaussian errors, empirical likelihood inferences had better coverage than the customary asymptotic ones, particularly for parameter values close to the boundary of the invertibility region.

Finite populations

Hartley & Rao (1968) provide one of the earliest NPMLE arguments, using the discrete likelihood (8.18). They also show how to optimize that likelihood over integer values to find the MLE. Hartley & Rao (1968) also provide what may be the very first MELE, maximizing a continuous version of the likelihood subject to a constraint on the mean. They do not consider nonparametric likelihood ratios, but show instead that the MELE closely approximates the regression estimator for which there are well-known variance estimates. They also consider a Bayesian formulation using a Dirichlet prior.

Chen & Qin (1993) present empirical likelihood for samples from a finite population. In addition to Theorem 8.2, they also present a consistent estimate of the variance of the MELE based on the jackknife. Under a superpopulation model with a continuous CDF for Y, Chen & Qin (1993) characterize the asymptotic behavior of the reweighted CDF of Y, using empirical likelihood weights based on known $\overline{X}$. Chen & Qin (1993) show that the MELE reproduces several well known estimates from survey sampling. A categorical X with known category frequencies gives rise to the post-stratified estimator of $\overline{Y}$. The MELE of the median of one variable, using as auxiliary information the known median of another variable, gives rise to a raking estimator.

Chen & Sitter (1999) formulate an empirical likelihood that respects design weights, and they use it to construct MELE's. Their statement of Theorem 8.3 does not put any conditions on Y_i. In a personal communication, Jiahua Chen indicates that the conditions on the Y_i should be the same as those on the X_i. Chen & Sitter (1999) show that the conditions in Theorem 8.3 are satisfied for sampling proportional to population size (pps) with replacement, for the Rao-Hartley-Cochran method (of pps without replacement), and for cluster sampling. Chen & Sitter (1999) also show how to define an MELE for sampling designs within strata, using side information. Zhong & Rao (2000) provide a central limit theorem for the MELE based on independent simple random samples within strata. They also show that the empirical likelihood ratio can be used to form confidence regions, if a correction generalizing $1 - f$ to the stratified case is applied.

Sitter & Wu (2000) consider estimating quadratic population quantities defined as $\sum_{i=1}^{N} \sum_{j=1}^{N} \phi(Y_i, Y_j)$ for some function ϕ. Variances and covariances are the motivating statistics. They modify the design effect likelihood (8.17) to take account of pairwise inclusion probabilities $\pi_{ij} = \Pr(Z_i Z_j = 1)$, and obtain range respecting estimators that incorporate side information.

Wu & Sitter (2001) consider a setting where the entire population $X_1, \ldots, X_N$ is known but only the sampled $y_1, \ldots, y_n$ are available. Then using a working model to link the mean and variance of Y_i to X_i, they develop estimators of $\bar{Y}$ that are consistent generally, and efficient if the working model holds.

Zhong, Chen & Rao (2001) consider combining multiple finite samples to estimate a common feature, such as a mean or CDF, when some of the samples have distorted observations.

MELE's for survey sampling have a lot in common with variance reduction methods in Monte Carlo. Hesterberg (1995b) presents the usual Monte Carlo variance reduction methods in terms of reweighted sample points.

Other dependencies

Longitudinal data are the subject of Diggle, Liang & Zeger (1994). They are usually analyzed by techniques in the companion papers Liang & Zeger (1986) and Zeger & Liang (1986). Bootstrap methods and references for spatial processes are discussed by Davison & Hinkley (1997, Chapter 8). A parametric model may be used, or the data can be resampled in spatial blocks. Loh (1996) considers empirical likelihood confidence regions for the mean based on Latin hypercube samples.

8.11 Exercises

Exercise 8.1 The GARCH(1,1) model is widely used for financial time series. Let e_i be a sequence of independent random variables with mean 0 and variance

1. The series values are $Y_i = e_i \sigma_i$ where the variance σ_i^2 evolves in time as

$$\sigma_i^2 = \alpha_0 + \alpha_1 Y_{i-1}^2 + \beta_1 \sigma_{i-1}^2.$$

This model captures the volatility clustering phenomenon often seen in financial data, where increases in variance have been seen to persist. It can also give rise to fatter than normal tails for the Y_i distribution, even if the e_i are normally distributed.

Let $\theta = (\alpha_0, \alpha_1, \beta)$ and suppose that e_i are independent $N(0, 1)$. Obtain conditional likelihood estimating equations for θ.

Exercise 8.2 Suppose that we suspect that the random variables e_i in Exercise 8.1 are not normally distributed but instead have some skewness. If Y_i are returns to holding an asset, negative skewness corresponds to downward price movements having a fatter tail than upward ones. Formulate estimating equations for the GARCH(1,1) model that support the construction of confidence intervals for the skewness.

Exercise 8.3 For the block approach to empirical likelihood, observation X_1 is always contained in block B_1. If T leaves a remainder of M when divided by L, then X_T appears in block B_n, but otherwise X_T does not appear in any block. Redefine the blocks, so that they have length M, have starting points separated by L units, and so that X_T is always used (in the last block) while X_1 may or may not be used at all.

Exercise 8.4 For the Campito tree ring data, the event of interest happened 145 times in 5395 yearly trials. The confidence interval for the event probability, taking account of dependence in the data, extends from 0.0214 to 0.0326. How much narrower (or wider) would the 95% confidence interval be for a problem with 145 occurrences in 5395 independent trials?

Exercise 8.5 What is the first year in the Campito tree series? (Hint: the year before 1 A.D. was 1 B.C., there being no year 0.)

Exercise 8.6 Suppose that $Z_1, \ldots, Z_{200} \sim N(\mu, \sigma^2)$, independently, with both μ and σ unknown. Statistician A has all the data and constructs an exact 95% confidence interval, of random length L_A, for μ using the $t_{(199)}$ distribution as usual. These same observations are then averaged in blocks of 10 and sent to Statistician B. This statistician gets $Y_j = (Z_{10(j-1)+1} + \cdots + Z_{10(j-1)+10})/10 \sim N(\mu, \sigma^2/10)$ for $j = 1, \ldots, 20$, and constructs an exact 95% confidence interval for μ of random length L_B, using the $t_{(19)}$ distribution. A is clearly better off than B. Find the mean, variance, and histogram of L_B/L_A, by simulation.

CHAPTER 9

Hybrids and connections

This chapter considers hybrid methods in which empirical likelihood is combined with other methods. There are a number of problems where a parametric likelihood is known or trusted for part, but not all, of the problem. In those cases, hybrid methods fill in the gaps with empirical likelihood. Similarly, an empirical likelihood can be combined with a Bayesian prior distribution, the bootstrap, and various jackknives. Bootstrap calibration of empirical likelihood is discussed elsewhere (Chapters 3.3 and 5.6.), as is a hybrid between empirical likelihood and permutation tests (Chapter 10.3).

There are also deep connections between empirical likelihood and other nonparametric methods of inference. This is not surprising. Loosely speaking, two methods that are asymptotically correct to some order might be expected to agree to at least that order. Connections to bootstraps, jackknives, and sieves are presented.

9.1 Product of parametric and empirical likelihoods

Consider a setting with two samples $X_1, \ldots, X_n$ and $Y_1, \ldots, Y_m$. Suppose that all $n + m$ observations are independent, and that we have a trusted parametric model in which $Y_i \sim g(y_i; \theta)$, but no such model is available for X_i. For example, Y_i may be from an instrument shown by experience to be normally distributed while the X_i may be from a newer kind of equipment. Similarly, the inter-arrival times Y_i for a queue may be known to have an exponential distribution, but the service times X_i may not belong to a known parametric family.

A natural approach is to form a likelihood that is nonparametric in the distribution F of the X_i but is parametric in the distribution G of the Y_i. That is,

$$L(F, \theta) = \prod_{i=1}^{n} F(\{X_i\}) \prod_{j=1}^{m} g(Y_j; \theta),$$

with likelihood ratio function

$$R(F, \theta) = \prod_{i=1}^{n} nw_i \prod_{j=1}^{m} \frac{g(Y_j; \theta)}{g(Y_j; \hat{\theta})},$$

where as usual $F = \sum_{i=1}^{n} w_i \delta_{X_i}$ for weights $w_i \geq 0$, $\sum_{i=1}^{n} w_i = 1$, and $\hat{\theta}$ is the parametric MLE of θ computed from the Y_i data.

Suppose that we are interested in a parameter ϕ defined through estimating equations

$$E\left(h(X,Y,\phi)\right) = \iint h(x,y,\phi)dG_\theta(y)dF(x) = 0.$$

If $h(X,Y,\phi)$ only involves X, or only involves Y, then ordinary empirical or parametric likelihood, respectively, is available. When both distributions are involved, define

$$\mathcal{R}(\phi) = \max_{F,\theta} R(F,\theta)$$

subject to

$$\sum_{i=1}^n w_i \int h(X_i, y, \phi)dG_\theta(y) = 0.$$

Under mild conditions (see Chapter 9.11), an asymptotic χ^2 calibration with degrees of freedom equal to the dimension of h is appropriate.

9.2 Parametric conditional likelihood

Consider independent (X_i, Y_i) pairs, for $i = 1, \ldots, n$, with $X_i \in \mathbb{R}^p$ and $Y_i \in \mathbb{R}^q$. Now suppose that we have a parametric density or mass function $g(y|x;\theta)$ for the conditional distribution of Y given X but no parametric likelihood for the marginal distribution of X. Then the hybrid likelihood is

$$L(F,\theta) = \prod_{i=1}^n F(\{X_i\})g(Y_i \mid X_i; \theta). \tag{9.1}$$

More generally, with $m-n \geq 0$ further observations with X_i but not Y_i measured, the likelihood is

$$L(F,\theta) = \prod_{i=1}^m F(\{X_i\}) \prod_{i=1}^n g(Y_i \mid X_i; \theta). \tag{9.2}$$

Exercise 9.1 considers observations with Y_i but not X_i measured.

It is natural to suppose that there is no known relationship between θ and F. If the statistic of interest only involves the X distribution, or only involves the distribution of Y given X, then we may use a marginal empirical likelihood of X or a parametric conditional likelihood of Y given X, respectively. But if the statistic of interest involves both distributions, then there is something to be gained by using (9.1) or (9.2).

Suppose, for example, that given $X = x$, the response Y has the $N(\beta_0 + \beta_1 x + \beta_2 x^2, \sigma^2)$ distribution. We might want to know the value $x_0 = -\beta_1/(2\beta_2)$. This x_0 represents a minimum of $E(Y \mid X = x)$ if $\beta_2 > 0$. It is a maximum if $\beta_2 < 0$. The average X value deviates from this optimum by an amount $\Delta = E(X) + \beta_1/(2\beta_2)$. This Δ represents the amount of the correction that needs to be applied to the average X value to reach the optimum. It depends on both the parametric and nonparametric parts of the data description.

An asymptotic χ^2 distribution has been shown for this hybrid likelihood ratio (see Chapter 9.11) for a problem incorporating some side information. That theorem has a scalar parameter of interest μ defined through a smooth estimating equation $E(h(X, \theta, \mu)) = 0$. There is no reason to suppose that the scalar setting is special here, and so it is reasonable to believe that the hybrid likelihood can be used to generate tests and confidence regions quite generally.

It is interesting to consider the reverse situation where the marginal likelihood is parametric and the conditional likelihood is empirical. Let $f(x; \theta)$ be the parametric density or mass function of X. Let $G_{0,x}(Y)$ be the conditional distribution of Y given $X = x$, and let $G_x(Y)$ be a candidate. Assuming no ties, the likelihood is

$$L(\theta, G_{X_1}, \ldots, G_{X_n}) = \prod_{i=1}^{n} f(X_i; \theta) G_{X_i}(Y_i).$$

Maximizing this likelihood is degenerate if there are no ties among the X_i. The NPMLE for θ matches the parametric one, but every G_{X_i} is a point mass at the corresponding value of Y_i. If there are a small number of distinct X_i with a large number of occurrences each, then NPMLE's for them are not degenerate. Failing that, a model linking G_x for different x values is required. If $G_x(y) = G(y - h(x))$, for a known function $h(x)$, or more generally $G_x(y) = G(y - h(x, \gamma))$ for a parameter vector γ, then the likelihood

$$L(\theta, G, \gamma) = \prod_{i=1}^{n} f(X_i; \theta) G(Y_i - h(X_i, \gamma))$$

is not necessarily degenerate.

9.3 Parametric models for data ranges

Another case where parametric and empirical likelihoods mix is where the data X_i are thought to follow a parametric model $f(x_i; \theta)$, over a subset of their range. For example, the random variable X_i may be thought to be normally distributed over some central values $[-M, M]$ but not necessarily in the tails. Or the X_i may be positive random variables known to have an exponential tail, $f(x; \theta) \propto \exp(-\theta x)$ for $x > x_0$. As in the previous sections, we might get a sample that is partly parametric and partly nonparametric, but in this case it is not known which likelihood applies to an observation until that observation becomes available.

Suppose that $X \sim f(x; \theta)$ for $x \in P_0$, but not necessarily for $x \notin P_0$. Introduce the shorthand $P_0(x)$ for $1_{x \in P_0}$. The appropriate likelihood hybrid is then

$$\prod_{i=1}^{n} [f(X_i; \theta)]^{P_0(X_i)} w_i^{1 - P_0(X_i)},$$

where $w_i \geq 0$, $w_i = 0$ when $X_i \in P_0$, taking $0^0 = 1$, and we impose the constraint

$$\int_{P_0} dF(x;\theta) + \sum_{i=1}^{n} w_i(1 - P_0(x_i)) = 1.$$

Again under mild conditions including smoothness in θ of the parametric distribution (see Chapter 9.11), the combined likelihood has the asymptotic χ^2 distribution that one would expect.

9.4 Empirical likelihood and Bayes

Let θ be a parameter with estimating equation $E(m(X, \theta)) = 0$. Suppose that we are willing to specify a prior density $\pi(\theta)$ for θ, but that we have no parametric family for the distribution F of X. The opposite situation, where a parametric model is given for θ but we are reluctant to specify a prior, is very commonly approached with flat non-informative prior distributions.

We will suppose that $\int_{\theta \in \Theta} \pi(\theta) d\theta = 1$. Then a natural procedure is to take the posterior distribution of θ to be

$$\mathcal{L}(\theta|X_1,\ldots,X_n) = \frac{\pi(\theta)\mathcal{R}(\theta)}{\int_{\theta \in \Theta} \pi(\theta)\mathcal{R}(\theta)d\theta}, \quad (9.3)$$

where $\mathcal{R}(\theta)$ is the profile empirical likelihood ratio function for θ. There is as yet little known about how well this proposal works. Some theory and simulations showing the asymptotic accuracy of posterior probability statements computed from $\mathcal{L}(\theta|X_1,\ldots,X_n)$, when θ is a univariate mean, are described in Chapter 9.11.

The process is simply to multiply the empirical likelihood by a prior distribution, and then renormalize to a proper density function. This appears to avoid putting a prior on the space of all distributions F, or even on the whole simplex of multinomial weights. The rationale behind the process is as follows: Empirical likelihood is approximately using a least favorable family for θ. This is a parametric family of the same dimension as θ. The prior distribution on θ induces a prior distribution on this same family. Multiplying the prior on the family by the likelihood on the family yields a posterior density on the family.

If nuisance parameters ν are defined jointly with θ through $E(m(X, \theta, \nu)) = 0$, then we place a prior $\pi(\theta, \nu)$ on θ and ν. The posterior distribution on θ and ν is then proportional to $\pi(\theta, \nu)\mathcal{R}(\theta, \nu)$, and the posterior distribution for θ is obtained by integrating out ν.

9.5 Bayesian bootstrap

To describe the Bayesian bootstrap, suppose at first that the distribution F_0 attaches probability 1 to the known finite set $\{z_1,\ldots,z_k\}$. This constraint will be lifted later. The z_j may be in $\mathbb{R}^d$ or even in more general spaces. Then there is a finite dimensional parametric space of candidate distributions F defined by the parameter vector $\omega = (\omega_1, \ldots, \omega_k)' \in \mathbb{S}_{k-1}$ with $\omega_j = F(\{z_j\})$. The unit probability simplex $\mathbb{S}_{k-1}$ is defined in equation (2.6), replacing n there by k.

The Bayesian bootstrap places a Dirichlet prior on θ. The Dirichlet prior is proportional to $1_{\omega \in \mathbb{S}_{k-1}} \prod_{j=1}^{k} \omega_j^{m_j}$. If the sample contains n_j observations equal to z_j, then the posterior distribution is also a Dirichlet, and is proportional to $1_{\omega \in \mathbb{S}_{k-1}} \prod_{j=1}^{k} \omega_j^{m_j + n_j}$. The choice $m_j = -1$ is particularly convenient. If there are any z_j with $n_j = 0$, then the posterior distribution is improper, having an infinite integral. That posterior can be interpreted as placing probability 1 on $\theta_j = 0$ for every j with $n_j = 0$. Then the posterior distribution is proportional to

$$1_{\omega \in \mathbb{S}_{k-1}} \prod_{j:n_j>0} \omega_j^{n_j - 1} \prod_{j:n_j=0} 1_{\omega_j=0}. \tag{9.4}$$

The unobserved z_j for which $n_j = 0$ do not appear in the posterior distribution, and this lifts the constraint that we have to know what they are.

The Bayesian bootstrap samples from the posterior distribution implied by (9.4). To generate a sample from the posterior distribution, draw n independent $U(0,1)$ random variables U_i, transform them into exponential random variables $Y_i = -\log(U_i)$, and then take $w_i = Y_i / \sum_{j=1}^{n} Y_j$. The sampled value of ω_j is then $\sum_{i:X_i=z_j} w_i$. For a statistic $T(F)$, the resampled value is $T(\sum_{i=1}^{n} w_i \delta_{X_i})$. The Bayesian bootstrap sample consists of B independently sampled values of T. The posterior probability of a set C is estimated by the fraction of the B resampled T values that happen to be in C, and a posterior moment is simply the average over B sampled values of the corresponding power of T.

The empirical likelihood is proportional to $\prod_{j=1}^{k} \omega_j^{n_j}$. Apart from the way unobserved z_j are handled, the empirical likelihood is obtained as a posterior distribution for the non-informative Dirichlet prior having $m_j = 0$. This is a nonparametric analogue of the familiar fact that the posterior is proportional to the likelihood, when a non-informative prior is used.

9.6 Least favorable families and nonparametric tilting

Empirical likelihood works with an n-dimensional family of distributions supported on the sample points. For data $X_i \in \mathbb{R}^d$, we maximize the empirical likelihood subject to a constraint like $\sum_i w_i X_i = \mu$. As μ varies through a d-dimensional space, a d-dimensional subfamily of the multinomial distributions arise as constrained maxima. This family may be indexed by μ, or by the Lagrange multiplier λ.

For a statistic $\theta = T(F) \in \mathbb{R}^p$, there is usually a reduction to a p-dimensional family of multinomial distributions. Similar p-dimensional subfamilies may also be defined through other discrepancies such as empirical entropy or the Euclidean likelihood.

The nonparametric tilting bootstrap draws samples from members of one of these lower dimensional subfamilies of multinomial distributions. For a scalar parameter θ such as the univariate mean, there is a univariate family of distributions. Denote the generic member of this family by F_θ and let $w_i(\theta) = F_\theta\{X_i\}$. Bootstrap samples can be drawn from these multinomial distributions by taking $X_i^{*b} = X_{J(i,b)}$, where for $i = 1, \ldots, n$ and $b = 1, \ldots, b$ the indices $J(i,b)$ are independent with $\Pr(J(i,b) = k) = w_k(\theta)$. Then $\hat\theta^{*b}$ is the value of $\hat\theta$ on the data $X_1^{*b}, \ldots, X_n^{*b}$. The values

$$\theta_L = \min\{\theta \mid \Pr(\hat\theta^* \geq \hat\theta \,;\, F_\theta) \geq \alpha/2\},$$
$$\theta_U = \max\{\theta \mid \Pr(\hat\theta^* \leq \hat\theta \,;\, F_\theta) \geq \alpha/2\},$$

are the lower and upper limits, respectively, of the nonparametric tilting approximate $100(1-\alpha)\%$ confidence interval.

It can be very laborious to sample from many members of a parametric family, searching for the desired endpoints. Importance sampling can be used to reweight data from one distribution F_θ to obtain unbiased expectations under another distribution $F_{\theta'}$. The nonparametric tilting bootstrap samples from $F_{\hat\theta}$ with every $w_i = 1/n$, and so the importance sampling weight is $\prod_{i=1}^{n} n w_{J(i,b)}(\theta)$ in bootstrap sample b. A further advantage of importance sampling is that the simulations for different values of θ are coupled, which makes it more likely that the search for endpoints can be done by looking for the point at which a monotone function is zero. Using the Kullback-Liebler family gives an especially convenient exponential tilting form for the importance sampling weight factors.

A least favorable parametric family is sometimes described as one in which the estimation problem is hardest, sometimes as one in which the estimation problem is as hard as in a parametric problem. Usually difficulty is measured through a discrepancy measure. So if a statistic is defined through $T(F)$ and θ_0 is the true value of T, then family might consist of distributions F_θ for which $T(F_\theta) = \theta$, and subject to this, a distance measure $D(F_{\theta_0}, F_\theta)$ is minimized.

Empirical likelihood, nonparametric tilting, and some other methods described in Chapter 9.11 all employ least favorable families. These methods use random, sample based families. The value of a least favorable family is that it has not made the statistical problem artificially easy.

It is possible to construct some parametric families in which inference is outlandishly hard, although still not least favorable. The $N(0,1)$ distribution is a member of the family

$$f(x;\theta) = (1-\epsilon)N(0,1) + \epsilon N(\theta/\epsilon, 1). \tag{9.5}$$

Here $f(x;\theta)$ has mean θ. For $\epsilon = 10^{-100}$ and any reasonable sample size, it would be very hard to estimate θ. Using multinomial families on the data rules out unreasonably hard cases like (9.5).

9.7 Bootstrap likelihood

Suppose that we compute an estimate $\widehat{T} = T(\widehat{F})$ of $\theta = T(F_0)$, using an IID sample $X_1, \ldots, X_n \in \mathbb{R}^d$. In a parametric family indexed by θ, the probability density function $f(\widehat{T}; \theta)$ could be interpreted as a partial likelihood for θ. The qualifier "partial" reflects that f does not give the joint density of $X_1, \ldots, X_i$, but just that of the function $\widehat{T}$ computed from them. In Chapter 13.3 a similar density is called a pseudo-likelihood.

The bootstrap likelihood uses two levels of resampling, some density estimation, and some regression smoothing to estimate $f(\widehat{T}; \theta)$ from the data. We will suppose that $T(F) \in \mathbb{R}$. For $r = 1, \ldots, R$, let $X_1^{*r}, \ldots, X_n^{*r}$ be a bootstrap sample of the data, with corresponding T value $\widehat{T}^{*r}$. Then for $r = 1, \ldots, R$ and $s = 1, \ldots, S$, let $X_1^{*rs}, \ldots, X_n^{*rs}$ be a bootstrap sample from $X_1^{*r}, \ldots, X_n^{*r}$ with corresponding T value $\widehat{T}^{*rs}$. A preliminary bootstrap likelihood at $\widehat{T}^{*r}$ is then estimated by a kernel density estimate

$$L(\widehat{T}^{*r}) = \hat{f}(\widehat{T}; \widehat{T}^{*r}) = \frac{1}{Sh_2} \sum_{s=1}^{S} K_2\left(\frac{\widehat{T}^{*rs} - \widehat{T}}{h_2}\right),$$

where K_2 is a kernel function (Chapter 5) and h_2 is a bandwidth.

It is possible to have $L(\widehat{T}^{*r}) \neq L(\widehat{T}^{*r'})$ even when $\widehat{T}^{*r} = \widehat{T}^{*r'}$. For this reason, and to interpolate, the bootstrap likelihood is defined through further smoothing, such as

$$L_B(\theta) = \hat{f}(\widehat{T}; \theta) = \frac{\sum_{r=1}^{R} K_1\left(\frac{\widehat{T}^{*r} - \theta}{h_1}\right) L(\widehat{T}^{*r})}{\sum_{r=1}^{R} K_1\left(\frac{\widehat{T}^{*r} - \theta}{h_1}\right)},$$

for the kernel K_1 and bandwidth h_1, or through some other scatterplot smoother.

The bootstrap likelihood has been shown to match the empirical likelihood, but only to first order. Much of the research on bootstrap likelihood aims at reducing the computational burden. See Chapter 9.11.

9.8 Bootstrapping from an NPMLE

The usual form of the bootstrap resamples from the empirical distribution F_n. In IID sampling the empirical distribution is the NPMLE. In settings with side information, F_n is not the NPMLE, and an attractive alternative is to use empirical likelihood to construct the NPMLE $\hat{F}$, and then resample from $\hat{F}$. When the side information is specified by $E(m(X, \theta, \nu)) = 0$ then the NPMLE of Chapter 3.10 is $\hat{F} = \sum_{i=1}^{n} w_i \delta_{X_i}$, where $w_i \geq 0$, $\sum_{i=1}^{n} w_i = 1, \sum_{i=1}^{n} w_i m(X_i, \theta, \nu) = 0$, and $\prod_{i=1}^{n} nw_i$ is maximized subject to these constraints. Similarly, when the data were obtained by biased sampling, then bootstrapping from the NPMLE becomes attractive.

Resampling from $\hat{F}$ is straightforward. Let $C_i = \sum_{j=1}^{i} w_j$, with the understanding that $C_0 = 0$. To generate a bootstrap sample draw nB independent

$U(0,1)$ random variables U_i^b, for $1 \le i \le n$ and $1 \le b \le B$. Turn these into resampled observations where $X_i^b = X_j$ whenever $C_{j-1} < U_i^b \le C_j$. This produces B bootstrap data sets $(X_1^b, \dots, X_n^b)$, for $b = 1, \dots, B$.

The Euclidean likelihood can also be used to define an NPMLE from which to resample. But sampling from the Euclidean likelihood NPMLE is hard to define in cases where some $w_i < 0$.

9.9 Jackknives

The jackknife is a leave-one-out method of forming confidence regions for a statistical quantity θ. Suppose that the true value is $\theta_0 = T(F_0) \in \mathbb{R}^p$ and the sample value is $T(\hat{F})$. For a candidate distribution $F = \sum_{i=1}^n w_i \delta_{X_i}$ where $w_i \ge 0$ and $\sum_{i=1}^n w_i = 1$, define $T(w_1, \dots, w_n) = \theta(F)$. The NPMLE is $\hat{\theta} = T(1/n, \dots, 1/n)$. Define $T_{-i} = \theta((1 + 1/n)\hat{F} - (1/n)\delta_{X_i})$, the value of θ on a hypothetical sample of the $n-1$ observations other than X_i. Now let

$$T_{-\bullet} = \frac{1}{n} \sum_{i=1}^n T_{-i}, \quad \text{and}$$

$$S = \sum_{i=1}^n (T_{-i} - T_{-\bullet})(T_{-i} - T_{-\bullet})'.$$

The value $(n-1)(T_{-\bullet} - T(\hat{F}))$ is often used as an estimate of the bias in $T(\hat{F})$, and the corresponding bias-corrected estimate of $T(F)$ is $nT(\hat{F}) - (n-1)T_{-\bullet}$. Also S or $(n-1)S/n$ can often be used to estimate the variance of $T(\hat{F})$ or of $T_{-\bullet}$. For $p = 1$, a simple 95% confidence interval for θ_0 may then be calculated as $T(\hat{F}) \pm 1.96\sqrt{S}$.

The infinitesimal jackknife is based on the linear approximation

$$T(w_1, \dots, w_n) = \hat{\theta} + \sum_{i=1}^n w_i T_i(\hat{F}),$$

where

$$T_i(F) = \lim_{\varepsilon \to 0} \frac{T((1-\varepsilon)F + \varepsilon \delta_{X_i}) - T(F)}{\varepsilon}. \tag{9.6}$$

In the infinitesimal jackknife, a variance estimate for $\hat{\theta} - \theta_0$ can be constructed as

$$\frac{1}{n^2} \sum_{i=1}^n T_i(\hat{F}) T_i(\hat{F})'.$$

The coefficient n^{-2} can be replaced by $1/(n(n-1))$ in order to get an unbiased variance estimate for the variance of linear statistics of the form $T(w_1, \dots, w_n) = \sum_{i=1}^n w_i Q(X_i)$. One of the algorithms for maximizing empirical likelihood (see

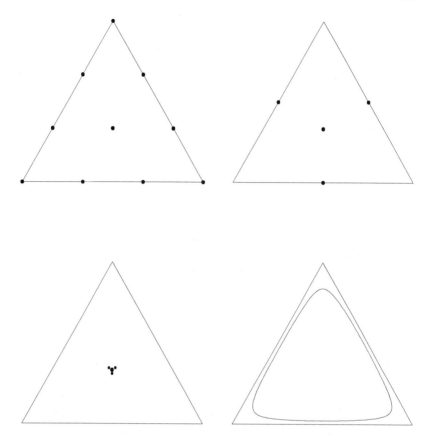

Figure 9.1 *The upper left plot shows the reweighings of the data used by the bootstrap, for $n = 3$. The upper right and lower left show the jackknife and infinitesimal jackknife reweightings, respectively. The lower right plot depicts a region used by empirical likelihood.*

Chapter 12.6) is an empirical likelihood test of whether the $T_i(\hat{F})$ have mean $\theta_0 - \hat{\theta}$. They have sample mean 0.

Figure 9.1 compares the points in the simplex used in nonparametric inferences, for $n = 3$. When the infinitesimal jackknife fails it is often because it only uses reweightings that are a negligible distance from the center $(1/n, \ldots, 1/n)$. It works with a linear approximation to T, and this may be inadequate if T is nonlinear enough. The jackknife has similar difficulties because the points it uses are only $O(1/n)$ away from the center. Even when they do not fail, approximations based on expansions around the MLE can work badly in some settings. Linearization inference in nonlinear least squares, Greenwood's formula in survival analysis, and Wald tests are often found to be inferior to methods that are not so strongly localized around the MLE.

The bootstrap and empirical likelihood achieve greater generality than the jackknives in part because they consider reweightings at a distance (or average distance) of $O(n^{-1/2})$ from the center. The difference is illustrated by the median, for which the jackknife and infinitesimal jackknife do not provide consistent variance estimates.

9.10 Sieves

In some nonparametric problems, there is an infinite dimensional family of estimators and an MLE does not exist. For example, consider the problem of estimating the density f from an IID sample of random variables $X_i \in [0, 1]$. The density f is a nonnegative function integrating to 1 over $[0, 1]$. The natural likelihood to use is $L(f) = \prod_{i=1}^n f(X_i)$, but $L(f)$ is unbounded over the family of densities, thus there is no NPMLE.

A sieve approach to this problem is to consider a regularized set of densities, such as densities that are piecewise constant on $[0, 1]$ with the allowed discontinuities at knots $t_j = j/m$ for $j = 1, \ldots, m-1$. Given a value of $m > 1$, the NPMLE is a histogram estimator. The NPMLE might not be unique if some X_i equals some t_j, but a unique choice can be forced by taking f to be continuous from the left. The sieve approach to the histogram estimator lets $m \to \infty$ at a suitable rate, as $n \to \infty$. Sieves are more general than the histogram estimator, and the controlling parameter m is not always integer valued. For example, a sieve could be constructed by taking all continuous densities with $\int_0^1 f(x)^2 dx < m$.

Empirical likelihood can be thought of as a sieve method. The family of candidate distributions is usually all those that reweight the sample. It may also include distributions that put some weight on a number of non-observed values. Two features of empirical likelihood distinguish it from sieve methods. First, the family of candidate distributions is random, as it depends on the sample. Second, the emphasis is shifted from maximum likelihood to likelihood ratios.

While sieves originated to handle an infinite dimensional parameter set, they may also be applied in problems with an infinite number of estimating equation constraints. Some examples of such problems appear in Chapter 10.

The sieved empirical likelihood (SEL) has been developed for some conditional moment restriction problems. Suppose that X and Z are jointly distributed random variables and that the estimating equation

$$E(m(Z, \theta) \mid X = x) = 0 \tag{9.7}$$

holds for all x. Then, of course, $E(m(Z, \theta)) = 0$ holds unconditionally, too, but (9.7) is a more stringent condition. For instance, in a regression model we might have $Z = (X', Y')'$, and $m(Z, \theta) = Y - g(X, \theta)$. Let $u(X)$ be a vector of functions of X. Then $E(u(X)(Y - g(X, \theta))) = 0$ and if u has enough component functions in it, we get an overdetermined set of unconditional estimating equations for θ.

A lower bound is known for the variance of an asymptotically unbiased estimator $\hat{\theta}$ from n IID observations:

$$n\text{Var}(\hat{\theta}) \geq V_0 \equiv \left(E(D(X)'\Omega(X)^{-1}D(X))\right)^{-1}$$

where $D(x) = E(\partial m(Z, \theta)/\partial \theta \mid X = x)$ and $\Omega(x) = E(m(Z, \theta)m(Z, \theta)' \mid X = x)$. Moreover, the solution to estimating equations $E(u(X)m(Z, \theta)) = 0$ attains this asymptotic variance bound for the (usually unknown) vector $u_0(x) = D(x)'\Omega(x)^{-1}$. The functions $u(X)$ are called instrumental variables and u_0 are the optimal instruments.

A sieved empirical likelihood approach imposes (9.7) at the n sample values x_i, using some smoothing. Let $K_{ij} = K((x_i - x_j)/h)$, for a kernel function K and bandwidth h, and let $K_{i\bullet} = \sum_{j=1}^{n} K_{ij}$. Define the conditional empirical log likelihood function $L_i(F) = \sum_{j=1}^{n} K_{ij} \log(w_{ij})$ for $F(\{Z_j\} \mid X_i) = w_{ij} \geq 0$ and $\sum_{j=1}^{n} w_{ij} = 1$. Maximizing L_i subject to $\sum_{j=1}^{n} w_{ij} m(Z_j, \theta) = 0$ is the same weighted empirical likelihood problem we solved in Chapter 8.7 for finite population sampling designs, with $K_{i\bullet}$ playing the role of D there. Let the solution be $\hat{F}_i$ with weights $\hat{w}_{ij}(\theta)$, and now define $\ell(\theta) = \sum_{i=1}^{n} \log \hat{w}_{ii}(\theta)$, and let $\hat{\theta}$ maximize $\ell(\theta)$.

Theorem 9.1 *Let (X_i, Z_i) be IID. Under some mild conditions that uniquely identify the true value θ_0, conditions that impose smoothness on m, and conditions on K and h, $\sqrt{n}(\hat{\theta} - \theta_0) \to N(0, V_0)$. Further, if $\tilde{\theta}$ maximizes $\ell(\theta)$ subject to s conditions $C(\theta) = 0$ where $\partial C/\partial \theta$ has rank s, then $-2(\ell(\tilde{\theta}) - \ell(\hat{\theta})) \to \chi^2_{(s)}$ as $n \to \infty$.*

Proof. Fan & Gijbels (1999) and Kitamura, Tripathi & Ahn (2000). □

More general random sieves have been proposed, particularly for certain regression problems. In linear regression models relating Y_i to a predictor X_i and a parameter β, there is a residual $e_i = Y_i - X_i'\beta$. In certain models, however, there is no unique e_i that can be computed from a given X_i, Y_i and β. A simple example without any X_i, is the following mixed effects model

$$Y_{ij} = \theta_j + \alpha_i + \epsilon_{ij}, \quad i = 1, \ldots, n, \quad j = 1, \ldots, k$$

for scalar parameters θ_j and independent errors α_i, ϵ_{ij}, with mean 0, $\text{Var}(\alpha_i) = \sigma_1^2$, and $\text{Var}(\epsilon_{ij}) = \sigma_2^2$. Given $\theta = (\theta_1, \ldots, \theta_k)$ and $Y_i = (Y_{i1}, \ldots, Y_{ik})$ we cannot compute $e_i = (\alpha_i, \epsilon_{i1}, \ldots, \epsilon_{ik})$ though we do know that $\alpha_i + \epsilon_{ij} = Y_{ij} - \theta_j$.

For a random sieve model, we can identify a set $B_i(\theta)$ known to contain e_i. For the random effect example

$$B_i(\theta) = \{(\alpha_i, \epsilon_{i1}, \ldots, \epsilon_{ik}) \mid \alpha_i + \epsilon_{ij} = Y_{ij} - \theta_j, j = 1, \ldots, k\}.$$

If $B_i(\theta)$ is not a finite set of points, the random sieve method selects a set of points $E_{ij} \in B_i(\theta)$, $j = 1, \ldots, n_i$. For larger n, the numbers n_i increase to give better coverage of B_i. Consider a family of distributions, with the generic member F putting probability $p_{ij} \geq 0$ on the observation (X_i, Y_i, E_{ij}) for $i = 1, \ldots, n$ and $j = 1, \ldots, n_i$ with $p_i = \sum_{j=1}^{n_i} p_{ij}$ and $\sum_{i=1}^{n} p_i = 1$. The sieved likelihood is the maximum of $\prod_{i=1}^{n} p_i$ over members of the family satisfying a set of constraints. For random effects the constraints might express that $E(\alpha_i) = 0$, $\mathrm{Var}(\alpha_i) = \sigma_1^2$, and that for $j = 1, \ldots, k$, $E(\epsilon_{ij}) = E(\epsilon_{ij} - \sigma_2^2) = E(\alpha_i \epsilon_{ij}) = 0$. Methods for choosing the E_{ij} are in articles cited in Chapter 9.11.

9.11 Bibliographic notes

Parametric-empirical hybrids

Qin (1994) presents the semi-empirical likelihood, in which an empirical likelihood is used for one sample and a parametric one is used for the other. He considers the example where X_i and Y_j are scalars and interest centers on $E(X) - E(Y)$. Qin (2000) considers multiplying a parametric conditional likelihood by an empirical marginal one, and presents a theorem from Qin (1992) establishing a $\chi^2_{(1)}$ limit for a statistic defined through a smooth estimating function. Qin (2000) simulates some examples that Imbens & Lancaster (1994) treated by generalized method of moments, and remarks that the likelihood formulation makes it easier to incorporate observations where X_i but not Y_i was measured. Those examples consider samples augmented by side information from the census.

Qin & Wong (1996) present another semi-empirical likelihood in which the data are parametric or not, depending on their values. They consider the case where the parametric model holds if $x \leq T_0$, and establish a χ^2 calibration for a univariate θ that enters the likelihood smoothly. Moeschberger & Klein (1985) consider a parametric model for a tail subject to censoring combined with a nonparametric model to the left of that tail.

Lazar (2000) studies the product of a prior density on the univariate mean and an empirical likelihood for that mean. She shows that the posterior distribution is asymptotically normal, as one would expect from an asymptotically quadratic log likelihood. As a result, we can expect in general that the computed posterior probabilities of intervals are asymptotically justified. The arguments parallel those of Monahan & Boos (1992) for parametric likelihoods. In simulations, accuracy can be measured by finding the distribution under sampling of the posterior probability attached to the set $(-\infty, \mu]$ where μ is the true mean. This posterior probability should have nearly a uniform distribution. Figure 1 of Lazar (2000) shows an example in which the accuracy is good for $n = 50$ but perhaps not for $n = 10$. Figure 2 shows an example where the accuracy appears good with $n = 20$.

Bootstrap connections

The Bayesian bootstrap was proposed by Rubin (1981). Newton & Raftery (1994) illustrate its use on a number of frequentist inference problems. The bootstrap likelihood was proposed by Davison, Hinkley & Worton (1992). Some computational improvements are given in Davison, Hinkley & Worton (1995) and Pawitan (2000).

Efron (1981) introduces the nonparametric/exponential tilting bootstrap. Further results for it appear in DiCiccio & Romano (1990), as described below under least favorable families. Efron emphasizes the Kullback-Liebler family but also presents a version using a nonparametric likelihood discrepancy, giving rise to the same family of distributions in the simplex that empirical likelihood uses.

Zhang (1999) resamples from an NPMLE constructed to incorporate side information in the form of estimating equations. He produces confidence bands and tests for the distribution function of a scalar random variable, and confidence intervals for the mean and variance of a scalar. Zhang (1997) constructs confidence bands for the quantile function by bootstrapping from an NPMLE. Hall & Presnell (1999b) term this the b-bootstrap, and they show its wide applicability and describe the asymptotic behavior.

Ren (2001) defines the leveraged bootstrap. The leveraged bootstrap takes IID samples of size m from an NPMLE. The bootstrap samples are IID, even though the original data was subject to interval censoring. Careful calibration makes up for the discrepancy.

Chuang & Lai (2000) describe a hybrid method in which they construct a parametric family on the simplex of observation weights and use the bootstrap in that family. They find good results this way for some hard inferential problems such as the analysis of group sequential trials, explosive time series, and Galton-Watson processes.

Least favorable families

The notion of a least favorable family is due to Stein (1956). He used it to reduce nonparametric problems to parametric ones that were at least as hard. Efron (1981) shows a least favorable family property of the one-dimensional multinomial sub-family. When θ is the mean, the Fisher information in the sub-family is the inverse of the sample variance, holding for $F_{\bar{X}}$ and even for other members F_θ.

DiCiccio & Romano (1990) show that one can construct a p-dimensional sub-family of multinomial distributions using any of various discrepancy measures. Once one has such a family, one can bootstrap within it, use the likelihood function in it, or use another distance function in it. They show that one-sided coverage errors are $O(1/n)$ for the nonparametric tilting bootstrap. They also show that the reduced family is least favorable for more general statistics than the mean. Hesterberg (1999) investigates exponential tilting bootstrap confidence intervals for nonlinear statistics, using importance sampling. He compares likelihood- and entropy-based intervals.

Empirical likelihood corresponds to using the likelihood ratio function in the least favorable family defined by the likelihood discrepancy. Lee & Young (1999) recommend defining a p-dimensional parametric family through the exponential discrepancy, and following up with a likelihood ratio test in that family. Corcoran et al. (1995) recommend using the exponential discrepancy (empirical entropy) to form a p-dimensional family, and then using the Wald test (a sandwich estimator) within that family. They report some simulations in which a χ^2 calibration is quite accurate for this combination. Davison & Hinkley (1997, Chapter 10) recommend a Rao test within a family formed by exponential discrepancies.

Where Efron (1981) considers sampling from various members of a data-determined least favorable family, the b-bootstrap and related papers above pick a single best-fitting member of that family and resample from it.

Other connections

The jackknife was introduced by Quenouille (1949) for correcting bias in estimates. Tukey (1958) showed that it can be used to construct variance estimates. The infinitesimal jackknife is due to Jaeckel (1972). Hesterberg (1995a) proposes a "butcher knife" formed by taking divided differences in (9.6) with $\epsilon = O(n^{-1/2})$. By varying the sample more than $O(1/n)$ this results in a more widely applicable jackknife. See also Shao & Tu (1995) for a discussion of jackknives that delete d of n observations.

Saddlepoint approximations allow one to construct approximate sampling densities for statistics. Monti & Ronchetti (1993) provide a formula that allows one to translate between empirical likelihoods and saddlepoint densities for statistics defined through estimating equations. Reid (1988) provides a survey of saddlepoint methods.

The connection between empirical likelihood and random sieves is given by Shen, Shi & Wong (1999), who also consider a random effects model like the one in Chapter 9.10 but incorporating predictors X_{ij}. Shen et al. (1999) describe methods for selecting a finite set of points within each $B_i(\theta)$.

The results in Theorem 9.1 are based on independent work by Fan & Gijbels (1999) and Kitamura et al. (2000). They combination of smoothing and empirical likelihood studied there was proposed by LeBlanc & Crowley (1995).

Fan, Zhang & Zhang (2001) describe a sieve-based approach to problems such as nonparametric regression, including testing whether a smooth function, or indeed an additive model, might be linear. They obtain more general Wilks's type results with $-r \log R$ approximately $\chi^2_{\nu(n)}$ with non-integer numbers of degrees of freedom $\nu(n) \to \infty$ as $n \to \infty$ and r not necessarily equal to 2. Their formulation has a parametric conditional likelihood with a parameter $\theta(x)$ estimated nonparametrically as a smooth function of x. Fan & Zhang (2000) make an extension to an empirical conditional likelihood for the nonparametric $\theta(x)$.

9.12 Exercises

Exercise 9.1 Suppose that there is no parametric model for the marginal distribution of X but there is a parametric density or mass function $f(y|x;\theta)$ for the conditional distribution of Y given X. Write a hybrid likelihood assuming that (X_i, Y_i) are observed for $i = 1,\ldots, n_1$, X_i only is observed for $i = n_1 + 1,\ldots, n_1 + n_2$, and Y_i only is observed for $i = n_1 + n_2 + 1,\ldots, n_1 + n_2 + n_3$. Assume that missingness of X_i or Y_i is non-informative in the sense of coarsening at random.

Exercise 9.2 The standard 95% confidence interval for the univariate sample mean is $\bar{X} \pm 1.96 s$ where $s^2 = (n-1)^{-1} \sum_{i=1}^{n}(X_i - \bar{X})^2$. What is the ratio of the length of the leave-one-out jackknife version of the 95% confidence interval to the length of the standard 95% confidence interval?

Exercise 9.3 The jackknife is known to fail for the median, and it seems to be because only a small number of T_{-i} values are possible for a given sample. But the jackknife provides a reliable confidence interval for $\Pr(X \le Q)$, so long as $0 < \Pr(X \le Q) < 1$. Describe how to employ the jackknife to construct a confidence set for the median. If necessary assume that the observations are IID from a distribution having a unique median and a positive density function on an open interval containing that median.

Exercise 9.4 Derive a bias estimate based in the infinitesimal jackknife.

CHAPTER 10

Challenges for EL

This chapter is devoted to problems where empirical likelihood has difficulties, and to ways of mitigating those difficulties. As a case in point, empirical likelihood inferences on the number of distinct possible values from a distribution is completely degenerate. The confidence interval only includes the number of distinct values in the sample, and that value is completely wrong for continuous distributions.

As a less outlandish example, the natural way to define empirical likelihood tests for symmetry or independence are degenerate. The root cause is that these conditions are equivalent to an infinite number of estimating equations. It is, however, possible to use a known point of symmetry, or known independence, as a side condition in inferences. In some settings this is degenerate, in some it reduces to ordinary empirical likelihood, and in others it gives something new. It is also possible to test for approximate symmetry or approximate independence, defined through a finite subset of the infinite set of constraints.

For some parametric likelihoods, the usual asymptotic theory does not hold. This can happen when the range of data values depends on the parameter, when the true value of the parameter is on the boundary of the set of possible values, or when the value of one parameter does not affect the predictions, if a second parameter is zero. Empirical likelihood based on the estimating equations from these likelihoods cannot be expected to have a χ^2 calibration.

10.1 Symmetry

The distribution F of $X \in \mathbb{R}$, is symmetric about a center c, if every interval (a, b) has the same probability as the interval $(c - b, c - a)$. In the familiar case where F has a density or mass function f, symmetry means that $f(c + x) = f(c - x)$, which we can rewrite as $f(x) = f(2c - x)$.

A natural approach to nonparametric inference under symmetry is to construct a family $\mathcal{F}_S$ of symmetric distributions that put positive probability on every observation. Such a family can be represented with $n + 1$ parameters: the center c of symmetry, and weights w_i attached to X_i, which by symmetry also attach to $\widetilde{X}_i = 2c - X_i$. We suppose that $\sum_{i=1}^n w_i = 1/2$. Under this setting the probability that F puts on x is $\sum_{i=1}^n w_i(1_{X_i = x} + 1_{\widetilde{X}_i = x})$, so that x is double counted if it is both a data point and the reflection of a data point.

First we consider whether nonparametric likelihood can help identify the center

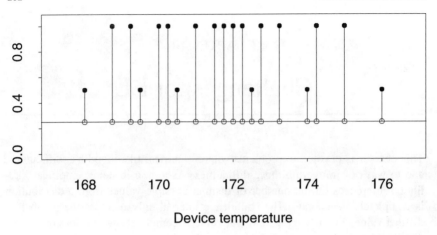

Figure 10.1 *The raw data are temperatures of six solid-state electronic devices, in degrees Celsius. The function shown is the empirical likelihood for the center of symmetry, assuming a symmetric distribution. The likelihood ratio takes the value 1/2 at each of the six data values. Source: Hahn & Meeker (1991, Chapter 13).*

c of symmetry. Define the points $c_{ij} = (X_i + X_j)/2$ for $1 \leq i \leq j \leq n$. For continuously distributed data, there are no ties among the c_{ij}. If c is not one of the c_{ij}, then the nonparametric likelihood is maximized at $w_i = 1/(2n)$ and takes the value $L_0 = (2n)^{-n}$. When $c = c_{ii} = X_i$, the likelihood $(2w_i)\prod_{j \neq i} w_j$ takes on a maximum value of $L_1 = 2(2n)^{-n}$. Finally, if $c = c_{ij}$ for $j \neq i$, then the likelihood $(w_i + w_j)^2 \prod_{k \notin \{i,j\}} w_k$ takes on maximum value $L_2 = 4(2n)^{-n}$.

Thus nonparametric likelihood arguments lead to a degenerate inference on c, under sampling from continuous F. The $n(n-1)/2$ points c_{ij} with $i < j$ maximize the likelihood, next come the n points $c_{ii} = X_i$ with a likelihood ratio of $1/2$, and finally every other point in the real line, with a likelihood ratio of $1/4$. Figure 10.1 shows this function for the operating temperatures of six solid-state electronic devices. Even the midpoint between the two largest X_i is a mode of the likelihood ratio function.

Although testing for symmetry and estimating a point of symmetry are degenerate, imposing a known symmetry as side information is not necessarily degenerate. In some cases this side information does not change empirical likelihood inferences, and in some it gives a new and nondegenerate method.

Suppose we know that F is symmetric about c. Then $F(\{x\}) = F(\{2c - x\})$ for all x. This symmetry can be imposed as a side constraint on F, to sharpen inferences on θ defined by $E(m(X, \theta)) = 0$, for some function m. We place weights w_i on X_i and weights w_i on $\tilde{X}_i = 2c - X_i$, where $w_i \geq 0$ and $\sum_{i=1}^n w_i = 1/2$.

SYMMETRY

The function m has even and odd parts defined by

$$m_{\mathrm{E}}(X,\theta) = \frac{1}{2}\Big[m(X,\theta) + m(2c - X,\theta)\Big], \quad \text{and}$$

$$m_{\mathrm{O}}(X,\theta) = \frac{1}{2}\Big[m(X,\theta) - m(2c - X,\theta)\Big].$$

Now $m = m_{\mathrm{E}} + m_{\mathrm{O}}$, and the odd part has weighted mean zero by symmetry. Thus the definition of θ may be replaced by $E(m_{\mathrm{E}}(X,\theta)) = 0$. Operationally, we replace m by m_{E} and apply empirical likelihood as usual. If $m = m_{\mathrm{E}}$ then empirical likelihood is unchanged by this operation, and so imposing symmetry this way does not make any difference. If $m = m_{\mathrm{O}}$ then the equations are degenerate, reducing to $E(0) = 0$ for all θ. An odd function is known to have mean zero under a symmetric distribution, and data are not required to draw that conclusion. If m has nonzero odd and even parts, then the result can be nondegenerate and different from the usual empirical likelihood.

The following three examples illustrate the possible cases. In each of them X is a real random variable known to be symmetric about a value c.

Example 10.1 (Mean under symmetry) For $x, \mu \in \mathbb{R}$, the estimating function $m(x, \mu) = x - \mu$, defines a univariate mean μ. Then $m_E(x, \mu) = c - \mu = 0$, telling us what we already know ($\mu = c$), and without making any use of the data.

Example 10.2 (Variance under symmetry) Because $E(X) = c$, the variance σ^2 of X is defined by the estimating equation $E((X - c)^2 - \sigma^2) = 0$, so that $m = (x - c)^2 - \sigma^2$. This m is even, and so empirical likelihood inferences are unchanged by replacing m by m_{E}.

Example 10.3 (Tail probability under symmetry) Suppose that we are wish to estimate a tail probability $\theta = \Pr(X > x)$. To avoid trivialities, assume that $x \neq c$, and $\min_i X_i < x < \max_i X_i$. The estimating function for θ is $m(X, \theta) = 1_{X > x} - \theta$. This m has nondegenerate even and odd parts

$$m_{\mathrm{E}}(X,\theta) = \frac{1}{2}\Big(1_{X>x} + 1_{X<2c-x}\Big) - \theta,$$

$$m_{\mathrm{O}}(X,\theta) = \frac{1}{2}\Big(1_{X>x} - 1_{X<2c-x}\Big) - \theta.$$

Thus, for this case, empirical likelihood inferences imposing symmetry are nondegenerate and different from empirical likelihood inferences that do not impose symmetry. Knowing that X is symmetric allows us to use data from one tail of the distribution to estimate a probability in the other tail. In practice, of course, we would have to be fairly sure of the symmetry to trust an estimate of one tail based on data from the other. For $x > c$, we could get the same result by replacing every X_i by $Z_i = \max(X_i, 2c - X_i)$, drawing inferences on $\Pr(Z > x) = 2\theta$, and dividing out the factor of 2 from this probability.

One consequence of symmetry of F is that odd (antisymmetric) functions of $X-c$ have mean zero. Suppose that $\phi(x) = -\phi(-x)$, so that ϕ is odd. If $\int |\phi(x-c)|dF(x) < \infty$, then $\int \phi(x-c)dF(x) = 0$. Conversely, this property can be used as a definition of symmetry about c, through specially chosen functions of the form

$$\phi_{a,b}(x) = 1_{a<x<b} - 1_{-b<x<-a}. \qquad (10.1)$$

We might replace symmetry by a weaker concept using only some finite number r of conditions $\int \phi_j(x-c)dF(x) = 0$, for $j = 1,\ldots,r$. Instead of testing for symmetry about c, we test $E(\phi_j(X-c)) = 0$ for $j = 1,\ldots,r$. If we have a specific value of c such as $c = 0$ in mind, then we may simply test whether these r functions all have mean 0. If we wish to test for symmetry about an unknown center, then all values of c that are not rejected form a conservative confidence region for the center of symmetry. If this region is empty, then we infer that F is not a symmetric distribution.

Here are some signed moment functions that might be used to approximate symmetry

$$M_j(z) = \text{sign}(z)|z|^{j-1}, \quad j = 1, 2, \ldots. \qquad (10.2)$$

A vanishing signed moment $E(M_j(X-c))$ corresponds to X having median c, when $j = 1$ and mean c, when $j = 2$. An alternative is to use a set of functions ϕ_{a_j,b_j} of the form (10.1) for a set of intervals (a_j, b_j).

For 672 National Football League (NFL) games of American professional football, the observed pointspread was compared to the pointspread established by professional book-makers. We will look at the values of $F - U - S$ where F is the actual number of points scored by the team favored to win, U is the number scored by the underdog team, and S is the published pointspread — the number of points by which the favorite was expected to win. The spreads S take values that are integer multiples of $1/2$ point. Figure 10.2 shows a histogram of these observed minus expected pointspreads. The data appear to be nearly normally distributed, with a center near zero and a surprisingly large standard deviation of 13.86 points. Thus about 95% of observed pointspreads came within plus or minus 28 points of the prediction.

It is plausible that data like these should be centered around 0. Even-money bets on whether the favorite beats the pointspread make most sense if the median discrepancy is close to zero. Otherwise bettors would catch on and exploit the difference. Because the data are nearly normally distributed, the mean may be a better estimate of the center of symmetry than the median. Also, if there are bets made that pay proportionally to the number of points by which the pointspread is beaten (or missed), then those bettors might pay attention to the mean difference between observed and predicted point spreads.

Figure 10.3 shows empirical likelihood curves for the mean and the median of the pointspread data. The sample mean is very close to zero, and the true mean seems to be within roughly one point of zero. The sample median is between -1

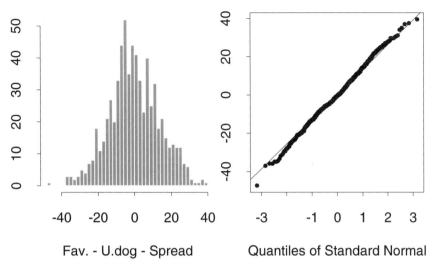

Figure 10.2 *For each of 672 NFL games, the betting-line pointspread has been subtracted from the observed one as described in the text. The data come from Stern (1991). The data are plotted as a histogram on the left, and in a normal QQ plot on the right.*

point and $-1/2$ point, and the true median appears to be between -2 points and $1/2$ point.

Now suppose that we consider a center of symmetry θ for which

$$0 = E(1_{X>\theta} - 1_{X<\theta}), \quad \text{and}$$
$$0 = E(X - \theta),$$

corresponding to signed moment functions M_1 and M_2 from equation (10.2) above. The empirical likelihood ratio function for θ is plotted in the lower panel of Figure 10.3. The unusual shape is easily explained. Consider a joint likelihood surface for the mean and median of the data. For a fixed value of the mean, that surface has step discontinuities as the median crosses observed data values. For a fixed value of the median, the surface is smooth as the mean varies. The function shown is a transect along the 45° line of the joint mean-median likelihood surface. This function takes jumps at the same points where the likelihood function for the median does. Between those jump points, it varies smoothly. The sawtooth shape of the likelihood curve gives a 95% confidence region that is a union of four disjoint intervals. The likelihood is relatively high at, and just to the left of, multiples of $1/2$ point. The smallest interval containing all four parts of the confidence interval for θ is narrower than either of the intervals for the mean or the median.

If we were not confident that the mean and median were at the same point, we could test this by examining the maximum value over θ of the empirical likelihood

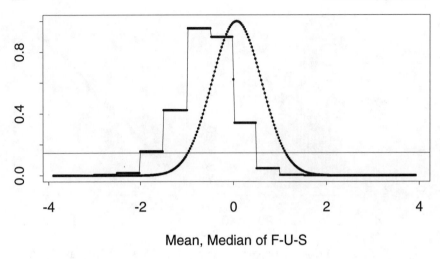

Mean, Median of F-U-S

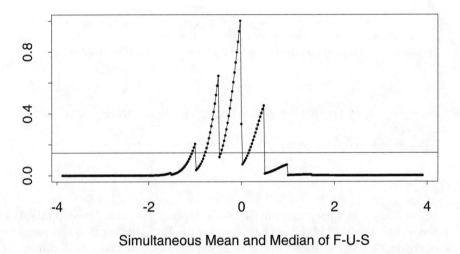

Simultaneous Mean and Median of F-U-S

Figure 10.3 *The top plot shows empirical likelihood curves for the mean and for the median of the pointspread data in Figure 10.2. The bottom plot shows the empirical likelihood for a center θ that is simultaneously the mean and the median of the distribution.*

for the hypothesis that θ is both the mean and median of X. At the MELE $\hat{\theta}$, the log empirical likelihood is -0.498. Using a $\chi^2_{(1)}$ calibration, we obtain a p-value of $\Pr(\chi^2_{(1)} \geq -2 \times (-0.498)) = 0.318$, so there is no reason to reject the presumed equality of the mean and median.

Further analysis shows that using the first three signed moments does not make much difference to the empirical likelihood ratio function, but that using the fourth and higher signed moments calls into question the symmetry of the data. Because these higher order signed moments have very heavy tails, dominated by the small number of extreme games, it does not seem wise to use them. If one were interested in imposing higher order symmetry, it would be preferable to use a set of bounded functions ϕ_j, perhaps of the form given in (10.1).

10.2 Independence

Suppose that $(X, Y) \in \mathbb{R}^2$ and that we want to test for independence of X and Y. As with the problem of symmetry, a direct formulation of empirical likelihood leads to a degenerate answer.

Suppose that the data are (X_i, Y_i) pairs, $i = 1, \ldots, n$, from a continuous joint distribution. Consider a distribution for X that has weight u_i on X_i and a distribution for Y with weight v_j on Y_j. Under independence of X and Y, the weight on the pair (X_i, Y_j) is $u_i v_j$. Without an assumption of independence, a distribution can put weight w_{ij} where w_{ij} is not necessarily of product form. To maximize the likelihood without assuming independence, put $w_{ij} = 0$ if $i \neq j$ and $w_{ii} = 1/n$. To maximize the likelihood ratio assuming independence put $u_i = v_j = 1/n$. As a result, the likelihood ratio for independence is $n^{-2n}/n^{-n} = n^{-n}$, without regard to the data. Thus the empirical likelihood test for independence is degenerate on continuous data.

If the (X_i, Y_i) pairs are from a discrete distribution, with finitely many possible values, the situation changes. Then the empirical likelihood test becomes a standard multinomial likelihood test of independence. Significance levels for that likelihood ratio test can be set using the $\chi^2_{(r-1)(s-1)}$ distribution where X takes r distinct values and Y takes s distinct values.

For the problem of symmetry, tests were degenerate, but some nondegenerate methods arose when symmetry was imposed as a side constraint. A similar effect occurs when independence is imposed as a side constraint.

Suppose that we know that $X \in \mathbb{R}^p$ and $Y \in \mathbb{R}^q$ are independent. Then there is no difference between n paired observations (X, Y) and two unpaired samples of X and Y. Because the data are effectively unpaired, there is no real need for the sample sizes to be equal. Suppose then that $X_i \sim F_0$ for $i = 1, \ldots, n$, and $Y_j \sim G_0$ for $j = 1, \ldots, m$ are independent. We are interested in θ defined by

$$E(h(X, Y, \theta)) = 0.$$

Here are example functions h that are of interest in applications,

$$h_1(X, Y, \theta) = X - Y - \theta, \quad \text{and}$$
$$h_2(X, Y, \theta) = 1_{X>Y} - \theta.$$

When $p = q$, the function h_1 can be used for inferences on $E(X) - E(Y)$. When $p = q = 1$, the function h_2 can be used for inferences on $\Pr(X > Y)$.

Let $F = \sum_{i=1}^n u_i \delta_{X_i}$ and $G = \sum_{j=1}^m v_j \delta_{Y_j}$ where $u_i \geq 0$, $v_j \geq 0$, and

$\sum_{i=1}^{n} u_i = \sum_{j=1}^{m} v_j = 1$. The likelihood of the pair (F,G) is $\prod_{i=1}^{n} u_i \prod_{j=1}^{m} v_j$. This is maximized by taking $u_i = 1/n$ and $v_j = 1/m$, so the likelihood ratio function is $\prod_{i=1}^{n} nu_i \prod_{j=1}^{m} mv_j$. The estimating equation is

$$\sum_{i=1}^{n}\sum_{j=1}^{m} u_i v_j h(X_i, Y_j, \theta) = 0.$$

The profile empirical likelihood ratio function for θ is

$$\mathcal{R}(\theta) = \max\Bigg\{ \prod_{i=1}^{n} nu_i \prod_{j=1}^{m} mv_j \mid \sum_{i=1}^{n}\sum_{j=1}^{m} u_i v_j h(X_i, Y_j, \theta) = 0,$$

$$u_i \geq 0, \sum_{i=1}^{n} u_i = 1, \ v_j \geq 0, \sum_{j=1}^{m} v_j = 1 \Bigg\}.$$

Under mild conditions $-2\log\mathcal{R}(\theta_0) \to \chi^2_{(d)}$ where d is the dimension of θ. A nonrigorous argument is given in Chapter 11.4 for the case of two samples with a scalar function $h(X, Y, \theta)$ and a scalar θ.

Independence corresponds to an infinite set of constraints. These take the form $E(\phi(X)\eta(Y)) = E(\phi(X))E(\eta(Y))$ for every ϕ, η pair with $E(|\phi(X)\eta(Y)|) < \infty$. If independence of X and Y is in doubt, and we want to test it, we can retain the original (X_i, Y_i) pairings and then for a list $j = 1, \ldots, r$ of functions ϕ_j and η_j test whether $E(\phi_j(X)\eta_j(Y)) = E(\phi_j(X))E(\eta_j(Y))$.

10.3 Comparison to permutation tests

A disadvantage of testing approximate symmetry or approximate independence is that symmetry or independence can be violated by a data distribution satisfying all of the estimating equations used in the approximate test. This motivates taking a large value of r, to get a test sensitive to more kinds of departures. For larger r, a larger $\chi^2_{(r)}$ threshold must be surpassed in order to reject independence. So large r can result in less power than small r. Judgment is required in order to select r conditions that cover either the likely departures, or if possible, the consequential departures, from symmetry or independence. A sieve method (Chapter 9.10) letting $r = r(n) \to \infty$ as $n \to \infty$ might provide an effective approach.

Permutation tests are commonly used for testing independence, and similar procedures are available for testing symmetry. For a permutation test of independence one starts with a statistic $T((X_i, Y_i), i = 1, \ldots, n)$ for which larger values represent greater departures from independence of X and Y. Then the Y_i are permuted with respect to the X_i, replacing Y_i by $Y_{\pi(i)}$ where $(\pi(1), \ldots, \pi(n))$ is a uniform random permutation of $(1, \ldots, n)$. A uniform random permutation takes each of the $n!$ possible permutations with probability $1/n!$. Then a histogram is made of the values $T((X_i, Y_{\pi(i)}), i = 1, \ldots, n)$ taken by T under a large number of independent random permutations. If the original T value is in the largest 5% of this histogram, then independence is rejected at the 5% level. When $n!$ is not too large then it is possible to consider all the permutations instead of sampling them.

As with empirical likelihood tests of approximate independence, these permutation tests can fail to detect those departures from independence which do not have a strong effect on T. The statistic T can be constructed as a composite of some number r of test statistics, each of which is sensitive to one aspect of dependence between X and Y. Once again there is a trade-off in the value of r.

Compared to permutation tests, empirical likelihood offers the advantage of using the data to combine the r test statistics. Consider testing $E(m_j(X_i, Y_i)) = 0$, simultaneously for all $j = 1, \ldots, r$, for real-valued estimating functions m_j. A permutation test could take

$$T = \sum_{j=1}^{r} \left(\sum_{i=1}^{n} m_j(X_i, Y_i) \right)^2$$

as its test criterion. But it might be better to weight the criteria, either a priori or based on the data. An empirical likelihood test based on $\mathcal{R}_0 = \max \prod_{i=1}^{n} nw_i$ subject to constraints $w_i \geq 0$, $\sum_{i=1}^{n} w_i = 1$, and $\sum_{i=1}^{n} w_i m_j(X_i, Y_i) = 0$ for $j = 1, \ldots, r$ uses the data to combine the individual tests.

The significance level for the test can be based on an asymptotic χ^2 distribution for $\mathcal{R}_0$ under the null hypothesis that X_i and Y_i are independent. We can also use the permutation distribution of $\mathcal{R}_0$ directly. For N random permutations $\pi(i)$ of $1, \ldots, n$, compute $\mathcal{R}_0$ on the data pairs $(X_i, Y_{\pi(i)})$, $i = 1, \ldots, n$. If the original $\mathcal{R}_0$ is smaller than αN of the permuted $\mathcal{R}_0$ values, then reject the null hypothesis at level α.

10.4 Convex hull condition

Empirical likelihood confidence regions for the mean are nested within the convex hull of the data. Their coverage level is necessarily smaller than that of the convex hull itself. This constraint is limiting when n is small, p is large, or the confidence level $1 - \alpha$ is high.

The Euclidean likelihood and some other members of the Cressie-Read family do not restrict the weights w_i to be nonnegative, thus allowing the reweighted mean $\sum_i w_i X_i$ to escape the convex hull of the data.

The empirical likelihood-t method also produces confidence regions for the mean that can escape from the convex hull. For testing whether $E(X) = \mu$, the constraint $\sum_{i=1}^{n} w_i X_i = \mu$ is replaced by

$$\left(\sum_{i=1}^{n} w_i (X_i - \mu_w)(X_i - \mu_w)' \right)^{-1/2} (\mu_w - \bar{X})$$

$$= \left(\frac{1}{n} \sum_{i=1}^{n} (X_i - \bar{X})(X_i - \bar{X})' \right)^{-1/2} (\bar{X} - \mu),$$

where $\mu_w = \sum_{i=1}^{n} w_i X_i$.

For $d = 1$, the new constraint equates two quantities. The first is the number of weighted standard deviations by which the reweighted mean differs from $\bar{X}$. The second is the number of unweighted standard deviations by which $\bar{X}$ differs from the candidate mean μ. Notice that the reweighted mean is not μ. If it were, the result could not escape the convex hull. For $d > 1$ a similar interpretation can be made in terms of analogous Mahalanobis distances $(u - \mu)'S^{-1}(u - \mu)$.

The empirical likelihood-t method is analogous to the bootstrap-t method. It replaces a quantity by a studentized version of that quantity. The empirical likelihood-t constraints can be applied to nonparametric likelihoods other than empirical likelihood. For empirical likelihood-t, a Bartlett correction is available. See Chapter 10.8.

10.5 Inequality and qualitative constraints

Many interesting hypotheses are expressed in terms of inequalities, not equalities. For example, if $\mu_1, \ldots, \mu_k$ are the means of k different populations, the constraints

$$\mu_{j+1} - \mu_j \geq 0, \quad j = 1, \ldots, k-1, \tag{10.3}$$

impose a qualitative constraint that μ_j is a nondecreasing function of the group labels. Estimating and testing of parameters subject to isotone constraints like (10.3) is known as order restricted inference.

Another qualitative constraint that might be useful is stochastic monotonicity. Suppose that $(X, Y) \in \mathbb{R}^2$ are jointly observed and we are sure that Y can only increase if X increases. Stochastic monotonicity corresponds to infinitely many constraints of the form $\Pr(Y \geq y | X = x) \geq \Pr(Y \geq y | X = x')$ for all y and all $x \geq x'$.

A standard technique for imposing unimodality constraints is to impose monotonicity constraints on either side of a mode $c \in \mathbb{R}$. When c is not known, then an outer loop searching over candidate modes is required.

A widely studied qualitative constraint for $X \in [0, \infty)$ is that f has a nonincreasing density. The NPMLE in this setting is known to be a mixture of $U[0, x)$ distributions, characterized as the smallest concave function lying above the empirical CDF.

Some results have been obtained for empirical likelihood ratios in inequality constrained problems. Suppose that a parameter $\theta \in \mathbb{R}^p$ is defined by

$$E(m(X, \theta)) = 0,$$

where $m(X, \theta) \in \mathbb{R}^p$, and consider the hypotheses

$$\mathcal{H}_0 : g_j(\theta) = 0, \quad j = 1, \ldots, k, \quad \text{and}$$
$$\mathcal{H}_1 : g_j(\theta) \geq 0, \quad j = 1, \ldots, k,$$

where $k \leq p$. The functions g_j may represent all the constraints of the qualitative feature we are interested in or a judiciously chosen subset of them.

Two statistical problems of interest are the testing of $\mathcal{H}_1$ and the testing of $\mathcal{H}_0$ assuming that $\mathcal{H}_1$ holds. Following parametric likelihood theory, we define

$$\mathcal{R}(\theta) = \max \left\{ \prod_{i=1}^{n} nw_i \mid \sum_{i=1}^{n} w_i m(X_i, \theta) = 0, w_i \geq 0, \sum_{i=1}^{n} w_i = 1 \right\}$$

and then

$$\mathcal{R}_0 = \max \{\mathcal{R}(\theta) \mid g_j(\theta) = 0, j = 1, \ldots, k\}, \quad \text{and}$$
$$\mathcal{R}_1 = \max \{\mathcal{R}(\theta) \mid g_j(\theta) \geq 0, j = 1, \ldots, k\}.$$

The hypothesis $\mathcal{H}_1$ is rejected if $\mathcal{R}_1$ is too small, and assuming $\mathcal{H}_1$, the further constraint $\mathcal{H}_0$ is rejected if $\mathcal{R}_0/\mathcal{R}_1$ is too small.

This poses two new problems for empirical likelihood. The first is how to maximize an empirical likelihood subject to inequality constraints. The second is how to calibrate the likelihood ratios. These issues are described in the references in Chapter 10.8. Here we summarize the calibration results.

The asymptotic distribution of $-2 \log \mathcal{R}_1$ is not necessarily chisquared. The null hypothesis is quite heterogenous. Perhaps the true value θ_0 has $g_j(\theta_0) = 0$ for some j, and $g_j(\theta_0) > 0$, for other j. There may be as many as 2^k different cases to consider. It is known that under mild conditions, for $Q > 0$,

$$\lim_{n \to \infty} \Pr(-2 \log \mathcal{R}_1 \leq Q) = \sum_{i=0}^{k} \gamma_i \Pr(\chi^2_{(i)} \leq Q),$$

where $\gamma_i = \gamma_i(\theta_0) \geq 0$ are weights that sum to 1, and $\chi^2_{(0)} = 0$. This limiting distribution of $-2 \log \mathcal{R}_1$ is known as a chi-bar squared, often written $\bar{\chi}^2$. A chi-bar squared distribution commonly applies to parametric likelihoods in the presence of inequality constraints. Similarly,

$$\lim_{n \to \infty} \Pr(-2 \log(\mathcal{R}_0/\mathcal{R}_1) \leq Q) = \sum_{i=0}^{k} \gamma_i \Pr(\chi^2_{(k-i)} \leq Q)$$

with the same $\gamma_i(\theta_0)$.

Each γ_i is a sum of products of probabilities that certain multivariate normal random variables have no negative components. The normal vectors have mean zero and it is possible to construct sample estimates of their covariance matrices.

10.6 Nonsmooth estimating equations

Theorem 3.4 allows us to construct tests and confidence regions for parameters defined through very general estimating equations. Theorem 3.6 justifies profiling out nuisance parameters, but only from smooth estimating equations.

Let θ and a nuisance parameter ν be defined through $E(m(X, \theta, \nu)) = 0$ for a nonsmooth estimating function $m(X, \theta, \nu)$. Then profiling out ν raises both theoretical and numerical challenges.

In this text we have encountered several problems of this type: the interquartile range defined through (3.21), a problem in regression tolerance intervals defined by (4.16) through (4.19), the sensitivity and specificity of logistic regression (Chapter 4.7), and the difference between medians of censored data.

Some results are known for problems in which the estimating equations are smoothed, with decreasing smoothness as $n \to \infty$. The theorem below requires no smoothing:

Theorem 10.1 *For $i = 1, \ldots, k$, let $X_{ij} \in [0, \infty)$ for $j = 1, \ldots, n_i$ be failure times from distribution F_i. Suppose that all F_i have a common median θ_0, and that F_i is a continuous distribution with density f_i where $f_i(\theta_0) > 0$. Let $Y_{ij} \in [0, \infty)$ be corresponding right censoring times from distribution G_i, where $G_i([\theta_0, \infty)) > 0$. Assume that all X_{ij} and Y_{ij} are independent. Let $Z_{ij} = \min(X_{ij}, Y_{ij})$ and $\delta_{ij} = 1_{X_{ij} \leq Y_{ij}}$, and define*

$$L(F_1, \ldots, F_k) = \prod_{i=1}^{k} \prod_{j=1}^{n_i} F(\{Z_{ij}\})^{\delta_{ij}} F((Z_{ij}, \infty))^{1-\delta_{ij}}$$

$$\mathcal{R}(\theta_1, \ldots, \theta_k) = \frac{\max \{L(F_1, \ldots, F_k) \mid F_i([\theta_i, \infty)) = 1/2, i = 1, \ldots, k\}}{\max_{F_1, \ldots, F_k} \{L(F_1, \ldots, F_k)\}},$$

and $\mathcal{R}(\theta) = \mathcal{R}(\theta, \theta, \ldots, \theta)$. Then $-2 \max_\theta \log \mathcal{R}(\theta) \to \chi^2_{(k-1)}$ as $\min n_i \to \infty$.

Proof. Naik-Nimbalkar & Rajarshi (1997). □

Now suppose that $X_1, \ldots, X_n \sim F$ are independent real random variables with CDF $\mathsf{F}(x) = F((-\infty, x])$. As usual, the order statistics are denoted by $X_{(1)} \leq \ldots X_{(n)}$. Statistics known as L-estimators can be written as $T_n = \sum_{i=1}^{n} c_{ni} X_{(i)}$ where $c_{ni} = \int_{(i-1)/n}^{i/n} dG(u)$. Here $G(u)$ can be a distribution function, or more generally, a weighted difference of distribution functions, $G(u) = c_+ G_+(u) - c_- G_-(u)$, where $G_\pm$ are distribution functions and $c_\pm \geq 0$. The population value of T is $T(F) = \int_0^1 \mathsf{F}^{-1}(u) dG(u)$, where $\mathsf{F}^{-1} = \inf\{x \mid \mathsf{F}(x) \geq u\}$.

For example, when $0 < \alpha < 1/2$, the α-trimmed mean has $dG(u) = (1 - 2\alpha)^{-1}$ if $\alpha/2 < u < 1 - \alpha/2$ and $dG(u) = 0$ otherwise. As a second example, the interquartile range has $c_\pm = 1$, $G_+(u) = 1_{u \leq 3/4}$ and $G_-(u) = 1_{u \leq 1/4}$.

The empirical likelihood function is

$$\mathcal{R}(\theta) = \max \left\{ \prod_{i=1}^{n} nw_i \mid T\left(\sum_{i=1}^{n} w_i \delta_{X_i}\right) = \theta, w_i \geq 0, \sum_{i=1}^{n} w_i = 1 \right\},$$

where we suppose that T uses F determined from F in the obvious way.

We suppose here that $dG(u) = g(u)$ where $\int_0^1 |g(u)| < \infty$. This rules out the interquartile range, though it can be closely approximated by an L-estimate satisfying this condition, using smoothed versions of $G_\pm = 1_{u \leq (2\pm 1)/4}$.

Theorem 10.2 *Let* $X_1, \ldots, X_n$ *be IID from* F_0. *Suppose that for some* $p \in (0, \infty)$, *some* $M \in (0, \infty)$, *and some* $d \in (1/6, 1/2)$ *both*

$$|g(u)| \leq M[u(1-u)]^{\frac{1}{p} - \frac{1}{2} + d}, \quad \text{and}$$

$$|\mathsf{F}^{-1}(u)| \leq M[u(1-u)]^{-\frac{1}{p}}$$

hold for all $0 < u < 1$. *Then* $-2 \log \mathcal{R}(T(F_0)) \to \chi^2_{(1)}$ *as* $n \to \infty$.

Proof. La Rocca (1995a) □

10.7 Adverse estimating equations and black boxes

In parametric likelihood settings there can be difficulty in passing from maximizing the log likelihood $\sum_{i=1}^{n} \log f(X_i; \theta)$, to solving $\sum_{i=1}^{n} m(X_i, \theta) = 0$. The set of solutions θ may include multiple global maxima, local maxima that are not global maxima, local minima, and saddlepoints. Such equations pose a challenge for empirical likelihood too, for solutions to $\sum_{i=1}^{n} w_i m(X_i, \theta) = 0$ might not correspond to unique global maxima of $\sum_{i=1}^{n} w_i \log f(X_i; \theta)$ as we would have hoped.

Parametric likelihood methods have developed the farthest for models in which solving the likelihood equation does indeed give the maximum likelihood estimate. The most widely used parametric models are those with log concave likelihoods, such as the normal distribution, the binomial, and the Poisson. When $\log f(X; \theta)$ is concave in θ then so is $\sum_{i=1}^{n} w_i \log f(X_i; \theta)$ and so empirical likelihood is on firmer ground in these cases, too.

Another big challenge comes from methods such as classification and regression trees (CART) and projection pursuit regression. These may be thought of as black boxes that take data in and produce answers. It would be a daunting computational challenge to find the weighting of the data closest to equality for which a classification tree predicted $y_0 \in \mathbb{R}$ given a feature vector X_0. One reason is that the method is nonsmooth. A small change in the weighting of the data can change the shape of the tree and the variables used to split the data. Bootstrap resampling is ideally suited to computations for black box estimators of this type. In addition to the computational challenge, the theoretical behavior of the resulting likelihood ratio would have to be at least as complicated as that leading to the chi-bar squared distribution in Chapter 10.5.

10.8 Bibliographic notes

Brown & Chen (1998) study empirical likelihood for an assumed common value of the mean and median. They prefer the Euclidean log likelihood for this problem because it allows negative weights and so produces a bounded log likelihood. Brown & Chen (1998) prefer a smoothed version of the log likelihood ratio. They discuss robustness and asymptotic normality of the combined mean/median estimator.

The signed moments are due to Qu (1995). He shows that a location estimator defined as an MELE using a small number of signed moments is nearly fully efficient compared to some parametric maximum likelihood estimators from symmetric distributions. He extends the signed moment arguments to multiple regression estimators with symmetrically distributed residuals.

The approach in Chapter 10.1 symmetrizes the estimating equations. By contrast, Jing (1995a) symmetrizes a data set around its sample mean, then constructs an empirical likelihood test of whether the resulting $2n$ observations have mean μ. He obtains a χ^2 limit but finds that Bartlett correctability does not hold.

Romano (1988) considers bootstrap versions of permutation tests for symmetry and independence.

Breiman, Friedman, Olshen & Stone (1984) describe classification and regression trees. Friedman & Stuetzle (1981) introduce projection pursuit regression.

Zhou (2000) considers a location family in which $X_{ij} - \mu_i \sim F$ for $i = 1,\ldots,k$ and $j = 1,\ldots,n_i$. As with symmetry and independence, this location family constraint corresponds to an infinite number of moment constraints.

Empirical likelihood-t was proposed by Baggerly (1999), who also considers an entropy version. Baggerly (1999) notes that one price to be paid for escaping the convex hull is a loss of transformation invariance. Baggerly (1999) reports some simulations in which the studentized empirical entropy method gives very good coverage accuracy in small samples. He also applies a Bartlett correction to those intervals. Bartlett correction of empirical entropy appears in simulations to make a worthwhile improvement in the coverage error, although it does not improve the error rate from $O(1/n)$. Baggerly speculates that this is because the empirical entropy distance gives relative error of $O(1/n)$ instead of an absolute error of that order, pointing to work by Jing, Feuerverger & Robinson (1994).

The discussion of empirical likelihood ratios under inequality constraints in Chapter 10.5 is based on El Barmi (1996). El Barmi gives expressions for the weights in the $\bar{\chi}^2$ distributions. El Barmi & Dykstra (1994) describe maximization of multinomial likelihoods subject to constraints expressed as the intersection of a finite number of convex sets. Hoff (2000) considers maximizing empirical likelihoods subject to stochastic monotonicity constraints linking a finite number of distributions.

Grenander (1956) constructs an NPMLE for a distribution known to have a monotone density on $[0, \infty)$. Groeneboom & Wellner (1992) characterize the result as the least concave majorant of the empirical CDF. Lindsay (1995) describes mixture-based methods for this and other statistical problems. Banerjee & Wellner (2000) present describe the asymptotics of likelihood ratios under monotonicity constraints. The results are characterized by $n^{1/3}$ rate asymptotics involving Brownian motion with quadratic drift. Azzalini & Hall (2000) show how qualitative information can be used to reduce variability. The result can be like having $O(n^{2/3})$ extra observations. Although this effect becomes negligible as $n \to \infty$, the benefit may be substantial at practical sample sizes.

Qin & Zhao (1997) prove a chisquared limit for the difference of two smoothed quantiles from two different populations, by extending the arguments of Chen & Hall (1993) for smoothing of a single quantile. The amount of smoothing applied to the two quantiles tends to zero as the sample sizes tend to infinity. Naik-Nimbalkar & Rajarshi (1997) prove Theorem 10.1 without requiring that the underlying statistics be smoothed.

La Rocca (1995a) proves Theorem 10.2, and also considers more general statistics defined as L estimators applied to values $h(X_{(i)})$ where h is a function of bounded variation. La Rocca (1995a) shows empirical likelihood ratio curves for trimmed means of Newcomb's passage time of light data, and presents some simulations. La Rocca (1996) obtains a χ^2 calibration for the difference between the trimmed means of two populations.

10.9 Exercises

Exercise 10.1 Suppose that X_i are independent random variables from a distribution with maximum value $\theta < \infty$. That is $\Pr(X \le \theta) = 1$, but $\Pr(X \le \theta - \epsilon) < 1$ for any $\epsilon > 0$. Let $\hat\theta = \max_{1 \le i \le n} X_i$. Describe how the empirical likelihood confidence region for θ fails. Invent and describe a nondegenerate approach to empirical likelihood for the sample maximum.

Exercise 10.2 Derive the values of L_1 and L_2 given in Chapter 10.1. For L_1, when $c = X_i$ what is the value that w_i takes in order to maximize the likelihood? For L_2, when $c = (X_i + X_j)/2$, what is the maximizing value of w_i?

Exercise 10.3 Let $(X, Y) \in \mathbb{R}^2$ be from a distribution F that is thought to be symmetric about the line $x \cos\theta + y \sin\theta = r$.

a) For $(x, y) \subset \mathbb{R}^2$, find its reflection $(\widetilde{x}, \widetilde{y})$ under this symmetry.

b) For $i = 1, \ldots, n$, suppose that (X_i, Y_i) are sampled from a continuous distribution. Consider the $n + 2$ dimensional family of distributions given by parameters θ, r, and weights w_i summing to $1/2$ that apply equally to (X_i, Y_i) and $(\widetilde{X}_i, \widetilde{Y}_i)$. For almost all lines, the maximum over w_i of $\prod_i w_i$ is $(2n)^{-n}$. Describe geometrically any lines that give a larger likelihood. How much larger is that likelihood? Do not bother with any kind of line that is not sure to arise in the sample for large enough n.

Exercise 10.4 Suppose that F is known to be symmetric about the line $x + y = 0$ (for which $\theta = \pi/4$ and $r = 0$ in the notation of the previous question). Find an example estimating equation $E(m(X, Y, \theta)) = 0$, for each of the following cases:

a) Imposing the symmetry constraint is degenerate.

b) Imposing the symmetry constraint leaves empirical likelihood unchanged.

c) Imposing the symmetry constraint with empirical likelihood gives something new.

Exercise 10.5 Construct a family of moment-like estimating equations to use for testing or imposing approximate symmetry of F about the line $x\cos\theta + y\sin\theta = r$.

Exercise 10.6 Now suppose that F is thought to be symmetric about the point (x_c, y_c). Construct a set of moment-like estimating equations for imposing approximate symmetry of this kind.

CHAPTER 11

Some proofs

This chapter contains some of the proofs of the empirical likelihood theorems. The arguments presented here are more difficult than the proofs that appear throughout the text. For the most advanced material the reader is referred to the published literature, as outlined in Chapter 13.

11.1 Lemmas

This section presents some lemmas needed to prove the ELT's. They are used to handle some technical details showing that the Lagrange multiplier λ is asymptotically small, and to show that higher order terms in some Taylor series can be neglected.

A distribution with nondegenerate variance matrix on $\mathbb{R}^p$ cannot put all of its probability on a half space defined as one side of a hyperplane through its mean. A stronger conclusion is that, for any distribution, there is some $\varepsilon > 0$ that provides a uniform lower bound on the probability of all possible half-spaces defined by hyperplanes through the mean.

Lemma 11.1 *Let F_0 be a distribution on $\mathbb{R}^p$ with mean μ_0 and finite variance matrix V_0 of full rank p. Let Θ be the set of unit vectors in $\mathbb{R}^p$. Then for $X \sim F_0$*

$$\inf_{\theta \in \Theta} \Pr\left((X - \mu_0)' \theta > 0\right) > 0.$$

Proof. Without loss of generality take $\mu_0 = 0$. Suppose that the infimum above is 0. Then there exists a sequence θ_n such that $\Pr(X'\theta_n > 0) < 1/n$. By compactness of Θ there is a convergent subsequence $\theta_n^* \to \theta^*$. Let $H = \{X \mid X'\theta^* > 0\}$. Then $1_{X'\theta_n^* > 0} \to 1_{X'\theta^* > 0}$ holds at any $X \in H$. Now by Lebesgue's dominated convergence theorem

$$\begin{aligned}
\Pr\left(X'\theta^* > 0\right) &= \int_H 1_{X'\theta^* > 0} dF_0(X) \\
&= \lim_{n \to \infty} \int_H 1_{X'\theta_n^* > 0} dF_0(X) \\
&\leq \lim_{n \to \infty} \Pr\left(X'\theta_n^* > 0\right) \\
&= 0.
\end{aligned}$$

Since $X'\theta^*$ has mean zero we must also have $\Pr(X'\theta^* < 0) = 0$ from which

$X'\theta^* = 0$ with probability 1. Then $\text{Var}(X'\theta^*) = 0$ contradicting the assumption on V_0. □

The next lemma shows that for random variables with a finite variance, the largest value in a sample of size n cannot grow to infinity as fast as $n^{1/2}$.

Lemma 11.2 *Let Y_i be independent random variables with a common distribution and $E(Y_i^2) < \infty$. Let $Z_n = \max_{1 \le i \le n} |Y_i|$. Then $Z_n = o(n^{1/2})$.*

Proof. Since $E(Y_i^2) < \infty$, we have $\sum_{i=1}^n \Pr(Y_i^2 > n) < \infty$. Therefore, by the Borel-Cantelli lemma there is probability 1 that $|Y_n| > n^{1/2}$ happens only for finitely many n. This implies that there are only finitely many n for which $Z_n > n^{1/2}$. A similar argument shows that for any $A > 0$, there are only finitely many n for which $Z_n > An^{1/2}$, and hence

$$\limsup_{n \to \infty} Z_n n^{-1/2} \le A$$

holds with probability 1. The probability 1 applies simultaneously over any countable set of values for A so $Z_n = o(n^{1/2})$. □

Similarly, a finite variance bounds how fast a sample third moment can diverge to infinity.

Lemma 11.3 *Let Y_i be independent random variables with common distribution and suppose that $E(Y_i^2) < \infty$. Then*

$$\frac{1}{n} \sum_{i=1}^n |Y_i|^3 = o(n^{1/2}).$$

Proof. Write

$$\frac{1}{n} \sum_{i=1}^n |Y_i|^3 \le \frac{Z_n}{n} \sum_{i=1}^n Y_i^2$$

where $Z_n = \max_{1 \le i \le n} |Y_i|$. The result follows by Lemma 11.2 applied to Z_n and the strong law of large numbers applied to the average of Y_i^2. □

The next lemma shows how quickly the probability that the maximum observation exceeds $n^{1/2}$ decreases to zero with increasing n.

Lemma 11.4 *Let Y_i be independent random variables with common distribution and suppose that $E(|Y_i|^3) < \infty$. Define $Z_n = \max_{1 \le i \le n} |Y_i|$ and let $A > 0$. Then*

$$\Pr(Z_n > An^{1/2}) = O(n^{-1/2}).$$

Proof.

$$n^{1/2}\Pr\left(Z_n > An^{1/2}\right) \le n^{3/2}\Pr\left(|Y_1| > An^{1/2}\right)$$
$$\le n^{3/2} E\left(|Y_1|^3\right)/(An^{1/2})^3$$
$$= A^{-3} E(|Y_1|^3)$$
$$< \infty.$$

□

11.2 Univariate and Vector ELT

The univariate ELT, Theorem 2.2, is a special case of the vector ELT, Theorem 3.2, corresponding to dimension $d = 1$. In this section we prove Theorem 3.2 of Chapter 3.

The Lagrange multiplier λ plays a key role in the proof. The proof goes in stages. First we show that $\lambda = O_p(n^{-1/2})$. Then, knowing $\lambda = O_p(n^{-1/2})$, we show that $\lambda = S^{-1}(\bar{X} - \mu_0) + o_p(n^{-1/2})$, for a certain sample covariance matrix S. Plugging this expression for λ into the profile empirical log likelihood ratio statistic, applying a central limit theorem, and verifying that some other terms are negligible completes the proof.

Proof. [Proof of Theorem 3.2, Vector ELT] Without loss of generality $q = p$. Otherwise we can replace X_i by a subset of q components having a variance of full rank. Convexity of $C_{r,n}$ follows easily by the same argument used in Chapter 2 to show that confidence regions for the univariate mean are intervals.

Let Θ denote the set of unit vectors in $\mathbb{R}^p$. By Lemma 11.1

$$\inf_{\theta \in \Theta} F_0\left(\{\theta'(X - \mu_0) > 0\}\right) > 0.$$

By a version of the Glivenko-Cantelli theorem for uniform convergence over half spaces

$$\sup_{\theta \in \Theta} |F_0(\{\theta'(X - \mu_0) > 0\}) - F_n(\{\theta'(X - \mu_0) > 0\})| \to 0$$

with probability 1 as $n \to \infty$. It follows that with probability tending to 1 that the mean μ_0 is inside the convex hull of X_i.

When the mean is inside the convex hull of the X_i, then there is a unique set of weights $w_i > 0$ with $\sum_{i=1}^n w_i = 1$ and $\sum_{i=1}^n w_i(X_i - \mu_0) = 0$ for which $\prod_{i=1}^n nw_i$ is maximized. By the arguments in Chapter 3 these maximizing weights may be written

$$w_i = \frac{1}{n}\frac{1}{1 + \lambda'(X_i - \mu_0)},$$

where the vector $\lambda = \lambda(\mu_0) \in \mathbb{R}^p$ satisfies p equations given by

$$g(\lambda) \equiv \frac{1}{n} \sum_{i=1}^{n} \frac{X_i - \mu_0}{1 + \lambda'(X_i - \mu_0)} = 0. \tag{11.1}$$

The next step is to bound the magnitude of λ. Let $\lambda = \|\lambda\|\theta$ where $\theta \in \Theta$ is a unit vector. Introduce

$$Y_i = \lambda'(X_i - \mu_0), \quad \text{and} \quad Z_n^\star = \max_{1 \le i \le n} \|X_i - \mu_0\|.$$

Substituting $1/(1+Y_i) = 1 - Y_i/(1+Y_i)$ into $\theta' g(\lambda) = 0$ and simplifying, we find that

$$\|\lambda\|\theta'\widetilde{S}\theta = \theta'(\bar{X} - \mu_0) \tag{11.2}$$

where

$$\widetilde{S} = \frac{1}{n} \sum_{i=1}^{n} \frac{(X_i - \mu_0)(X_i - \mu_0)'}{1 + Y_i}. \tag{11.3}$$

Let

$$S = \frac{1}{n} \sum_{i=1}^{n} (X_i - \mu_0)(X_i - \mu_0)'.$$

Every $w_i > 0$, so $1 + Y_i > 0$ and therefore

$$\|\lambda\|\theta'S\theta \le \|\lambda\|\theta'\widetilde{S}\theta(1 + \max_i Y_i)$$
$$\le \|\lambda\|\theta'\widetilde{S}\theta(1 + \|\lambda\|Z_n^\star)$$
$$= \theta'(\bar{X} - \mu_0)(1 + \|\lambda\|Z_n^\star),$$

by (11.2) and so

$$\|\lambda\|\left(\theta'S\theta - Z_n^\star\theta'(\bar{X} - \mu_0)\right) \le \theta'(\bar{X} - \mu_0).$$

Now $\sigma_1 + o_p(1) \ge \theta'S\theta \ge \sigma_p + o_p(1)$, where $\sigma_1 \ge \sigma_p > 0$ are the largest and smallest eigenvalues of $\text{Var}(X_i)$. Also by Lemma 11.2, $Z_n^\star = o(n^{1/2})$. The central limit theorem applied to the vector $\bar{X} - \mu_0$ implies that $\theta'(\bar{X} - \mu_0) = O_p(n^{-1/2})$. It follows that

$$\|\lambda\|(\theta'S\theta + o_p(1)) = O_p(n^{-1/2}),$$

and hence

$$\|\lambda\| = O_p(n^{-1/2}).$$

Having established an order bound for $\|\lambda\|$, we have from Lemma 11.2 that

$$\max_{1 \le i \le n} |Y_i| = O_p(n^{-1/2}) o(n^{1/2}) = o_p(1). \tag{11.4}$$

Now

$$0 = \frac{1}{n}\sum_{i=1}^{n}(X_i - \mu_0)\left(1 - Y_i + Y_i^2/(1 - Y_i)\right)$$

$$= \bar{X} - \mu_0 - S\lambda + \frac{1}{n}\sum_{i=1}^{n}\frac{(X_i - \mu_0)Y_i^2}{1 - Y_i}. \quad (11.5)$$

The final term in (11.5) above has a norm bounded by

$$\frac{1}{n}\sum_{i=1}^{n}\|X_i - \mu_0\|^3 \|\lambda\|^2 |1 - Y_i|^{-1} = o(n^{1/2})O_p(n^{-1})O_p(1) = o_p(n^{-1/2}),$$

using Lemma 11.1 from which we find

$$\lambda = S^{-1}(\bar{X} - \mu_0) + \beta,$$

where

$$\beta = o_p(n^{-1/2}).$$

By (11.4), we may write

$$\log(1 + Y_i) = Y_i - \frac{1}{2}Y_i^2 + \eta_i,$$

where for some finite $B > 0$

$$\Pr(|\eta_i| \leq B|Y_i|^3, \quad 1 \leq i \leq n) \to 1,$$

as $n \to \infty$.

Now we may write

$$-2\log\mathcal{R}(\mu_0) = -2\sum_{i=1}^{n}\log(nw_i)$$

$$= 2\sum_{i=1}^{n}\log(1 + Y_i)$$

$$= 2\sum_{i=1}^{n}Y_i - \sum_{i=1}^{n}Y_i^2 + 2\sum_{i=1}^{n}\eta_i$$

$$= 2n\lambda'(\bar{X} - \mu_0) - n\lambda'S\lambda + 2\sum_{i=1}^{n}\eta_i$$

$$= n(\bar{X} - \mu_0)S^{-1}(\bar{X} - \mu_0) - n\beta'S^{-1}\beta + 2\sum_{i=1}^{n}\eta_i.$$

In the limit as $n \to \infty$,

$$n(\bar{X} - \mu_0)S^{-1}(\bar{X} - \mu_0) \to \chi^2_{(p)}$$

in distribution,
$$n\beta' S^{-1}\beta = no_p(n^{-1/2})O_p(1)o_p(n^{-1/2}) = o_p(1),$$
and
$$\left|\sum_{i=1}^{n}\eta_i\right| \le B\,\|\lambda\|^3 \sum_{i=1}^{n}\|X_i - \mu_0\|^2 = O_p(n^{-3/2})o_p(n^{3/2}) = o_p(1).$$

Therefore $-2\log \mathcal{R}(\mu_0) \to \chi^2_{(p)}$ in distribution. $\square$

11.3 Triangular array ELT

In this section we prove Theorem 4.1 of Chapter 4.3. The proof follows the same lines as the proof of the Vector ELT in Chapter 11.2.

Proof. [Proof of Theorem 4.1, Triangular Array ELT] Without loss of generality we may take $\mu_n = 0$ and $\sigma_{1n} = 1$, and then seek the asymptotic distribution of $-2\log \mathcal{R}(0)$. This is equivalent to reformulating the problem with $\sigma_{1n}^{-1/2}(Z_{in} - \mu_n)$ in place of Z_{in}. To simplify the notation, we drop the second subscript n, using Z_i for Z_{in}. Let
$$\hat{V}_n = \frac{1}{n}\sum_{i=1}^{n} Z_i Z_i'.$$

By assumption (4.4) we may assume that the convex hull of Z_i contains the origin. It then follows by Lagrange multiplier arguments that
$$\mathcal{R}(0) = \prod_{i=1}^{n}\frac{1}{1+\lambda' Z_i}$$
where $\lambda = \lambda(0)$ is uniquely determined by
$$\sum_{i=1}^{n}\frac{Z_i}{1+\lambda' Z_i} = 0.$$

Write $\lambda = \|\lambda\|\theta$ where $\theta'\theta = 1$. Let
$$Y_i = \lambda' Z_i, \quad \text{and} \quad Z^* = \max_{1 \le i \le n}\|Z_i\|.$$

By an argument used in the proof of the Vector ELT, we obtain
$$\|\lambda\|\left(\theta'\hat{V}\theta - Z_n^*\theta'\bar{Z}\right) \le \theta'\bar{Z}.$$

From assumption (4.5) the variance of each entry in $\hat{V}_n$ tends to zero and so by Chebychev's inequality $\hat{V}_n - V_n = o_p(1)$. Then by assumption (4.6) on V_n, we obtain
$$c + o_p(1) \le \theta'\hat{V}\theta \le 1 + o_p(1),$$

where $c > 0$ is the constant in that assumption.

Assumption (4.5) also implies that $Z_n^\star = o(n^{1/2})$, which in turn implies Lindeberg's condition for Z_i. The central limit theorem applied to the vector $\bar{Z}$ implies that $\theta'\bar{Z} = O_p(n^{-1/2})$. It follows that

$$\|\lambda\| \left(\theta'\hat{V}\theta - o_p(1)\right) = O_p(n^{-1/2}),$$

and hence $|\lambda| = O_p(n^{-1/2})$.

The rest of the proof follows the same argument as used for the Vector ELT, and so $-2\log\mathcal{R}(0) \to \chi^2_{(p)}$ in distribution. □

11.4 Multi-sample ELT

This section presents a nonrigorous argument that empirical likelihood inferences have a χ^2 calibration in multi-sample settings. For simplicity we consider the case of two samples and a single estimating equation for a scalar parameter. A rigorous argument would have to impose explicit moment conditions in order to bound the Lagrange multiplier, along the lines of the proof in Chapter 11.2 of the vector ELT.

Let $X_1, \ldots, X_n \in \mathbb{R}^p \sim F_0$ and $Y_1, \ldots, Y_m \in \mathbb{R}^q \sim G_0$, with all observations independent. Let $\theta \in \mathbb{R}^t$ be defined by $E(h(X, Y, \theta)) = 0$, where $h(x, y, \theta) \in \mathbb{R}^t$. For example when $p = q = t$, $h(X, Y, \theta) = X - Y - \theta$ defines θ as the difference in means $E(X) - E(Y)$. When $p = q = t = 1$, the function $h(X, Y, \theta) = 1_{X>Y} - \theta$ defines θ as the probability that X is larger than Y. We might be interested in testing $\theta = 0$ in the first case and $\theta = 1/2$ in the second case.

The ANOVA setting of Chapter 4.4 applies to group means and functions of group means. The setting here is more general in that the expectation in the estimating equation is with respect to both distributions jointly.

The argument below provides a $\chi^2_{(t)}$ limit, under some mild assumptions. To simplify the presentation $t = 1$ is used, but the argument goes through for general t with natural modifications. We assume that $\min(n, m) \to \infty$, and that $0 < E(h(X, Y, \theta_0)^2) < \infty$. We also rule out cases such as those where $h(X, Y, \theta_0)$ can be written in the form $\phi(X)\eta(Y)$ with $E(\phi(X)) = E(\eta(Y)) = 0$. In such a case, independence of X and Y implies that $E(h(X, Y, \theta_0)) = 0$, and there is no need to infer it from data. The extra assumption we will make is that either $E(E(h(X, Y, \theta_0)|X)^2) > 0$ or $E(E(h(X, Y, \theta_0)|Y)^2) > 0$.

The empirical likelihood ratio may be written

$$R(F, G) = \prod_{i=1}^{n} nu_i \prod_{j=1}^{m} mv_j$$

where F puts weight u_i on X_i and G puts weight v_j on Y_j. We will assume that $u_i \geq 0$, $\sum_{i=1}^{n} u_i = 1$, $v_j \geq 0$, $\sum_{j=1}^{m} v_j = 1$, so that the empirical likelihood is a product of two multinomials, one on each sample. The profile empirical likelihood ratio function is

$$\mathcal{R}(\theta) = \max\left\{ R(F,G) \mid \sum_{i=1}^{n}\sum_{j=1}^{m} u_i v_j H_{ij}(\theta) = 0 \right\}$$

where $H_{ij}(\theta) = H_{ij} = h(X_i, Y_j, \theta)$, and the simplex constraints on u_i and v_j have been left out to shorten the expression.

Using Lagrange multipliers we find that

$$u_i = \frac{1}{n + \lambda \sum_{r=1}^{m} v_r H_{ir}}$$

$$v_j = \frac{1}{m + \lambda \sum_{s=1}^{n} u_s H_{sj}},$$

where λ is defined by $\sum_{i=1}^{n}\sum_{j=1}^{m} u_i v_j H_{ij} = 0$. Introduce the terms

$$\bar{H}_{i\bullet} = \frac{1}{m}\sum_{j=1}^{m} H_{ij}, \quad \widetilde{H}_{i\bullet} = \sum_{r=1}^{m} v_r H_{rj}$$

$$\bar{H}_{\bullet j} = \frac{1}{n}\sum_{i=1}^{n} H_{ij}, \quad \widetilde{H}_{\bullet j} = \sum_{s=1}^{n} u_s H_{sj},$$

and $\bar{H}_{\bullet\bullet} = (nm)^{-1}\sum_{i=1}^{n}\sum_{j=1}^{m} H_{ij}$. Then

$$u_i = \frac{1}{n}\left[1 - \left(\frac{\lambda}{n}\widetilde{H}_{i\bullet}\right) + \left(\frac{\lambda}{n}\widetilde{H}_{i\bullet}\right)^2 - \left(\frac{\lambda}{n}\widetilde{H}_{i\bullet}\right)^3 + \cdots\right], \text{ and}$$

$$v_j = \frac{1}{m}\left[1 - \left(\frac{\lambda}{m}\widetilde{H}_{\bullet j}\right) + \left(\frac{\lambda}{m}\widetilde{H}_{\bullet j}\right)^2 - \left(\frac{\lambda}{m}\widetilde{H}_{\bullet j}\right)^3 + \cdots\right].$$

Substituting these values into $\sum_{i=1}^{n}\sum_{j=1}^{m} u_i v_j H_{ij} = 0$, we get

$$0 = \bar{H}_{\bullet\bullet} - \lambda\left[\frac{1}{n^2 m}\sum_{i=1}^{n}\sum_{j=1}^{m} H_{ij}\widetilde{H}_{i\bullet} + \frac{1}{nm^2}\sum_{i=1}^{n}\sum_{j=1}^{m} H_{ij}\widetilde{H}_{\bullet j}\right]$$

$$+ \lambda^2\left[\frac{1}{n^3 m}\sum_{i=1}^{n}\sum_{j=1}^{m} H_{ij}\widetilde{H}_{i\bullet}^2 + \frac{1}{nm^3}\sum_{i=1}^{n}\sum_{j=1}^{m} H_{ij}\widetilde{H}_{\bullet j}^2 + \right.$$

$$\left.\frac{1}{n^2 m^2}\sum_{i=1}^{n}\sum_{j=1}^{m} H_{ij}\widetilde{H}_{i\bullet}\widetilde{H}_{\bullet j}\right] + \cdots.$$

MULTI-SAMPLE ELT

Ignoring higher terms in λ, we find $\lambda \doteq D^{-1}\bar{H}_{\bullet\bullet}$ where

$$D = \frac{1}{n^2m^2}\sum_{i=1}^{n}\sum_{j=1}^{m}H_{ij}\sum_{r=1}^{m}H_{ir} + \frac{1}{n^2m^2}\sum_{i=1}^{n}\sum_{j=1}^{m}H_{ij}\sum_{s=1}^{n}H_{sj}.$$

In finding this D, the term

$$\widetilde{H}_{i\bullet} = \bar{H}_{i\bullet} - \frac{\lambda}{m^2}\sum_{r=1}^{m}H_{ir}\widetilde{H}_{\bullet r}$$

has been replaced by $\bar{H}_{i\bullet}$, and $\widetilde{H}_{j\bullet}$ has been replaced by $\bar{H}_{j\bullet}$, with the differences being absorbed into the coefficient of λ^2.

Now keeping terms up to order λ^2 in the profile empirical log likelihood function, we find

$$-2\log\mathcal{R}(\theta_0) = 2\sum_{i=1}^{n}\log\left(1 + \frac{\lambda}{n}\widetilde{H}_{i\bullet}\right) + 2\sum_{j=1}^{m}\log\left(1 + \frac{\lambda}{m}\widetilde{H}_{\bullet j}\right)$$

$$\doteq 2\sum_{i=1}^{n}\left(\frac{\lambda}{n}\widetilde{H}_{i\bullet} - \frac{1}{2}\left(\frac{\lambda}{n}\widetilde{H}_{i\bullet}\right)^2\right)$$

$$+ 2\sum_{j=1}^{m}\left(\frac{\lambda}{m}\widetilde{H}_{\bullet j} - \frac{1}{2}\left(\frac{\lambda}{m}\widetilde{H}_{\bullet j}\right)^2\right).$$

Replacing $\widetilde{H}$'s by corresponding $\bar{H}$'s and keeping terms to order λ^2, we get

$$-2\log\mathcal{R}(\theta_0) \doteq 2\sum_{i=1}^{n}\frac{\lambda}{n}\bar{H}_{i\bullet} - \frac{2\lambda^2}{m^2}\sum_{r=1}^{m}\bar{H}_{\bullet r}^2 - \sum_{i=1}^{n}\left(\frac{\lambda}{n}\bar{H}_{i\bullet}\right)^2$$

$$+ 2\sum_{j=1}^{m}\frac{\lambda}{m}\bar{H}_{\bullet j} - \frac{2\lambda^2}{n^2}\sum_{s=1}^{n}\bar{H}_{s\bullet}^2 - \sum_{j=1}^{m}\left(\frac{\lambda}{m}\bar{H}_{\bullet j}\right)^2$$

$$= 4\lambda\bar{H}_{\bullet\bullet} - 3\lambda^2\left(\frac{1}{n^2}\sum_{i=1}^{n}\bar{H}_{i\bullet}^2 + \frac{1}{m^2}\sum_{j=1}^{m}\bar{H}_{\bullet j}^2\right)$$

$$\doteq \bar{H}_{\bullet\bullet}^2\left(4D^{-1} - 3KD^{-2}\right),$$

where

$$K = \frac{1}{n^2}\sum_{i=1}^{n}\bar{H}_{i\bullet}^2 + \frac{1}{m^2}\sum_{j=1}^{m}\bar{H}_{\bullet j}^2.$$

To complete the argument, we need to show that $\bar{H}_{\bullet\bullet}$, suitably scaled, is asymptotically normally distributed and that the coefficient $4D^{-1} - 3KD^{-2}$ of $\bar{H}_{\bullet\bullet}^2$ is a consistent estimator of the suitably scaled variance. To make this argument, we introduce an ANOVA decomposition $h(X, Y, \theta_0) = A(X) + B(Y) + C(X, Y)$. Here $A(X) = E(h(X, Y, \theta_0) \mid X)$, $B(Y)$ is defined similarly and C is found by subtraction. We have $E(C(X,Y) \mid X) = E(C(X,Y) \mid Y) = 0$, and $E(A(X)) = E(B(X)) = 0$. Let $\sigma_A^2 = E(A(X)^2)$, $\sigma_B^2 = E(B(X)^2)$, and $\sigma_C^2 = E(C(X,Y)^2)$. These components are uncorrelated: $E(AB) = E(AC) = E(BC) = 0$.

Now

$$\bar{H}_{..} = \frac{1}{nm} \sum_{i=1}^{n} \sum_{j=1}^{m} A_i + B_j + C_{ij}$$

has mean 0 and variance

$$\frac{\sigma_A^2}{n} + \frac{\sigma_B^2}{m} + \frac{\sigma_C^2}{nm}.$$

The expected value of D is

$$E(D) = \frac{\sigma_A^2}{n} + \frac{\sigma_B^2}{m},$$

and the expected value of K is

$$E(K) = \frac{\sigma_A^2}{n} + \frac{\sigma_B^2}{m} + \frac{\sigma_A^2 + \sigma_B^2 + 2\sigma_C^2}{mn}.$$

Under mild moment conditions D and K approach their expectations with small relative errors, as $m, n \to \infty$, and so

$$\left(4D^{-1} - 3KD^{-2}\right) V\left(\bar{H}_{..}\right)$$
$$\doteq \left(\frac{\sigma_A^2}{n} + \frac{\sigma_B^2}{m} + \frac{\sigma_C^2}{nm}\right) \left(\frac{\sigma_A^2}{n} + \frac{\sigma_B^2}{m} - 3\frac{\sigma_A^2 + \sigma_B^2 + 2\sigma_C^2}{mn}\right) \left(\frac{\sigma_A^2}{n} + \frac{\sigma_B^2}{m}\right)^{-2}$$
$$\to 1,$$

as $\min(n, m) \to \infty$. This final limit also relies on the condition that at least one of σ_A^2 and σ_B^2 is positive. If $\sigma_A^2 = \sigma_B^2 = 0$, then $E(h(X, Y, \theta_0)) = 0$ follows, as remarked above, from the independence of X and Y.

From this it follows that empirical likelihood on two independent samples has an asymptotic $\chi_{(1)}^2$ distribution, for inferences on θ defined by $E(h(X, Y, \theta)) = 0$. We needed to have $\min(n, m) \to \infty$, but there was no need to have them grow at the same rate. Also, the second moment $E(h(X, Y, \theta)^2)$ must be finite to allow the ANOVA decomposition, and positive to get a central limit theorem for $\bar{H}_{..}$.

11.5 Bibliographic notes

The proof of the vector ELT is taken from Owen (1990*b*). The triangular array ELT is from Owen (1991). A univariate ELT was proved in Owen (1988*b*) for the mean, as well as for some M-estimates and statistics defined as (Fréchet) differentiable functions of F.

The basic strategy of forcing λ to be small and then making Taylor expansions is also used by Jing Qin in his dissertation (Qin 1992) and in subsequent papers.

For smooth estimating equations Qin (1992) assumes that $\|m(x,\theta)\|^3 \le M(x)$ uniformly in a neighborhood of θ_0 with $E(M(X)) < \infty$. Along with a few other conditions, having $-\log \mathcal{R}(\theta)$ smaller than a certain multiple of $n^{1/3}$ forces $\|\theta - \theta_0\| \le n^{-1/3}$. Then Taylor expansions of the log likelihood can be made in (λ, θ) around $(0, \theta_0)$ jointly.

CHAPTER 12

Algorithms

This chapter describes how to approach the optimization problems posed by empirical likelihood. The emphasis is on computing the empirical likelihood, for statistics defined through estimating equations, with nuisance parameters and side constraints. Much of the discussion carries over to other nonparametric likelihoods.

Several optimization methods are presented. Choosing a method entails some trade-offs. The Newton-Lagrange algorithm appears to be fastest. A nested search algorithm is slower but more reliable. The most reliable algorithm tried so far is sequential quadratic programming, using the NPSOL code (see Chapter 12.7). It also handles censoring and truncation in a direct way. But NPSOL is slower than special purpose code could be, because for problems without censoring or truncation, the Hessian is very sparse and NPSOL does not exploit that sparsity. Sequential linearization methods appear to be both fast and reliable, but they only compute an approximation to the desired likelihood. An approach based on range space methods has the potential to be as reliable as sequential quadratic programming, but much faster.

There is also a choice of which likelihood to use. The empirical likelihood respects range restrictions on parameters and is transformation invariant, but only provides positive likelihood if 0 is in the convex hull of $m(X_i, \theta)$. This convex hull constraint may be unduly restrictive when n is small, or the number of parameters is large. Alternatives such as the Euclidean likelihood can escape the convex hull, but do not obey range constraints. The empirical likelihood-t approach from Chapter 10.4 has some attractive properties, but it has not as yet been thoroughly explored. Empirical entropy methods do not escape the convex hull, but they do allow distributions with weight 0 on one or more data points to contribute to the confidence region.

Numerical searches typically require starting values in order to succeed. Usually the MLE $\hat{\theta}$ corresponds to solving the estimating equations with equal weight on every observation, and for most statistics of interest we can find $\hat{\theta}$ by some pre-existing algorithm. The profiles and hypothesis tests of interest are best found through a sequence of optimization problems starting at $\hat{\theta}$. There can also be a need to rescale the problem. A cubic regression of Y on $X/1000$ may succeed where a cubic regression on X fails. Similarly, a B-spline basis is a better choice than the truncated power spline basis when one wants a spline.

For this text, it was desired to use exact likelihood computations instead of the approximate ones, and NPSOL was used, except for some problems with quantiles that allow special methods. In the author's experience NPSOL has never failed to compute the empirical likelihood, if $\hat{\theta}$ has been available, though sometimes the problem must be rescaled. NPSOL often succeeds without $\hat{\theta}$. For small n it is best to use NPSOL directly, while for larger n, speed considerations lead to the use of NPSOL in the outer problem of the nested algorithm described in Chapter 12.3. While NPSOL was reliable without using explicit second derivatives, an algorithm using those derivatives might be faster or might be less sensitive to starting values and scaling.

12.1 Statistical tasks

The inferential problems to be addressed are testing hypotheses, computing maximum empirical likelihood estimates (MELE's), profiling likelihood ratio functions, and finding confidence sets.

The basic chore in all of these tasks is to maximize $R(F) = \sum_{i=1}^{n} \log(nw_i)$ subject to constraints

$$w_i \geq 0,$$
$$\sum_{i=1}^{n} w_i = 1,$$
$$\sum_{i=1}^{n} w_i m(X_i, \theta, \eta) = 0,$$
$$C(\theta, \eta, \gamma, \delta) = 0.$$

The second last equation describes the estimating equations, defined in terms of observations X_i, the parameter vector θ, and another parameter vector η not previously discussed. For autoregressive models described in Chapter 8.2, m also depends on the index i, but we suppress that dependence here for simplicity of exposition. There are also side constraints on θ and η expressed through the vector-valued function C of θ, η, and two new parameter vectors γ and δ.

It is assumed that the estimating function m can be differentiated at least once with respect to the components of θ, but not necessarily with respect to η. Optimization with respect to components of θ may then be approached by standard smooth function optimization methods, but components of η are either not to be optimized over, or must be optimized over by other means, such as grid searches. For example, a tail probability could be a part of θ, while the corresponding quantile would be part of η. As a more complicated example, a conditional quantile may be expressed through

$$m(X, \theta, \eta) = I(X_{\eta_1} = \eta_2)(I(X_{\eta_3} \leq \eta_4) - \theta_1),$$

where to avoid deeply nested subscripts, 1_A has been written as $I(A)$. Conditionally on component η_1 of $X \in \mathbb{R}^d$ taking the value η_2, component η_3 of X

has quantile η_4 with corresponding tail probability θ_1. We might optimize over θ_1 with η fixed, or optimize over η_4 with θ_1 and the rest of η fixed. For a given data set there may not be a need to optimize over η_2, or especially over the indices η_1 and η_3.

The function C describes constraints on the parameters θ and η, introducing two new parameter vectors γ and δ. We assume that C is differentiable at least once with respect to γ (and θ), but not necessarily with respect to δ (or η). The reason to keep C separate from m is that it does not depend on X_i. The function C has to be computed only once where m has to be computed n times. If n is large or C is expensive, the time saving can be large. An example of a side constraint is that a linear combination of θ values takes a particular value:

$$C = \sum_{j=1}^{p} \theta_j \gamma_j - \gamma_{p+1} = 0.$$

If γ_{p+1} is supposed to be a response variable for observation i then we can use $\gamma_1, \ldots, \gamma_p$ and $\delta_1 = i$, and encode the constraint as $\sum_{j=1}^{p} \theta_j \gamma_j - Y_{\delta_1}$.

A hypothesis test may be conducted by solving the problem above with some part of the list $(\theta, \eta, \gamma, \delta)$ fixed at defining values and the rest of it free to vary. The same problem arises in computing an MELE. To profile one or more members of the list $(\theta, \eta, \gamma, \delta)$, we conduct a sequence of hypothesis tests, in which some parameter components are frozen permanently at defining values, others change as we step through the sequence of hypothesis tests but are fixed in any one test, while the remainder are optimized over in each hypothesis test.

A confidence region is most easily found by computing a profile, and observing where the profile likelihood ratio function is above a threshold used for the confidence region. For a confidence interval the problem may also be approached by alternately maximizing and minimizing θ_j (or γ_j) subject to constraints including $\sum_{i=1}^{n} \log(nw_i) \geq \log(r_0)$.

To compute a univariate profile of θ_j, η_j, γ_j, or δ_j, we start at the MLE (or MELE) and solve a sequence of problems in which the parameter increases by steps. Then we return to the starting point and solve a sequence of problems in which the parameter decreases in a sequence of steps. The steps continue until either the desired range is covered, or the likelihood has fallen below a threshold.

Suppose that θ_1 has been profiled by maximizing empirical likelihood with θ_1 fixed at values $\theta_1(g)$ for a grid indexed by $g = 1, \ldots, G$. The profile trace contains the record of concomitant values

$$(\theta(g), \eta(g), \gamma(g), \delta(g), \mathcal{R}(g)), \quad g = 1, \ldots, G,$$

and possibly other quantities of interest, as computed during the optimizations.

For some parameters η_j and δ_j, such as observation component indices, a profile is not statistically meaningful. In other settings a profile can be used to generate the MLE of a curve. For example, if θ_1 is a tail probability corresponding to a quantile η_1 then profiling η_1 generates the empirical CDF in the profile trace

$(\theta_1(g), \eta_1(g))$. Of course, the CDF is easy to compute directly. But if we seek a CDF subject to some other estimating equation constraints, then computing it by profiling the quantile is attractive.

A bivariate profile may be computed by a sequence of problems in which the point (θ_j, θ_k) varies over a grid in the plane. The boundaries of this grid may be chosen after inspecting the two uni-parameter profiles to find the ranges of interest. It is advantageous to use a grid that contains the MLE $(\hat{\theta}_j, \hat{\theta}_k)$, which we take as the origin of the grid. Starting at the origin, we compute the empirical likelihood function outward in each of the four axis directions, with the solution at each grid point supplying starting values for the next one out. For every other cell in the grid define its predecessor as the cell that is one unit towards the MLE in the j direction and one unit towards the MLE in the k direction. These other cells can be visited in any order at all, so long as each cell is visited after its predecessor has been visited. This iteration allows starting values to be taken from the predecessors. If we anticipate a unimodal likelihood, then some speed can be gained by not computing at a cell whose predecessor had a very small likelihood.

The iteration above is how the bivariate profiles for this text were computed. Profiles of (γ_j, γ_k) or (θ_j, γ_k) are similar to the ones described above. Profiles involving statistically meaningful quantities η_j or δ_k might be approached this same way. Having computed the empirical log likelihood on a grid, the resulting values can be used in contour plots and perspective plots.

Some other strategies may be necessary to follow very twisty or narrow confidence regions, especially those from undetermined parameter vectors. It might also be faster to find and follow the contours of interest directly.

12.2 Smooth optimization

This section surveys numerical optimization of smooth functions. The texts cited in the bibliographic notes of Chapter 12.7 provide more detail. Parameters like η and δ are assumed to be fixed during any smooth optimization. Accordingly, we absorb η and δ into the definitions of m and C.

Basic smooth optimization methods are variations of Newton's method. They typically converge to local minima. If, however, the function is known to be convex, then a local minimum is a global one, and if the function is strictly convex, then a local minimum is the unique global minimum.

Let $f(x)$ be a real-valued function over $x \in \mathbb{R}^p$, and suppose that f is twice continuously differentiable. The gradient of f is

$$g(x) = \frac{\partial}{\partial x} f(x),$$

interpreted here as a column vector, and the Hessian of f is

$$H(x) = \frac{\partial^2}{\partial x \partial x'} f(x),$$

a p by p matrix of second order partial derivatives of f.

SMOOTH OPTIMIZATION

The basic method for optimizing f is Newton's method. Starting with a value x_0, the Newton iteration takes

$$x_{k+1} = x_k - H(x_k)^{-1} g(x_k), \quad k \geq 0.$$

To simplify the notation, write $H_k = H(x_k)$ and $g_k = g(x_k)$. The typical behavior for Newton's method is quadratic convergence to a solution x_∞ of the equations $g(x) = 0$, if x_0 is close enough to x_∞. Quadratic convergence means that $\limsup_{k \to \infty} \|x_{k+1} - x_\infty\| / \|x_k - x_\infty\|^2 \in (0, \infty)$. Roughly speaking, the number of correct bits or digits in x_{k+1} is twice the number in x_k. It is not uncommon for Newton's method to require only four or five iterations to converge in double precision.

The simplest version of Newton's method is not reliable. One problem is that Newton's method converges quadratically to a solution x_∞ that could be a local minimum, a local maximum, or a saddlepoint of f. A second problem is that if H_k is nearly singular, then a very large step can be taken and the algorithm might never recover.

The simplest usable versions of Newton's method apply some checks to the value $x_k - H_k^{-1} g_k$ before accepting it as x_{k+1}. A step-halving approach takes $x_{k+1} = x_{k+1,r} = x_k - 2^{-r} H_k^{-1} g_k$ where r is the smallest nonnegative integer for which $f(x_{k+1,r}) < f(x_k)$. A variant is to take $x_{k+1,r} = x_k - [H_k + \lambda_r D_k]^{-1} g_k$, where λ_r is an increasing sequence of nonnegative numbers, and D is a positive definite matrix such as the identity or perhaps the diagonal of H_k. This generalizes the Levenberg-Marquardt method, familiar in nonlinear least squares. As r increases, the step $[H_k + \lambda_r D_k]^{-1} g_k$ becomes shorter, and if $D = I$ it becomes more nearly parallel to the steepest descent direction $-g_k$.

More sophisticated step reduction approaches insist on a sufficient decrease in f. The decrease in f must be comparable to that predicted through a quadratic Taylor approximation $f \doteq f_k + (x - x_k)' g_k + (x - x_k)' H_k (x - x_k)/2$. If the decrease is not comparable, then the Taylor approximation is called into question for the vector $x_{k+1,r}$, and a shorter vector $x_{k+1,r+1}$ is tried.

When the Hessian is expensive to compute or difficult to program, then quasi-Newton methods are often used. Quasi-Newton methods do not compute the Hessian, but approximate it instead by using the computed values of g_k. If x_{k+1} is close to x_k then $g_{k+1} - g_k$ is approximately $H'_{k+1}(x_{k+1} - x_k)$. Starting with an approximation H_0 such as the identity matrix, each new gradient value g_{k+1} that is observed is used to make an update to H_{k+1} so that

$$(x_{k+1} - x_k)' H'_{k+1}(x_{k+1} - x_k) = (x_{k+1} - x_k)'(g_{k+1} - g_k),$$

making H_{k+1} consistent with the gradient information from step $k+1$.

In some cases the Hessian is simply too large to store, whether exactly or approximately. If sparse methods are not available, then the method of conjugate gradients is a candidate.

Constrained optimization is a much deeper topic than unconstrained optimization. A brief sketch of the subject is given below. The reader may consult references in Chapter 12.7 for more details. We only consider general smooth nonlinear equality constraints, as these are the kind that arise in most empirical likelihood problems. Special methods exist to exploit simpler cases such as bounds on parameters or linear constraints. Other specialized methods have been developed for inequality constraints.

Minimization of $f(x)$ subject to nonlinear equality constraints $c(x) = 0$, is usually organized around the Lagrangian

$$G(x, \lambda) = f(x) + \lambda' c(x).$$

The constrained optimum is described by values x and λ with $\partial G/\partial x = 0$ and $\partial G/\partial \lambda = 0$. If we knew the Lagrange multiplier λ, then the solution would be a stationary point x with $\partial G(x, \lambda)/\partial x = 0$. The stationary point is not necessarily a minimum of $G(\cdot, \lambda)$.

Augmented Lagrangian methods replace G by $G_\rho = G + (\rho/2)c(x)'c(x)$ for some $\rho > 0$. Some versions use a different ρ_i for each constraint in $c(x)$. For large enough ρ, the constrained optimum is a local minimum of G_ρ. The augmented Lagrangian works better than simplistic penalty methods that seek to minimize $P_\rho = f(x) + (\rho/2)c(x)'c(x)$, because these latter require $\rho \to \infty$ and the Hessian of P_ρ becomes badly conditioned for large ρ.

Sequential quadratic programming (SQP) methods may be applied to the augmented Lagrangian function. In SQP, given λ and ρ, the function $G_\rho(x, \lambda)$ is approximated by a quadratic in x, and the constraint $c(x)$ is approximated as a linear function in x. A solution is sought minimizing the quadratic function subject to linear constraints.

Practical implementations of SQP must have means for detecting defective subproblems, such as those that are unbounded below due to an indefinite Hessian. Augmented Lagrangian methods must also have methods for updating ρ and λ. The convergence rate for x is limited by that of λ.

The NPSOL software uses sequential quadratic programming with an augmented Lagrangian. The user of the code needs to provide starting values, routines to compute f and c, and ideally, routines to compute gradients of f and c with respect to x.

12.3 Estimating equation methods

Here we consider optimization of empirical likelihood subject to a constraint fixing some but not all of the parameters in the estimating equation. This formulation is the most commonly used one in this text. The parameters that are free to vary in the optimization may be nuisance parameters that had to be introduced in order to define the parameters of interest. Other times all the parameters are of interest, but each one becomes free to vary when one or more others are being profiled.

Suppose that the data are $X_i \in \mathbb{R}^d$, for $i = 1, \dots, n$. Suppose that the estimating equation of interest is $E(m(X, \theta, \nu)) = 0$. The nonsmooth parameter η from Chapter 12.1 has been absorbed into the definition of m. The parameter θ

ESTIMATING EQUATION METHODS

from there has been split into two pieces. The parameters fixed for the inference have remained in $\theta \in \mathbb{R}^p$, while the freely varying parameters are now taken to be components of $\nu \in \mathbb{R}^q$. Assume that $m(X, \theta, \nu) \in \mathbb{R}^{p+q}$.

We wish to compute

$$\mathcal{R}(\theta) = \max_{\nu} \mathcal{R}(\theta, \nu).$$

where

$$\mathcal{R}(\theta, \nu) = \max \left\{ \prod_{i=1}^{n} nw_i \mid \sum_{i=1}^{n} w_i m(X_i, \theta, \nu) = 0, w_i \geq 0, \sum_{i=1}^{n} w_i = 1 \right\}.$$

The value of $\mathcal{R}(\theta, \nu)$ can be found by solving the numerical problem for simple hypotheses, by iterated least squares as described in Chapter 3.14. The convex dual representation from that chapter allows us to write

$$\log \mathcal{R}(\theta) = \max_{\nu} \min_{\lambda} \mathbb{L}_*(\lambda, \theta, \nu), \qquad (12.1)$$

where

$$\mathbb{L}_*(\lambda, \theta, \nu) = -\sum_{i=1}^{n} \log_*(1 + \lambda' m(X_i, \theta, \nu)), \qquad (12.2)$$

and

$$\log_*(z) = \begin{cases} \log(z), & \text{if } z \geq \varepsilon \\ \log(\varepsilon) - 1.5 + 2z/\varepsilon - z^2/(2\varepsilon^2), & \text{if } z \leq \varepsilon, \end{cases} \qquad (12.3)$$

for some small $\varepsilon > 0$.

The point of replacing log by $\log_*$ is that it frees us from imposing the constraints $w_i \geq 0$. The function $\mathbb{L}$ is twice continuously differentiable in λ, and for each ν, it is convex in λ. If we know that $\max_i nw_i \leq A$ at the solution, then we can choose $\varepsilon = A^{-1}$ and get the same solution using $\log_*$ without constraints on λ as we would get using log with constraints $w_i = (1/n)(1+\lambda' m(X_i, \theta, \nu))^{-1} \geq 0$. In Chapter 3.14 we used $\varepsilon = 1/n$, because $w_i \leq 1$ at the solution. Larger values of ε might improve the condition of the Hessian with respect to λ of $\mathbb{L}_*$, without affecting the asymptotic validity of the empirical likelihood inferences. See the discussion around equation (3.39).

A nested algorithm uses a literal interpretation of the maximin formulation (12.1). The inner stage is to minimize $\mathbb{L}_*(\lambda, \theta, \nu)$ over λ for fixed values of θ and ν. Let $\lambda(\theta, \nu)$ be the minimizing value of λ. The outer stage is to maximize

$$\mathbb{M}_*(\nu, \theta) \equiv \mathbb{L}_*(\lambda(\theta, \nu), \theta, \nu)$$

over ν. The parameter θ remains fixed.

At the inner stage of the nested algorithm we may required the first and second partial derivatives:

$$\mathbb{L}_*^\lambda \equiv \frac{\partial}{\partial \lambda} \mathbb{L}_* \quad \text{and} \quad \mathbb{L}_*^{\lambda\lambda} \equiv \frac{\partial^2}{\partial \lambda \partial \lambda'} \mathbb{L}_*.$$

At the outer stage, we may need

$$\mathbb{M}_\star^\nu(\nu,\theta) = \frac{\partial}{\partial \nu}\mathbb{L}_\star(\lambda(\theta,\nu),\theta,\nu), \quad \text{and}$$

$$\mathbb{M}_\star^{\nu\nu}(\nu,\theta) = \frac{\partial^2}{\partial \nu \partial \nu'}\mathbb{L}_\star(\lambda(\theta,\nu),\theta,\nu).$$

Expressions for all of these partial derivatives are given in Chapter 12.4.

Ideally $\mathbb{M}_\star^{\nu\nu}$ should be negative definite, or at least negative semidefinite. This indeed holds for large enough n, well-posed estimating equations, and values of ν and θ that are not extreme. But $\mathbb{M}_\star$ can fail to have a negative definite Hessian in practice. The outer optimization in the nested algorithm is more difficult than the inner one because of this feature.

Sophisticated optimization code takes account of the possibility that the Hessian might not be definite. For minimization problems, such as minimizing $-\mathbb{M}_\star$, the target function should have positive curvature along all linear paths. Directions of negative curvature tend to be promising because they may lead further downhill. For empirical likelihood problems we do not have to worry about the objective function $-\mathbb{M}_\star$ being unbounded below, so the situation is simpler than for some other optimization problems.

Figure 12.1 illustrates this problem. The example is taken from Huber's robust location and scale estimator as applied to the passage time of light data in Chapter 3.12. Suppose that θ is the scale parameter and ν is the location parameter. As θ is reduced from the MLE it becomes harder to find the optimizing ν. Imagine a hiker starting at the MLE in Figure 12.1 on a hill with contours equal to the empirical likelihood function. In each step in the profiling of θ, the hiker is instructed to step one meter south, and then move to the highest point possible on the present line of latitude. At first there is a local maximum in ν at each θ. The hiker can follow a ridge. Eventually, however, the hiker encounters a θ for which there is a local minimum of ν between two local maxima corresponding to two ridges.

The profile trace in Figure 12.1 was found by NPSOL. For small values of the scale parameter θ there are two ridges in ν. NPSOL found the higher ridge, though perhaps it overstepped before finding it. The ideal path may be one where the profile trace $\nu(\theta)$ has a discontinuity in it. For very small values of θ there are a great many ridges, but these correspond to exceedingly small $\mathcal{R}$ values that are not in any reasonable confidence set.

The expression (12.5) for $\mathbb{M}_\star^{\nu\nu}$ is a sum of two terms, one negative definite, and one not. For problems of maximizing a Gaussian log likelihood for nonlinear least squares, a similar phenomenon occurs. The Gauss-Newton algorithm for nonnegative least squares simply uses the definite term and ignores the other one. A similar algorithm may improve the outer loop in the nested algorithm, but there will still be a difficulty if the gradient $\mathbb{M}_\star^\nu$ is zero at a non-optimal value of ν. For this reason, a sophisticated algorithm like NPSOL is recommended for the outer optimization.

ESTIMATING EQUATION METHODS

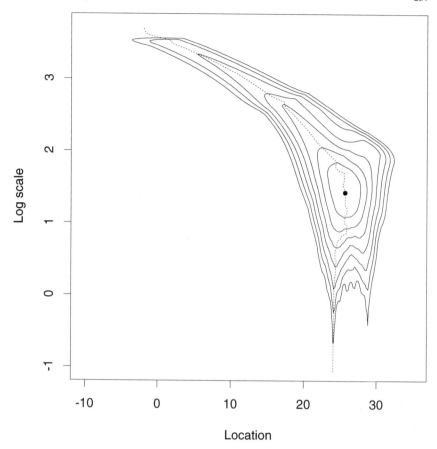

Figure 12.1 *The contours are for the empirical likelihood of μ and $\log \sigma$ for Huber's robust location andscale estimates applied to the passage time of light data. This example was discussed in Chapter 3.12. Contour levels correspond to $\log \mathcal{R}(\mu, \sigma)$ from -1 to -7 by steps of -1. The solid point is at $(\hat{\mu}, \log \hat{\sigma})$. The dotted path contains points of the form $(\mu(\sigma), \log \sigma)$ where $\mu(\sigma_0)$ maximizes empirical likelihood subject to $\sigma = \sigma_0$.*

A non-nested approach is to solve the equations

$$\begin{pmatrix} \mathbb{L}_*^\lambda \\ \mathbb{L}_*^\nu \end{pmatrix} = 0 \in \mathbb{R}^{p+q}$$

jointly for λ and ν. Here $\mathbb{L}_*^\nu$ is the partial derivative of $\mathbb{L}_*$ with respect to ν, given in Chapter 12.4.

The Newton-Lagrange algorithm is to apply Newton's method for solving these equations. This algorithm will converge quadratically, if started close enough to the solution. The advantage in this approach is that it can be faster than a nested method. A nested method may expend a lot of effort optimizing λ for values of ν that are not nearly optimal. When quadratic convergence holds, it is typical to require 4 or 5 iterations of a non-nested algorithm. By comparison, a nested algorithm would require 4 or 5 outer iterations and 16 to 25 inner iterations. The speed advantage is smaller than this argument would suggest because the estimating equations $m(X_i, \theta, \nu)$ and their derivatives do not need to be recomputed at each inner iteration.

The Hessian matrix for this Newton-Lagrange approach is

$$\begin{pmatrix} \mathbb{L}_\star^{\lambda\lambda} & \mathbb{L}_\star^{\lambda\nu} \\ \mathbb{L}_\star^{\nu\lambda} & \mathbb{L}_\star^{\nu\nu} \end{pmatrix},$$

once again using partial derivatives from Chapter 12.4.

Step reduction for the Newton-Lagrange method is quite difficult. It is harder to measure progress towards a constrained optimum than towards an unconstrained one. A step that improves the objective function might also increase the violation of one or more constraints. A hybrid algorithm can give good results. The hybrid starts with Newton-like methods to solve the equations above, and then if the problem appears ill conditioned, it switches over to the nested algorithm.

12.4 Partial derivatives

Here we present the partial derivatives required by the optimization algorithms of Chapter 12.3 with $\mathbb{L}_\star(\lambda, \theta, \nu)$ defined at (12.2). Some optimization software does not require second derivatives. The second derivatives have more complicated expressions than the first derivatives. They are presented here because analytic second derivatives can lead to more reliable optimization.

The partial derivatives of $\mathbb{L}_\star(\lambda, \theta, \nu)$ require derivatives of the function $\log_\star(z)$. The first two derivatives of $\log_\star(z)$ are

$$\log_\star^{(1)}(z) = \frac{d}{dz} \log_\star(z) = \begin{cases} 1/z, & \text{if } z \geq \varepsilon \\ 2/\varepsilon - z/\varepsilon^2, & \text{if } z \leq \varepsilon, \end{cases}$$

and

$$\log_\star^{(2)}(z) = \frac{d^2}{dz^2} \log_\star(z) = \begin{cases} -1/z^2, & \text{if } z \geq \varepsilon \\ -1/\varepsilon^2, & \text{if } z \leq \varepsilon. \end{cases}$$

For $w_i \leq \varepsilon^{-1}/n$, we have $z_i = 1 + \lambda' m(X_i, \theta, \nu) \geq \varepsilon/n$, and

$$\log_\star^{(1)}(z_i) = nw_i$$
$$\log_\star^{(2)}(z_i) = -(nw_i)^2.$$

To make some expressions shorter and more intuitive we define $\widehat{w}_i = \widehat{w}_i(\theta, \nu)$

and $\widetilde{w}_i = \widetilde{w}_i(\theta, \nu)$ through

$$\log_*^{(1)}(1 + \lambda' m(X_i, \theta, \nu)) = n\widehat{w}_i \qquad (12.4)$$
$$\log_*^{(2)}(1 + \lambda' m(X_i, \theta, \nu)) = -(n\widetilde{w}_i)^2.$$

Ordinarily $w_i \leq \varepsilon^{-1}/n$, and then $\widehat{w}_i = \widetilde{w}_i = w_i$.

The partial derivatives of $\mathbb{L}_*$ that we need are:

$$\mathbb{L}_*^\lambda = \frac{\partial}{\partial \lambda}\mathbb{L}_*(\lambda, \theta, \nu), \qquad \mathbb{L}_*^{\lambda\lambda} = \frac{\partial^2}{\partial \lambda \partial \lambda'}\mathbb{L}_*(\lambda, \theta, \nu),$$

$$\mathbb{L}_*^\nu = \frac{\partial}{\partial \nu}\mathbb{L}_*(\lambda, \theta, \nu), \qquad \mathbb{L}_*^{\nu\nu} = \frac{\partial^2}{\partial \nu \partial \nu'}\mathbb{L}_*(\lambda, \theta, \nu),$$

and

$$\mathbb{L}_*^{\nu\lambda} = \frac{\partial^2}{\partial \lambda \partial \nu'}\mathbb{L}_*(\lambda, \theta, \nu).$$

These are of dimension p, $p \times p$, q, $q \times q$, and $q \times p$, respectively. Of course, $\mathbb{L}_*^{\lambda\nu}$ is the transpose of $\mathbb{L}_*^{\nu\lambda}$. By straightforward calculus,

$$\mathbb{L}_*^\lambda = -\sum_{i=1}^n \log_*^{(1)}(1 + \lambda' m(X_i, \theta, \nu)) m(X_i, \theta, \nu)$$
$$= -\sum_{i=1}^n n\widehat{w}_i m(X_i, \theta, \nu)$$

and

$$\mathbb{L}_*^{\lambda\lambda} = -\sum_{i=1}^n \log_*^{(2)}(1 + \lambda' m(X_i, \theta, \nu)) m(X_i, \theta, \nu) m(X_i, \theta, \nu)'$$
$$= \sum_{i=1}^n (n\widetilde{w}_i)^2 m(X_i, \theta, \nu) m(X_i, \theta, \nu)'.$$

This Hessian is positive semidefinite and grows proportionally to n, assuming that $\widetilde{w}_i$ is roughly $1/n$. It is typical for the curvature in a parametric log likelihood ratio to grow proportionally to n. Here λ is a Lagrange multiplier and does not correspond to a tangible parameter of the data distribution. It is however the parameter in the dual likelihood. The fact that the curvature grows proportionally to n is a consequence of the scaling chosen when the problem was set up in equation (3.32).

For derivatives with respect to ν we introduce the $(p+q) \times q$ matrix

$$m_\nu(X_i, \theta, \nu) = \frac{\partial}{\partial \nu} m(X_i, \theta, \nu)$$

of partial derivatives of the $p + q$ estimating equations with respect to the q free parameters. Now

$$\mathbb{L}_\star^\nu = -\sum_{i=1}^n \log_\star^{(1)}(1+\lambda' m(X_i,\theta,\nu))m_\nu(X_i,\theta,\nu)'\lambda$$

$$= -\sum_{i=1}^n n\widehat{w}_i m_\nu(X_i,\theta,\nu)'\lambda,$$

and

$$\mathbb{L}_\star^{\nu\lambda} = -\sum_{i=1}^n \log_\star^{(1)}(1+\lambda' m(X_i,\theta,\nu))m_\nu(X_i,\theta,\nu)$$

$$-\sum_{i=1}^n \log_\star^{(2)}(1+\lambda' m(X_i,\theta,\nu))m_\nu(X_i,\theta,\nu)'\lambda m(X_i,\theta,\nu)'$$

$$= -\sum_{i=1}^n n\widehat{w}_i m_\nu(X_i,\theta,\nu)$$

$$+\sum_{i=1}^n (n\widetilde{w}_i)^2 m_\nu(X_i,\theta,\nu)'\lambda m(X_i,\theta,\nu)'.$$

Differentiating twice with respect to ν requires second derivatives of m with respect to ν. There is a $(p+q)\times q\times q$ array of such derivatives. To keep within the usual matrix notation, we introduce an index $j=1,\ldots,p+q$ for the estimating equations. That is,

$$\lambda' m(X_i,\theta,\nu) = \sum_{j=1}^{p+q} \lambda_j m^j(X_i,\theta,\nu).$$

Introduce

$$m_\nu^j \equiv \frac{\partial}{\partial \nu} m^j(X_i,\theta,\nu), \quad m_{\nu\nu}^j \equiv \frac{\partial^2}{\partial \nu \partial \nu'} m^j(X_i,\theta,\nu),$$

of dimensions q and $q\times q$, respectively. Now

$$\mathbb{L}_\star^{\nu\nu} = -\sum_{i=1}^n \log_\star^{(1)}(1+\lambda' m(X_i,\theta,\nu)) \sum_{j=1}^{p+q} m_{\nu\nu}^j(X_i,\theta,\nu)$$

$$-\sum_{i=1}^n \log_\star^{(2)}(1+\lambda' m(X_i,\theta,\nu)) \left(\sum_{j=1}^{p+q} \lambda_j m_\nu^j\right)\left(\sum_{j=1}^{p+q} \lambda_j m_\nu^j\right)'$$

$$= -\sum_{i=1}^n n\widehat{w}_i \sum_{j=1}^{p+q} \lambda_j m_{\nu\nu}^j(X_i,\theta,\nu)$$

$$+\sum_{i=1}^n (n\widetilde{w}_i)^2 m_\nu' \lambda \lambda' m_\nu.$$

Next we develop expressions for the derivatives of $\mathbb{M}_\star$. We need

$$\frac{\partial \lambda}{\partial \nu} = \frac{\partial}{\partial \nu}\lambda(\theta,\nu),$$

a $(p+q) \times q$ matrix of partial derivatives. The chain rule gives

$$\mathbb{M}_\star^\nu(\nu,\theta) = \left(\frac{\partial \lambda}{\partial \nu}\right)' \mathbb{L}_\star^\lambda + \mathbb{L}_\star^\nu = \mathbb{L}_\star^\nu,$$

because $\mathbb{L}_\star^\lambda = 0$ at the minimizer $\lambda(\theta,\nu)$. Similarly,

$$\mathbb{M}_\star^{\nu\nu}(\nu,\theta) = \left(\frac{\partial \lambda}{\partial \nu}\right)' \mathbb{L}_\star^{\lambda\lambda}\left(\frac{\partial \lambda}{\partial \nu}\right) + 2\left(\frac{\partial \lambda}{\partial \nu}\right)' \mathbb{L}_\star^{\lambda\nu} + \mathbb{L}_\star^{\nu\nu}.$$

From a local quadratic approximation to $\mathbb{L}_\star$ we may find that

$$\frac{\partial \lambda}{\partial \nu} = -\left(\mathbb{L}_\star^{\lambda\lambda}\right)^{-1}\mathbb{L}_\star^{\lambda\nu},$$

so that

$$\mathbb{M}_\star^{\nu\nu}(\nu,\theta) = \mathbb{L}_\star^{\nu\nu} - \mathbb{L}_\star^{\nu\lambda}\left(\mathbb{L}_\star^{\lambda\lambda}\right)^{-1}\mathbb{L}_\star^{\lambda\nu}. \tag{12.5}$$

Optimization over ν is easier if $\mathbb{M}_\star^{\nu\nu}$ is negative definite. The second term in (12.5) is negative semidefinite, but the first term $\mathbb{L}_\star^{\nu\nu}$ might not be. For $\lambda = O_p(n^{-1/2})$ and $n\widehat{w}_i$ and $n\widetilde{w}_i$ both near 1, we expect $\mathbb{L}_\star^{\nu\nu} = O_p(n^{-1/2})$, and $\mathbb{L}_\star^{\nu\lambda} = \mathbb{L}_\star^{\lambda\lambda} = O(n)$. Then $\mathbb{M}_\star^{\nu\nu}$ is $O(n)$ and dominated by a negative semidefinite term.

12.5 Primal problem

The primal problem is to maximize $\sum_{i=1}^n \log(nw_i)$, subject to the constraints $w_i \geq 0$, $\sum_{i=1}^n w_i = 1$, and $\sum_{i=1}^n w_i m(X_i, \theta, \nu) = 0$, without first encoding the n dimensional vector of weights through a Lagrange multiplier. The free variables in the primal problem are $u_i = nw_i$, $i = 1, \ldots, n$, and $\nu \in \mathbb{R}^q$. We work with $u_i = nw_i$ because they are of order 1 as n increases. As with the dual problem, we can replace $\log(u_i)$ by $\log_\star(u_i)$ instead of or in addition to imposing the constraints $u_i \geq 0$. The Lagrangian is

$$\mathbb{P}_\star(u,\theta,\nu,\lambda,\gamma) = \sum_{i=1}^n \log_\star(u_i) + \lambda'\sum_{i=1}^n u_i m(X_i,\theta,\nu) + \gamma\left(\sum_{i=1}^n u_i - n\right),$$

for $u \in \mathbb{R}^n$, $\lambda \in \mathbb{R}^r$, with $r = p+q$, $\theta \in \mathbb{R}^p$ fixed, $\nu \in \mathbb{R}^q$, and $\gamma \in \mathbb{R}$. Now instead of expressing u_i in terms of λ, we work directly with $n + r + q + 1$ free variables in u, λ, ν, and γ.

The gradient and Hessian may be written in obvious notation as

$$\begin{pmatrix} \mathbb{P}^u_\star \\ \mathbb{P}^\lambda_\star \\ \mathbb{P}^\nu_\star \\ \mathbb{P}^\gamma_\star \end{pmatrix} \quad \text{and} \quad \begin{pmatrix} \mathbb{P}^{uu}_\star & \mathbb{P}^{u\lambda}_\star & \mathbb{P}^{u\nu}_\star & \mathbb{P}^{u\gamma}_\star \\ \mathbb{P}^{\lambda u}_\star & \mathbb{P}^{\lambda\lambda}_\star & \mathbb{P}^{\lambda\nu}_\star & \mathbb{P}^{\lambda\gamma}_\star \\ \mathbb{P}^{\nu u}_\star & \mathbb{P}^{\nu\lambda}_\star & \mathbb{P}^{\nu\nu}_\star & \mathbb{P}^{\nu\gamma}_\star \\ \mathbb{P}^{\gamma u}_\star & \mathbb{P}^{\gamma\lambda}_\star & \mathbb{P}^{\gamma\nu}_\star & \mathbb{P}^{\gamma\gamma}_\star \end{pmatrix}$$

of dimensions

$$\begin{pmatrix} n \times 1 \\ r \times 1 \\ q \times 1 \\ 1 \times 1 \end{pmatrix} \quad \text{and} \quad \begin{pmatrix} n \times n & n \times r & n \times q & n \times 1 \\ r \times n & r \times r & r \times q & r \times 1 \\ q \times n & q \times r & q \times q & q \times 1 \\ 1 \times n & 1 \times r & 1 \times q & 1 \times 1 \end{pmatrix}.$$

In most problems n is much larger than $r + q + 1$, and so the components $\mathbb{P}^{uu}_\star$ comprise most of the Hessian.

The derivatives with respect to u_i are

$$\frac{\partial}{\partial u_i} \mathbb{P}_\star = \log^{(1)}_\star(u_i) + \lambda' m(X_i, \theta, \nu) + \gamma$$
$$= -\widehat{u}_i + \lambda' m(X_i, \theta, \nu) + \gamma,$$

where $\widehat{u}_i = n\widehat{w}_i$, defined at (12.4), and

$$\frac{\partial^2}{\partial u_i \partial u_j} \mathbb{P}_\star = \log^{(2)}_\star(u_i) 1_{i=j} = -\widetilde{u}_i^2 1_{i=j},$$

where $\widetilde{u}_i = n\widetilde{w}_i$, also defined at (12.4). Notice particularly that the n by n Hessian matrix $\mathbb{P}^{uu}_\star$ is diagonal and negative definite. Therefore, the Hessian of $\mathbb{P}_\star$ is very sparse when $n \gg q + r + 1$.

The derivatives with respect to λ are

$$\mathbb{P}^\lambda_\star = \sum_{i=1}^{n} u_i m(X_i, \theta, \nu) \quad \text{and} \quad \mathbb{P}^{\lambda\lambda}_\star = 0.$$

The derivatives with respect to ν are

$$\mathbb{P}^\nu_\star = \left(\sum_{i=1}^{n} u_i m_\nu(X_i, \theta, \nu)' \right) \lambda$$

and

$$\mathbb{P}^{\nu\nu}_\star = \sum_{i=1}^{n} u_i \sum_{j=1}^{p+q} \lambda_j m^j_{\nu\nu}(X_i, \theta, \nu).$$

The derivatives with respect to γ are

$$\mathbb{P}^\gamma_\star = \sum_{i=1}^{n} u_i - n \quad \text{and} \quad \mathbb{P}^{\gamma\gamma}_\star = 0.$$

PRIMAL PROBLEM

The cross partial derivatives $\mathbb{P}_*^{\lambda\gamma}$, and $\mathbb{P}_*^{\nu\gamma}$ are zero. Also

$$\mathbb{P}_*^{u\gamma} = \begin{pmatrix} 1 \\ 1 \\ \vdots \\ 1 \end{pmatrix}, \quad \mathbb{P}_*^{u\lambda} = \begin{pmatrix} m(X_1,\theta,\nu)' \\ m(X_2,\theta,\nu)' \\ \vdots \\ m(X_n,\theta,\nu)' \end{pmatrix}, \quad \mathbb{P}_*^{u\nu} = \begin{pmatrix} \lambda' m_\nu(X_1,\theta,\nu) \\ \lambda' m_\nu(X_2,\theta,\nu) \\ \vdots \\ \lambda' m_\nu(X_n,\theta,\nu) \end{pmatrix},$$

and finally,

$$\mathbb{P}_*^{\lambda\nu} = \sum_{i=1}^n u_i m_\nu(X_i,\theta,\nu).$$

The disadvantage of the primal formulation is that it requires a much larger Hessian matrix. Direct approaches based on computing and solving this matrix require $O(n^3)$ time per iteration and $O(n^2)$ storage. Methods that exploit sparsity can remove both issues at the expense of some additional algorithmic complexity.

The primal formulation has the advantage of simpler expressions for required partial derivatives. It may also be possible for a primal problem to find a path through the space of u_i variables that goes around some difficult point in the space of ν and λ values.

The Hessian in the primal formulation is sparse because the upper left n by n submatrix in it is diagonal. Partition the Hessian of $\mathbb{P}_*$ as

$$\begin{pmatrix} \mathbb{P}_*^{uu} & \mathbb{P}_*^{uv} \\ \mathbb{P}_*^{vu} & \mathbb{P}_*^{vv} \end{pmatrix},$$

where v denotes the variables other than u, namely λ, ν, and γ. Similarly, let $\mathbb{P}_*^v$ contain the gradient of $\mathbb{P}_*$ with respect to all variables other than u. Then the Newton step solves

$$\begin{pmatrix} \mathbb{P}_*^{uu} & \mathbb{P}_*^{uv} \\ \mathbb{P}_*^{vu} & \mathbb{P}_*^{vv} \end{pmatrix} \begin{pmatrix} u \\ v \end{pmatrix} = - \begin{pmatrix} \mathbb{P}_*^u \\ \mathbb{P}_*^v \end{pmatrix}$$

for vectors u and v. Because $\mathbb{P}_*^{uu}$ is diagonal, it is advantageous to rewrite the first row as

$$\mathbb{P}_*^{uu} u = - \left(\mathbb{P}_*^u + \mathbb{P}_*^{uv} v \right).$$

Thus, once v is found, it is easy to find u. To find v, write out the second row of equations, and substitute for u, obtaining

$$\left(\mathbb{P}_*^{vv} - \mathbb{P}_*^{vu} \left(\mathbb{P}_*^{uu} \right)^{-1} \mathbb{P}_*^{uv} \right) v = - \left(\mathbb{P}_*^v - \mathbb{P}_*^{vu} \left(\mathbb{P}_*^{uu} \right)^{-1} \mathbb{P}_*^u \right).$$

The equations to solve for v are not usually sparse, but there are only $q + r + 1 = p + 2q + 1$ of them. This approach to factoring sparse Hessians is known as the range space technique. Of course, step reduction strategies are required, as is a merit function trading off likelihood optimization and constraint satisfaction.

12.6 Sequential linearization

Since the optimization problem for a mean is so simple, a strategy based on approximating the statistic of interest by a mean can be effective. This approach is available for statistics not conveniently expressed in terms of estimating equations, such as U-statistics. Let us suppose that a statistic θ is defined through a function $\theta = T(F)$, for the distribution F putting weight w_i on observation X_i. We will suppose at first that any process for maximizing over nuisance parameters is embedded into $T(F)$. We will use the same symbol T to denote the function of $w_1, \ldots, w_n$ defined by

$$T(w_1, \ldots, w_n) \equiv T\left(\sum_{i=1}^{n} w_i \delta_{X_i}\right).$$

The NPMLE is $\hat{\theta} = T(\hat{F})$. Suppose that T is smooth, so that we may write

$$T(w_1, \ldots, w_n) \doteq \hat{\theta} + \sum_{i=1}^{n} w_i T_i(\hat{F}),$$

where

$$T_i(F) = \lim_{\varepsilon \to 0} \frac{T((1-\varepsilon)F + \varepsilon \delta_{X_i}) - T(F)}{\varepsilon} \qquad (12.6)$$

measures the effect of small increases in the weight w_i.

With linearization, we write

$$\mathcal{R}_L(\theta) = \max\left\{\prod_{i=1}^{n} nw_i \ \Big|\ \sum_{i=1}^{n} w_i \left(\hat{\theta} + T_i(\hat{F}) - \theta\right) = 0, w_i \geq 0, \sum_{i=1}^{n} w_i = 1\right\}.$$

This linearized profile empirical likelihood ratio function can be computed using the algorithm for the vector mean described in Chapter 3.14. Empirical likelihood inferences based on profiling $\mathcal{R}_L(\theta)$ use the same $T_i(\hat{F})$ quantities as are used in the infinitesimal jackknife (IJ) described in Chapter 9.9. The difference is that the IJ simply uses them to construct a variance estimate. Turning the IJ variance estimate into a confidence ellipsoid amounts to using the Euclidean likelihood for the mean of $T_i(\hat{F})$.

Having computed the optimizing w_i we can evaluate $T(w_1, \ldots, w_n)$. In general $T(w_1, \ldots, w_n) \neq \theta$, because the linearization was only an approximation. We can however re-linearize around these values w_i and repeat the process. Suppose that $w_i^{(k)}$ are the weights after k linearizations have been done, starting with $w_i^{(0)} = 1/n$. The weights $w_i^{(k+1)}$ are found by maximizing $\prod_{i=1}^{n} nw_i^{(k+1)}$ subject

SEQUENTIAL LINEARIZATION

to

$$0 = \sum_{i=1}^{n} w_i^{(k+1)} \left(\hat{\theta}_k + T_i(\hat{F}^{(k)}) - \theta \right),$$

$$0 \le w_i^{(k+1)}, \quad \text{and}$$

$$1 = \sum_{i=1}^{n} w_i^{(k+1)},$$

where $\hat{F}^{(k)}$ uses weights $w_i^{(k)}$, and $\hat{\theta}_k = T(\hat{F}^{(k)})$. The resulting maximum is denoted by $\mathcal{R}_{L,k+1}(\theta)$.

It is tempting to use $\mathcal{R}_{L,\infty}(\theta) = \lim_{k \to \infty} \mathcal{R}_{L,k}(\theta)$, but very often the sequence does not converge, especially for values of θ far from $\hat{\theta}$. Fortunately, good results can be obtained from a fixed number k of iterations. Let $\varphi \in \mathbb{R}^p$ be a fixed vector and define $\varphi_n = n^{-1/2} \varphi$. Then, under smoothness conditions, including existence of $k+2$ continuous derivatives for $T(w_1, \ldots, w_n)$,

$$-2 \log \mathcal{R}_{L,k}(\theta_0 + \varphi_n) = -2 \log \mathcal{R}(\theta_0 + \varphi_n)(1 + O_p(n^{-k/2})), \quad (12.7)$$

where θ_0 is the true value of θ. The two log likelihood ratios agree within a relative error of $O_p(n^{-k/2})$, over the region where the likelihood is not negligible. For $k \ge 1$, there is an asymptotic χ^2 calibration for $-2 \log \mathcal{R}_{L,k}(\theta_0)$. When $k \ge 2$, the result is qualitatively different from the simple linearization used in the infinitesimal jackknife. When the empirical likelihood is Bartlett correctable, then so is $\mathcal{R}_{L,k}$ for $k \ge 4$. See Chapter 12.7.

Now consider the problem of finding the derivatives $T_i(F)$. When F has an explicit representation in terms of w_i, it may be possible to compute $T_i(F)$ by analytic differentiation. Otherwise, divided differences of the form (12.6) for small $\varepsilon > 0$ may be used. This requires $n+1$ function evaluations per iteration, or fewer if there are ties among the X_i. A central divided difference approximation

$$T_i(F) \doteq \frac{T((1+\varepsilon_i)F - \varepsilon_i \delta_{X_i}) - T((1-\varepsilon_i)F + \varepsilon_i \delta_{X_i})}{2\varepsilon_i}, \quad (12.8)$$

should provide more accurate T_i at the expense of nearly doubling the cost of differentiation. If T is not defined for negative observation weights, then we require $\varepsilon_i < (1 - F(\{X_i\}))^{-1} F(\{X_i\})$ in order to use (12.8).

For some complicated statistics, $T(F)$ is only available for weights $w_i = n_i/N$, where n_i are integers and $N = \sum_{i=1}^{n} n_i$. This corresponds to making an artificial data set in which observation X_i appears n_i times. At least for $k = 0$ it is straightforward to find $T_i(F^{(k)})$. One can define T_i using equation (12.6) with $\varepsilon = -1/n$, or with $\varepsilon = 1/(n+1)$, by deleting observation X_i, or by including it twice, respectively. For $k \ge 1$, the distribution $\hat{F}^{(k)}$ may not correspond to one with an integer n_i copies of X_i, and a large value of N may be required for a good approximation.

Now consider applying linearization to a statistic $\theta \in \mathbb{R}^p$, defined through estimating equations involving a nuisance parameter $\nu \in \mathbb{R}^q$. Let $\theta(w_1, \ldots, w_n)$ and $\nu(w_1, \ldots, w_n)$ be jointly defined by

$$\sum_{i=1}^{n} w_i m(X_i, \theta, \nu) = 0,$$

where $m(X, \theta, \nu)$ is continuously differentiable in (θ, ν). The linearization of θ about $w^k = (w_1^k, \ldots, w_n^k)$ is written

$$\theta(w) \doteq \theta(w^k) + \sum_{i=1}^{n} w_i T_i(w^k),$$

with (12.6) expressed in weight notation as

$$T_j(w^k) = \lim_{\varepsilon \to 0} \frac{T((1-\varepsilon)w^k + \varepsilon \delta_{X_j}) - T(w^k)}{\varepsilon}.$$

Let $w_i^{k,\varepsilon} = (1-\varepsilon)w_i^k + \varepsilon 1_{i=j}$, and let θ^ε and ν^ε be defined through these weights. Let m_θ and m_ν be the $(p+q) \times p$ and $(p+q) \times q$ matrices of partial derivatives of m with respect to θ and ν. Then, ignoring terms of order ε^2 and higher,

$$0 = \sum_{i=1}^{n} w_i^{k,\varepsilon} m(X_i, \theta^\varepsilon, \nu^\varepsilon)$$

$$\doteq \sum_{i=1}^{n} w_i^{k,\varepsilon} \left[m(X_i, \theta, \nu) + \begin{pmatrix} m_\theta(X_i, \theta, \nu) & m_\nu(X_i, \theta, \nu) \end{pmatrix} \begin{pmatrix} \theta^\varepsilon - \theta \\ \nu^\varepsilon - \nu \end{pmatrix} \right]$$

$$\doteq \varepsilon m(X_j, \theta, \nu) + \left[\sum_{i=1}^{n} w_i^k \begin{pmatrix} m_\theta(X_i, \theta, \nu) & m_\nu(X_i, \theta, \nu) \end{pmatrix} \right] \begin{pmatrix} \theta^\varepsilon - \theta \\ \nu^\varepsilon - \nu \end{pmatrix},$$

from which

$$\begin{pmatrix} T_j(w^k) \\ N_j(w^k) \end{pmatrix} = -J(w^k)^{-1} m(X_j, \theta, \nu), \qquad (12.9)$$

where

$$J(w^k) = \sum_{i=1}^{n} w_i^k \begin{pmatrix} m_\theta(X_i, \theta, \nu) & m_\nu(X_i, \theta, \nu) \end{pmatrix} \qquad (12.10)$$

and $N_j(w^k)$ is the linearization term for ν, which we need not compute. Notice that the matrix J in (12.10) does not depend on j. Thus it only needs to be factored once for each set of w_i^k. The matrix is invertible under the assumption that the estimating equations define θ and ν.

For parameters defined through estimating equations with nuisance parameters, sequential linearization can be carried out for some number k of iterations as described above. The linearization (12.9) is recomputed at each iteration.

12.7 Bibliographic notes

Standard references on numerical optimization include Gill et al. (1981) and Fletcher (1987). Gill et al. (1981) include a very comprehensive discussion of constrained optimization. NPSOL is described in Gill, Murray, Saunders & Wright (1999). Coleman (1984) describes large sparse optimization problems, including range space methods.

Owen (1990b) describes how to compute empirical likelihood for a vector mean, corresponding to the inner loop of the nested algorithm. See Chapter 3.14. The nested algorithm was presented in Owen (1990a). and Owen (1990b) Newton's method was proposed by Hall & La Scala (1990). The primal problem was solved using NPSOL by Owen (1991), for regression problems, and by Kolaczyk (1994), for generalized linear models. Computation of profile empirical likelihood ratios for constraints was discussed in Owen (1992). Owen (1990b) reports that iterated linearizations can fail to converge. Wood et al. (1996) consider linearization inferences and were the first to point out the value of stopping after a fixed number of iterations. Equation (12.7) is based on a theorem in Wood et al. (1996).

Semi-infinite programming is a technique for handling optimizations with an infinite number of constraints, only a finite number of which are binding. Lesperance & Kalbfleisch (1992), use semi-infinite programming on a nonparametric likelihood ratio problem arising from mixture sampling.

Owen (1988a) describes an approach to profiling a pair of parameters by following a spiral path along a grid of points, starting at the MLE. Bates & Watts (1988) describe an approach to profiling a sum of squares function in a parameter $(\theta_1, \theta_2) \in \mathbb{R}^2$ using profile traces. The two profile traces intersect a desired contour at four points. The contour is also normal to the profile trace at the intersections. Thus the traces provide eight pieces of information on the desired contour, four locations and four slopes. They use a spline that agrees with these eight pieces of information.

CHAPTER 13

Higher order asymptotics

This chapter covers some theory of higher order asymptotics. The results described here show how to adjust empirical likelihood confidence intervals to have good one-sided coverage errors for one-dimensional parameters, and how to obtain similar effects for multidimensional parameters. Bartlett corrections are also presented, as are some large deviations results.

13.1 Bartlett correction

The profile empirical likelihood ratio statistic commonly has an asymptotic chi-squared distribution: $-2\log \mathcal{R}(\theta_0) \to \chi^2_{(p)}$ in distribution as $n \to \infty$, where p is the dimension of θ. A Bartlett correction is made by replacing the threshold $\chi^{2,1-\alpha}_{(p)}$, by a scale multiple $(1 + an^{-1})\chi^{2,1-\alpha}_{(p)}$. In parametric settings the value a can often be chosen to make the mean of minus twice the log likelihood ratio more nearly equal to p. Surprisingly, one Bartlett correction a improves the asymptotic error rate for all coverage levels $1 - \alpha$.

A Bartlett-corrected empirical likelihood confidence region takes either of the following asymptotically equivalent forms

$$\left\{ \theta \mid -2\log \mathcal{R}(\theta) \leq \left(1 + \frac{a}{n}\right) \chi^{2,1-\alpha}_{(p)} \right\}, \quad \text{or}$$

$$\left\{ \theta \mid -2\log \mathcal{R}(\theta) \leq \left(1 - \frac{a}{n}\right)^{-1} \chi^{2,1-\alpha}_{(p)} \right\}.$$

The appropriate value of a is typically unknown. We will consider $\theta \in \mathbb{R}^p$ obtained as a smooth function $h(\mu)$ of the mean of a random variable $X \in \mathbb{R}^d$, where $d \geq p$. For smooth functions of means, the value a can be expressed in terms of moments of X and derivatives of h. Plugging in sample versions of the moments and evaluating the derivatives at sample moments leads to a $n^{1/2}$ consistent estimator $\hat{a}$. This value can be substituted for a, giving regions such as

$$\left\{ \theta \mid -2\log \mathcal{R}(\theta) \leq \left(1 + \frac{\hat{a}}{n}\right) \chi^{2,1-\alpha}_{(p)} \right\},$$

with the same asymptotic order of coverage accuracy as those based on a.

The proof of Bartlett correctability requires Cramér's condition

$$\limsup_{\|t\|\to\infty} |E(\exp(it'X))| < \infty, \qquad (13.1)$$

which rules out some distributions supported on lattices like the integers. It also requires finiteness of some moments of X, and the existence of sufficiently many derivatives of h. These conditions are used to justify Edgeworth expansions, from which the Bartlett correctability follows. Under these conditions, Bartlett-corrected confidence regions have coverage $1 - \alpha + O(n^{-2})$, instead of $1 - \alpha + O(n^{-1})$, as holds for uncorrected regions.

The Bartlett correction for a univariate mean $\theta = E(X)$ is

$$a = \frac{1}{2}\frac{\mu_4}{\mu_2^2} - \frac{1}{3}\frac{\mu_3^2}{\mu_2^3} = \frac{\kappa+3}{2} - \frac{\gamma^2}{3}$$

where $\mu_k = E((X - E(X))^k)$, for integers $k \geq 2$. The skewness γ and kurtosis κ are defined on page 3. The natural sample estimates use $\hat{\mu}_k = (1/n)\sum_{i=1}^{n}(X_i - \bar{X})^k$. As a point of reference, $a = 1.5$ for univariate normal distributions. For heavy tailed distributions having $\kappa > 0$, a larger Bartlett correction is applied, while nonzero skewness reduces the Bartlett correction. If κ exists, then $\kappa \geq \gamma^2 - 2$, from which we know that $a \geq 1/2 + \gamma^2/6 \geq 1/2$.

A vector mean, $\theta = E(X) \in \mathbb{R}^d$, has $p = d$. Let $Y = V_0^{-1/2}(X - \mu_0)$, where $\mu_0 = \int x dF_0(x)$, and $V_0 = \int (x - \mu_0)(x - \mu_0)' dF_0(x)$ is assumed to be nonsingular. Introduce component superscripts $Y = (Y^1, \ldots, Y^p)$, and define

$$\mu_{jk\ell} = E\left(Y^j Y^k Y^\ell\right), \quad \text{and} \quad \mu_{jk\ell m} = E\left(Y^j Y^k Y^\ell Y^m\right). \quad (13.2)$$

Then the Bartlett correction is given by

$$a = \frac{5}{3}\sum_{j=1}^{p}\sum_{k=1}^{p}\sum_{\ell=1}^{p}(\mu_{jk\ell})^2 - 2\sum_{j=1}^{p}\sum_{k=1}^{p}\sum_{\ell=1}^{p}\mu_{jj\ell}\mu_{kk\ell} + \frac{1}{2}\sum_{j=1}^{p}\sum_{k=1}^{p}\mu_{jjkk}.$$

For a normally distributed vector X, the vector Y has the standard normal distribution with all $\mu_{jkl} = 0$, and $\mu_{jjkk} = 1$, so that $a = p(p-1) + 3p/2 = p^2 + p/2 = d^2 + d/2$.

The Bartlett correction for the mean can be computed in $O(nd^3)$ time. If θ is defined through p estimating equations $E(m(X, \theta)) = 0$, then we may apply the Bartlett correction for the mean of $Z_i = m(X_i, \theta)$ in order to determine whether θ should be in the confidence region.

13.2 Bartlett correction and smooth functions of means

Now consider Bartlett correction for a smooth function of a vector mean. As before, let Y be a standardized version of X, and define the moments μ_{jkl} and μ_{jklm} through (13.2). Suppose that $\theta = h(\mu) = (h_1(\mu), \ldots, h_p(\mu))' \in \mathbb{R}^p$, where $\mu = E(X)$. Now for $i = 1, \ldots, p$, let $\psi^i(\nu) = h_i(V_0^{1/2}\nu)$, and define the partial derivatives

$$\psi^i_{j_1, \ldots, j_r} = \left.\frac{\partial^r \psi^i(\nu)}{\partial \nu^{j_1} \cdots \partial \nu^{j_r}}\right|_{\nu = V_0^{-1/2}\mu_0}.$$

Let Ψ be the $p \times d$ matrix with elements ψ_j^i and define the matrices

$$Q = (\Psi\Psi')^{-1}, \quad M = \Psi'Q\Psi, \quad \text{and} \quad N = \Psi'Q.$$

Then the Bartlett correction is

$$a = \frac{1}{p}\left(\frac{5}{3}a_1 - 2a_2 + \frac{1}{2}a_3 - a_4 + \frac{1}{4}a_5\right),$$

where

$$a_1 = \sum_{j,k,\ell,m,n,o} \mu_{jk\ell}\mu_{mno} M^{jm} M^{kn} M^{\ell o},$$

$$a_2 = \sum_{j,k,\ell,m,n,o} \mu_{jk\ell}\mu_{mno} M^{jk} M^{\ell m} M^{no},$$

$$a_3 = \sum_{j,k,\ell,m} \mu_{jk\ell m} M^{jk} M^{\ell m},$$

$$a_4 = \sum_{j,k,\ell,m,n,u} \mu_{jk\ell} N^{ju} \psi_{mn}^u (I-M)^{mk}(I-M)^{n\ell},$$

$$a_5 = \sum_{\substack{j,k,\ell \\ m,u,v}} Q^{uv}\psi_{jk}^u \psi_{\ell m}^v \left[(I-M)^{jk}(I-M)^{\ell m} + 2(I-M)^{j\ell}(I-M)^{km}\right],$$

indicating the elements of M, N, and Q with superscripts, and summing every index over its entire range.

The moments required by the Bartlett correction for a smooth function of means are dominated by the $\mu_{jk\ell m}$ in the a_3 term. These take $O(nd^4)$ time to compute. The cost of the six-fold summations used in the Bartlett correction does not grow with n, but assuming that $p \leq d$, this cost is $O(d^6)$. For d as large as 10 or 20 this cost could be significant. For the vector mean itself, the cost is of the lower order $O(nd^3)$, suggesting that there may be a practical computational savings to Bartlett correcting estimating equations instead of smooth functions of means.

13.3 Pseudo-likelihood theory

Empirical likelihood provides a data-determined shape for confidence regions. For statistics that are smooth functions of means, this shape is asymptotically ellipsoidal. For a fixed sample size and confidence level, the confidence regions are not exactly ellipsoidal, and in some instances are far from ellipsoidal. For a mean, the confidence regions are elongated in the directions of greater skewness.

The coverage errors in empirical likelihood are typically $O(n^{-1})$, and this rate can also be achieved by ellipsoidal confidence regions. Pseudo-likelihood theory shows that the shape of the empirical likelihood confidence regions is informative. The contours of the profile empirical likelihood ratio for θ are close to contours of the probability density function of an approximate pivotal statistic.

The case $p = 1$ is simplest. Let $\theta = h(\mu)$, where h is a smooth function from $\mathbb{R}^d$ to $\mathbb{R}$ and $\mu = \int x dF(x)$. The true value is $\theta_0 = h(\mu_0)$, estimated by $\hat{\theta} = h(\bar{X})$. Let

$$\hat{\eta}_0 = \frac{\sqrt{n}(\hat{\theta} - \theta_0)}{\hat{\sigma}}, \tag{13.3}$$

where $\hat{\sigma}^2$ is a $n^{1/2}$ consistent estimator of $\sigma^2 = \text{Var}(n^{1/2}(\hat{\theta} - \theta_0))$. Let f_η be the probability density function of $\hat{\eta}_0$. For a properly chosen constant ψ described below, the set

$$C_n^{1-\alpha} = \left\{ \theta + \psi\sigma n^{-1} \mid -2\log\mathcal{R}(\theta) \leq \chi_{(1)}^{2,1-\alpha} \right\}$$

closely matches the set

$$P_\eta^{1-\alpha} = \left\{ \theta \mid f_\eta\left(\frac{\sqrt{n}(\hat{\theta} - \theta)}{\hat{\sigma}}\right) \geq v^{1-\alpha} \right\}$$

where $v^{1-\alpha}$ satisfies

$$\int_{P_\eta^{1-\alpha}} f_\eta(z) dz = 1 - \alpha.$$

Both $C_n^{1-\alpha}$ and $P_\eta^{1-\alpha}$ are asymptotically intervals, and their endpoints differ only by $O(n^{-3/2})$. If one shifts the contours (here endpoints) of the empirical likelihood function by $\psi\sigma n^{-1}$, then to high accuracy they match the contours of the density of $\hat{\eta}_0$.

For $\theta \in \mathbb{R}^p$, with $p \geq 1$, let $\hat{V}$ denote a $n^{1/2}$ consistent estimator of $V = \text{Var}(n^{1/2}(\hat{\theta} - \theta_0))$, and let

$$\hat{\eta}_0 = \sqrt{n}\left(V^{1/2}\hat{V}^{-1}V^{1/2}\right)^{1/2} V^{-1/2}\left(\hat{\theta} - \theta_0\right). \tag{13.4}$$

Definition (13.4) reduces to (13.3) when $p = 1$. Now let

$$C_n^{1-\alpha} = \left\{ \theta + \psi V^{1/2} n^{-1} \mid -2\log\mathcal{R}(\theta) \leq \chi_{(1)}^{2,1-\alpha} \right\},$$

for a properly chosen vector ψ (see Chapter 13.6), and let

$$P_\eta^{1-\alpha} = \left\{ \theta \mid f_\eta\left(\sqrt{n}\left(V^{1/2}\hat{V}^{-1}V^{1/2}\right)^{1/2} V^{-1/2}\left(\hat{\theta} - \theta\right)\right) \geq v^{1-\alpha} \right\},$$

with $v^{1-\alpha}$ chosen as before. Then the boundaries of $C_n^{1-\alpha}$ and $P_\eta^{1-\alpha}$ agree to $O(n^{-3/2})$.

The interpretation is that empirical likelihood contours are very close to those we would construct if we knew the density of $\hat{\eta}_0$. They have the same shape, size, and orientation as the contours of the density of $\hat{\eta}_0$, but they need to be shifted by a bias correction of order $1/n$ to make the match accurate. The term pseudo-

SIGNED ROOT CORRECTIONS

likelihood is used because it is the density of an observed statistic that is being studied, as opposed to the probability or density of the data.

For a multivariate mean, $\psi = (\psi_1, \ldots, \psi_p)'$ where

$$\psi_j = -\frac{1}{2} \sum_{r=1}^{p} \mu_{jrr},$$

and μ_{jrr} is defined at (13.2). For more general location adjustments, see Chapter 13.6.

Pseudo-likelihood arguments suggest shifting the confidence regions for θ by an amount of order $1/n$. The resulting confidence regions can also be Bartlett corrected, although a different Bartlett constant is required than for unshifted confidence regions.

13.4 Signed root corrections

Here we consider parameters defined as smooth function of means: $X \in \mathbb{R}^d$, $\mu = E(X)$, and $\theta = \theta(\mu) \in \mathbb{R}^p$, for $p \leq d$. The random vector X is assumed to satisfy Cramér's condition, to have sufficiently many finite moments, and to have a variance matrix of full rank d. The function $\theta(\mu)$ is assumed to have sufficiently many derivatives, and to have a gradient of full rank p at the true mean μ_0.

For smooth functions of means, the coverage error is typically $O(n^{-1})$ for empirical likelihood confidence regions. For a scalar-valued parameter θ, this $O(n^{-1})$ coverage error arises as two one-sided coverage errors of size $O(n^{-1/2})$ that nearly cancel each other. Bartlett correction reduces the two-sided error to $O(n^{-2})$, but leaves the one-sided error at $O(n^{-1/2})$. In some applications it is desirable to get smaller one-sided coverage errors, even if this means that some excluded parameter values have higher likelihood than some that are included.

This same phenomenon holds in parametric likelihood. Confidence intervals found by thresholding the likelihood function typically have coverage errors of order $O(n^{-1})$ for two-sided inferences but $O(n^{-1/2})$ for one-sided inferences. Bartlett correction improves the two-sided error rate but not the one-sided one. When $p = 1$, one solution in both parametric and empirical likelihoods is to consider the signed root of the log likelihood ratio statistic. For an empirical likelihood, let

$$R_0 = n^{-1/2} \operatorname{sign}(\hat{\theta} - \theta_0) \sqrt{-2 \log \mathcal{R}(\theta_0)},$$

so that the standard $\chi^2_{(1)}$ calibration is based on $n^{1/2} R_0 \overset{\cdot}{\sim} N(0,1)$. Sharper inferences can be obtained by comparing

$$\frac{n^{1/2}(R_0 - \hat{a}/n)}{1 + \hat{b}/(2n)} \tag{13.5}$$

to $N(0,1)$, where $\hat{a}$ and $\hat{b}$ are mean and variance adjustments to the signed root R_0. The adjusted signed root (13.5) has the $N(0,1)$ distribution to $O(n^{-3/2})$, and so it can be used to construct one-sided confidence intervals with coverage error $O(n^{-3/2})$.

For a univariate mean, the signed root correction uses sample analogs of

$$a = -\frac{\gamma}{6}, \quad \text{and} \quad b = \frac{5}{6}\kappa - \frac{31}{36}\gamma^2 + \frac{3}{2},$$

where γ and κ are the skewness and kurtosis of X. Corrections for more general smooth scalar functions of means are known to exist, but formulas for them have not been developed. The proper formula for b is likely to be awkward. It may be possible to use a form of bootstrap calibration to find $\hat{a}$ and $\hat{b}$.

For general $p \geq 1$, the signed root is constructed as a vector $R_0 \in \mathbb{R}^p$ such that

$$R_0'R_0 = -\frac{2}{n}\log \mathcal{R}(\theta_0).$$

The vector $n^{1/2}R_0$ has the $N(0, I_p)$ distribution to order $O(n^{-1/2})$, but after a mean shift the approximation improves to $O(n^{-1})$. Computing R_0 and the mean correction is awkward. A simpler computation with the same order of accuracy is to take the confidence region

$$\left\{\theta \mid -2\log \mathcal{R}(\theta) + \widehat{C}(\theta) \leq \chi^{2,1-\alpha}_{(p)}\right\}, \tag{13.6}$$

where $\widehat{C}$ is described below. The region (13.6) has coverage error $O(n^{-1})$, just as ordinary empirical likelihood does, but when specialized to $p = 1$ it has one-sided errors of order $O(n^{-1})$.

Suppose that the i'th data point is X_i with components X_i^j for $j = 1, \ldots, d$, and that $\theta = \theta(\mu)$ has components θ^u for $u = 1, \ldots, p$. The expression for the adjustment quantity $\widehat{C}$ depends on moments

$$\bar{X}^j = \frac{1}{n}\sum_{i=1}^n X_i^j, \quad \widehat{\Omega}^{jk} = \frac{1}{n}\sum_{i=1}^n (X_i^j - \bar{X}^j)(X_i^k - \bar{X}^k), \quad \text{and}$$

$$\widehat{\alpha}^{jk\ell} = \frac{1}{n}\sum_{i=1}^n \sum_{r,s,t=1}^d (\widehat{\Omega}^{-1/2})^{jr}(\widehat{\Omega}^{-1/2})^{ks}(\widehat{\Omega}^{-1/2})^{\ell t}$$
$$\times (X_i^r - \bar{X}^r)(X_i^s - \bar{X}^s)(X_i^t - \bar{X}^t),$$

where $\widehat{\Omega}$ is the $d \times d$ matrix of $\widehat{\Omega}^{jk}$ values, on the partial derivatives

$$\widehat{\theta}^u_j = \left.\frac{\partial \theta^u(\mu)}{\partial \mu^j}\right|_{\mu=\bar{X}}, \quad \widehat{\theta}^u_{jk} = \left.\frac{\partial^2 \theta^u(\mu)}{\partial \mu^j \partial \mu^k}\right|_{\mu=\bar{X}}$$

and on the related matrices

$$\widehat{\Theta} = (\widehat{\theta}^u_j), \quad \widehat{Q} = (\widehat{\Theta}\widehat{\Omega}\widehat{\Theta}')^{-1},$$
$$\widehat{N} = \widehat{\Omega}^{1/2}\widehat{\Theta}'\widehat{Q}, \quad \widehat{M} = \widehat{N}\widehat{\Theta}\widehat{\Omega}^{1/2}.$$

LARGE DEVIATIONS

The dimension of $\widehat{\Theta}$ is $q \times d$. In terms of these quantities, we may write

$$\widehat{C}(\theta) = \sum_{j,k,\ell,u,v} \left(\frac{1}{3} \widehat{\alpha}^{jk\ell} \widehat{M}^{jk} \widehat{N}^{lu} - \widehat{Q}^{uv} \widehat{\theta}^{v}_{jk}(I - \widehat{M})^{jk} \right) \left(\widehat{\theta} - \theta \right)^{u}. \quad (13.7)$$

Equation (13.7) requires a summation over five variables. Exercise 13.2 shows that for a vector mean, as one might use in conjunction with estimating equations, that the sum is simpler and only over three variables. For a scalar mean $\widehat{C}(\mu_0) = (1/3)(\bar{X} - \mu_0)\hat{\mu}_3/\hat{\mu}_2^2$ for sample second and third central moments $\hat{\mu}_2$ and $\hat{\mu}_3$.

13.5 Large deviations

The power of tests that $\theta = \theta_0 \in \mathbb{R}^p$ is usually investigated for Pitman alternatives of the form $\theta_0 + n^{-1/2}\tau$. This is a critical distance. Alternatives that are $o(n^{-1/2})$ cannot be distinguished from θ_0 in large samples, the power to reject Pitman alternatives usually tends to a limit (depending on τ), and for alternatives that are not $O(n^{-1/2})$ away from the null, the power typically approaches 1 as $n \to \infty$.

The asymptotic power at Pitman alternatives is usually the same for empirical, Euclidean, and the other likelihoods. A sharper distinction between tests can be drawn by considering a fixed alternative $\theta_1 \neq \theta_0$. For such alternatives, tests can be constructed at the level $\alpha_n = \exp(-n\eta)$ with power $1 - \beta_n = 1 - \exp(-n\gamma)$. That is, both type I and type II errors converge to zero exponentially fast in n. Because the errors are so small, they correspond to large deviations of the underlying test statistics.

For multinomial samples $X_1, \ldots, X_n \in \{z_1, \ldots, z_k\}$ the likelihood ratio test is optimal under the generalized Neyman-Pearson (GNP) criterion. Among tests that satisfy $\limsup_{n \to \infty} n^{-1} \log(\alpha_n) \leq -\eta$ for some $\eta > 0$, it minimizes $\limsup_{n \to \infty} n^{-1} \log(\beta_n)$. This property is referred to as universal optimality, because it holds for very general hypotheses and for very general alternatives to F.

For samples from a possibly continuous distribution on $\mathbb{R}^d$, the original argument establishing the GNP for multinomial likelihood breaks down. But other arguments can be used to show a kind of GNP optimality for empirical likelihood based tests of composite hypotheses under sampling from continuous distributions.

Here we quote the large deviations optimality result for empirical likelihood. This discussion leaves out measure theoretic details. For those, see the reference in Chapter 13.6.

Suppose that $X_1, \ldots, X_n \in \mathbb{R}^d$ are independent random vectors with common distribution F. For $\theta \in \Theta \subseteq \mathbb{R}^p$, let $m : \mathbb{R}^d \times \mathbb{R}^p \longrightarrow \mathbb{R}^q$ be an estimating function placing the constraint $E(m(X, \theta)) = \int m(X, \theta) dF(X) = 0$ on F.

For the rest of this section, θ is a freely varying nuisance parameter, and the rest of the hypothesis is specified in m. As a concrete example, to test the hypothesis that $X \in \mathbb{R}$ has variance 7.2, we take

$$m(X, \theta) = \begin{pmatrix} X - \theta \\ (X - \theta)^2 - 7.2 \end{pmatrix} \in \mathbb{R}^2, \quad \theta \in \Theta \subset \mathbb{R}.$$

Let $\mathcal{F}$ be the set of probability distributions on $\mathbb{R}^d$. Let

$$\mathcal{F}_\theta = \left\{ F \in \mathcal{F} \mid \int m(X, \theta) dF(X) = 0 \right\}, \quad \text{and} \quad \mathcal{H}_0 = \bigcup_{\theta \in \Theta} \mathcal{F}_\theta.$$

The empirical likelihood test of the hypothesis $F \in \mathcal{H}_0$, which specifies that $E(m(X, \theta)) = 0$ for some $\theta \in \Theta$, is based on

$$\mathcal{R}(\mathcal{H}_0) = \sup_{\theta \in \Theta} \max \left\{ \prod_{i=1}^n nw_i \mid \sum_{i=1}^n w_i m(X_i, \theta) = 0, w_i \geq 0, \sum_{i=1}^n w_i = 1 \right\}.$$

The empirical likelihood test rejects $\mathcal{H}_0$ if and only if $\mathcal{R}(\mathcal{H}_0)$ is small. Because this test depends on the data only through the empirical distribution function $\hat{F}$, we may write the rejection region itself as a set $\Lambda = \Lambda_n \subset \mathcal{F}$ of distributions. The empirical likelihood test rejects $\mathcal{H}_0$ if and only if $\hat{F} \in \Lambda_n$.

Let $\Omega = \Omega_n \subset \mathcal{F}$ be the rejection region for some other test. We introduce the concept of a Lévy-ball around a set of distributions, to define a regularity condition on Ω. For distributions F and G, let 1 be a vector of d ones. Then the Lévy distance between F and G is

$$\rho_{F,G} = \inf \left\{ \epsilon > 0 \mid F((-\infty, X - \epsilon 1]) - \epsilon \leq G(X) \leq F((-\infty, X + \epsilon 1]) + \epsilon \right\}.$$

Suppose that $X \sim F$ and that $Y = Y(X)$. For $\rho_{F,G}$ to be small, it must hold that $\|Y(X) - X\|$ is small with probability close to 1 under $X \sim F$. We can keep G within a small Lévy distance of F by making large changes $Y(X) - X$ to a small (under F) fraction of the X values, and/or small changes to a large fraction of the X values.

The rejection region Ω_n is a set of empirical distributions. The Lévy ball of radius $\delta > 0$ around Ω is

$$\Omega^\delta = \bigcup_{F \in \Omega} \left\{ G \mid \rho_{F,G} < \delta \right\}.$$

The test sequence Ω_n is regular for $\mathcal{H}_0$ if

$$\lim_{\delta \to 0} \sup_{F \in \mathcal{H}_0} \limsup_{n \to \infty} \frac{1}{n} \log \Pr(\hat{F} \in \Omega_n^\delta; F) \leq \sup_{F \in \mathcal{H}_0} \limsup_{n \to \infty} \frac{1}{n} \log \Pr(\hat{F} \in \Omega_n; F).$$

For example, the test Ω_n that rejects $\mathcal{H}_0$ if and only if $\mathcal{R}(\mathcal{H}_0)$ is a rational number is not regular.

Theorem 13.1 *Suppose that*

    A: *For all $F \in \mathcal{H}_0$, $\Pr(\sup_{\theta \in \Theta} \|m(X, \theta)\| = \infty; F) = 0$,*

    B: *$m(x, \theta)$ is continuous in θ for all x,*

C: *and let* $\Lambda = \{\hat{F} \mid \frac{1}{n}\log \mathcal{R}(\mathcal{H}_0) \leq -\eta\}$ *for some* $\eta > 0$.
Then

$$\sup_{F \in \mathcal{H}_0} \limsup_{n \to \infty} \frac{1}{n} \log \Pr(\hat{F} \in \Lambda; F) \leq -\eta. \tag{13.8}$$

If (13.8) *holds with* Λ *replaced by any regular (for* $\mathcal{H}_0$*) test* Ω*, then for all* $F \in \mathcal{F}$*,*

$$\limsup_{n \to \infty} \frac{1}{n} \log \Pr(\hat{F} \notin \Omega_n; F) \geq \limsup_{n \to \infty} \frac{1}{n} \log \Pr(\hat{F} \notin \Lambda_n; F). \tag{13.9}$$

Equation (13.9) *also holds for a test* Ω*, not necessarily regular for* $\mathcal{H}_0$*, if that test satisfies* (13.8) *with* Λ *replaced by* Ω^δ *for some* $\delta > 0$.

Proof. Kitamura (2001). □

In this theorem, the empirical likelihood test rejects $\mathcal{H}_0$ if and only if $\mathcal{R}(\mathcal{H}_0) \leq \exp(-n\eta)$. Result (13.8) compares the type I error α_n to $\exp(-n\eta)$. Result (13.9) shows that any regular test for $\mathcal{H}_0$ that also meets condition (13.8) has, asymptotically, a type II error at least as large as the empirical likelihood test has. Result (13.9) is universal, applying to any sampling distribution F, any regular test Ω, and very general hypotheses $\mathcal{H}_0$.

Condition A requires a bound in θ on $m(X, \theta)$ to hold with probability 1. Such a bound may require m to be a bounded function, or Θ to be a bounded domain.

13.6 Bibliographic notes

Adjustments

Most of the results on higher order asymptotics presented here were based on Edgeworth expansions for smooth functions of means. Bhattacharya & Ghosh (1978) established the validity of these expansions assuming Cramér's condition and that certain moments are finite.

The first Bartlett correction for empirical likelihood was published by DiCiccio et al. (1991) for smooth functions of means. The material on Bartlett correction for the mean is based on Hall & La Scala (1990).

Zhang (1996b) showed that the Bartlett correction for the univariate mean can be applied for $\theta \in \mathbb{R}$ defined through the estimating function $m(X, \theta) \in \mathbb{R}$. Lazar & Mykland (1999) showed that Bartlett correctability does not hold for empirical profile likelihoods obtained by maximizing over some nuisance parameters. This is the principle way in which empirical likelihood shows different behavior from parametric likelihoods. Mykland (1999) traces the discrepancy to a condition on the fifth moment of the signed square root of the empirical log likelihood.

Chen & Hall (1993) studied Bartlett correction of quantiles, and Chen (1993) established a Bartlett correction for regression with nonrandom predictors. Jing & Wood (1996) show that exponential empirical likelihood (empirical entropy) is not Bartlett correctable. Baggerly (1998) shows that empirical likelihood is the only member of the Cressie-Read family to be Bartlett correctable. Corcoran

(1998) constructed other Bartlett correctable nonparametric likelihoods given by (3.39).

The pseudo-likelihood theory for empirical likelihood was established by Hall (1990). The presentation here also makes use of the account in Hall & La Scala (1990).

Signed root corrections to empirical likelihood were established by DiCiccio & Romano (1989). Signed root corrections for parametric likelihoods are given by DiCiccio (1984) and by Barndorff-Nielsen (1986). The additive correction (13.7) is from DiCiccio & Romano (1989). McCullagh (1984) presents a similar correction for parametric likelihood. DiCiccio & Monti (2001) present adjustments for a scalar parameter based on third and fourth derivatives of the empirical log likelihood.

Large deviations

Dembo & Zeitouni (1998) is a monograph on large deviations. The generalized Neyman-Pearson result for finite multinomials, established by Hoeffding (1965), is of large deviations type. Tusnády (1977) extends Hoeffding's idea to more general distributions, by using finite partitions of the sample space that get finer as n increases. He finds that a likelihood ratio test on the partitions is asymptotically optimal (in Bahadur's sense). No construction is given for the partition. His methods require that the partition have $o(n/\log(n))$ elements.

Kitamura (2001) proves Theorem 13.1. He reports some simulations comparing empirical likelihood and three versions of generalized method of moments: 2-step, 10-step, and continuous updating (Euclidean likelihood). The problem had $n = 200$ observations and 3 parameters. A simulation at the null hypothesis showed that every method gave confidence regions that undercovered the true parameter. After an adjustment to make coverage 95%, the power was compared in simulations that varied the parameters at 8 places along 4 line segments through the null. Of 32 simulations empirical likelihood had the greatest power 22 times, 2-step updating did this 5 times, 10-step updating 7 times, and continuous updating never had the greatest power. There were two cases in which empirical likelihood and 2-step updating tied at power 1.0. As might be expected for a large deviations result like Theorem 13.1, empirical likelihood's power ranking was best at hypotheses farther from the null. Where any of simulated methods achieved power over 80%, empirical likelihood had the greatest power.

13.7 Exercises

Exercise 13.1 The Bartlett constant for a normal distribution is $d^2 + d/2$. An alternative to the scaled F critical value

$$r_F^{1-\alpha} = \frac{d(n-1)}{n-d} F_{d,n-d}^{1-\alpha}$$

EXERCISES

is to employ the normal theory Bartlett correction

$$r_{NB,1}^{1-\alpha} = \left[1 + \frac{d^2 + d/2}{n}\right] \chi_{(d)}^{2,1-\alpha}$$

or

$$r_{NB,2}^{1-\alpha} = \left[1 - \frac{d^2 + d/2}{n}\right]^{-1} \chi_{(d)}^{2,1-\alpha}.$$

As n increases, the Bartlett correction becomes smaller and so empirical likelihood with a normal theory Bartlett correction is asymptotically properly calibrated. For small n, the Bartlett correction appropriate to a normal distribution might be competitive with the F calibration. For $\alpha \in \{0.5, 0.1, 0.05, 0.01\}$ and $d \in \{1, 2, , 5, 10, 20, 50, 100, 200, 500\}$ find the values of n for which the scaled F critical value is larger than the first normal theory Bartlett correction. Find those values of $n > d^2 + d/2$ for which the scaled F critical value is larger than the second normal theory Bartlett correction.

Exercise 13.2 Show that for a vector mean $\theta = \mu$, formula (13.7) simplifies to

$$\widehat{C}(\theta) = \frac{1}{3}\widehat{\alpha}^{jjl} B^{lu} \left(\widehat{\theta} - \theta\right)^u, \qquad (13.10)$$

where $B = S^{-1/2}$ for $S = (1/n)\sum_{i=1}^{n}(X_i - \bar{X})(X_i - \bar{X})'$.

Appendix

This appendix presents some background on parametric likelihood inference and bootstrap methods. Both methods appeal to asymptotic characterizations of errors and other quantities, and so some asymptotic notions are reviewed first.

A.1 Order and stochastic order notation

The notations $O(\cdot)$, $o(\cdot)$, $O_p(\cdot)$ and $o_p(\cdot)$ are used to describe the asymptotic magnitude of statistical quantities.

Consider two sequences of real numbers a_n and b_n where n ranges through integers larger than 1. We say that $a_n = O(b_n)$ if there exists $B < \infty$ with

$$\limsup_{n \to \infty} \frac{|a_n|}{|b_n|} \leq B. \qquad (A.1)$$

Put another way, for some B, there are only finitely many n with $a_n > Bb_n$.

The expression $p_n = q_n + O(b_n)$ means that $p_n - q_n = O(b_n)$. Similarly, when $o(\cdot)$, $O_p(\cdot)$, or $o_p(\cdot)$ as defined below, are added to the right side of an equation, it means that the right side minus the left side is of that order. Here we consider limits indexed by n tending to ∞, but these notions also apply in other limits, such as quantities indexed by $\epsilon \to 0^+$.

For (A.1), it is sufficient to have $\lim_{n \to \infty} |a_n|/|b_n| \leq B$. The more general expression (A.1) also covers cases where $|a_n/b_n|$ is eventually bounded without converging to a limit.

As an example, suppose that $a_n = \gamma^2 n^{-1} + \kappa n^{-2} + n^{-3}$, where γ and κ are finite. Then $a_n = O(n^{-1})$. Here γ and κ are unknowns, such as population skewness and kurtosis, that could vary from one problem instance to another. Suppose that sometimes $\gamma = 0$. Then it is still true that $a_n = O(n^{-1})$, but now it is also true that $a_n = O(n^{-2})$. It is important to remember that a quantity of one order could also be of a smaller order. If $\gamma = \kappa = 0$, then $a_n = O(n^{-3})$, but there are no γ and κ values for which $a_n = O(n^{-4})$.

As a second example, consider $a_n = \mu^2 + n\sigma^2$ where μ and σ are finite. Now $a_n = O(n)$, and when $\sigma = 0$, then $a_n = O(1)$. In this example, a_n is not going to infinity faster than n, if at all, whereas in the previous example a_n is going to zero no slower than n^{-1}.

The notation $a_n = o(b_n)$ means

$$\limsup_{n \to \infty} \frac{|a_n|}{|b_n|} = 0. \tag{A.2}$$

The quantity $\gamma^2 n^{-1} + \kappa n^{-2} + n^{-3}$ is $o(n^{-1/2})$ and also $O(n^{-1})$, but it is not $o(n^{-1})$ unless $\gamma = 0$.

The o_p and O_p notations are used to describe bounds on a quantity whose magnitude is random. Suppose, for example, that Y_n are independent exponential random variables with mean 1. The quantity $Z_n = n^{-1} Y_n$ seems to be tending to zero like $1/n$, but Z_n is not $O(1/n)$. There is no finite B such that quantity $Z_n/(1/n) = Y_n$ is larger than B at most a finite number of times.

We say that $X_n = O_p(Y_n)$ if for any $\epsilon > 0$ there is a $B = B_\epsilon < \infty$ such that

$$\limsup_{n \to \infty} \Pr\left(|X_n| > B|Y_n|\right) < \epsilon. \tag{A.3}$$

Condition (A.3) allows $\Pr(|X_n| > B|Y_n|) \geq \epsilon$ to hold for at most a finite number of n. In the previous example it is indeed true that $Z_n = O_p(1/n)$. The order $O_p(\cdot)$ applies to quantities that are $O(\cdot)$ apart from exceptions that we can make as improbable as we please for large n.

Finally, $X_n = o_p(Y_n)$ if for any $\epsilon > 0$

$$\lim_{n \to \infty} \Pr\left(|X_n| > \epsilon|Y_n|\right) = 0. \tag{A.4}$$

For example, let $X_n = Y_n/n^{1+\delta}$ for $\delta > 0$ and Y_n independent exponential random variables with mean 1. Then $X_n = o_p(n^{-1})$.

A.2 Parametric models

When we make some calculations on data, we usually know that there is some uncertainty in our answer. The data could have come out differently, and then our answer almost certainly would have been different. But we also usually feel that there are reasonable limits to how different the answer might have been.

Probability models are widely used to provide a notion of the true value of a quantity that can be different from the value we have computed. Such models also allow us to quantify the uncertainty in our answers, help us to decide what to compute from the data, and even help us decide how to gather our data.

In a probability model we suppose that our data are the observed values, say $x_1, \ldots, x_n \in \mathbb{R}^d$ of corresponding random variables $X_1, \ldots, X_n \in \mathbb{R}^d$. In a parametric probability model, the joint distribution of $X_1, \ldots, X_n$ takes a known form, involving the data and a parameter vector $\theta \in \Theta \subseteq \mathbb{R}^p$. By contrast non-parametric models do not assume that a finite dimensional parameter indexes the distribution of the data.

Here we will consider only models in which X_i are independent and identically distributed and the sample size n is not random. Then specifying the distribution

PARAMETRIC MODELS

of X_1 specifies the distribution of the whole sample. Parametric models and likelihood methods extend to more general settings.

As an example, if X_i take nonnegative integer values, then one such model, the Poisson distribution, has

$$p(x;\theta) = \Pr(X = x;\theta) = \frac{e^{-\theta}\theta^x}{x!}, \quad x = 0, 1, \ldots \quad (A.5)$$

for $\theta > 0$, while for X_i taking nonnegative real values, the exponential distribution has

$$\Pr(X \le x;\theta) = \begin{cases} 1 - \exp(-x\theta), & 0 \le x < \infty \\ 0, & x < 0, \end{cases}$$

for $\theta > 0$, so the probability density function of X is $f(x;\theta) = \theta\exp(-x\theta)1_{x \ge 0}$.

When we compute a value from the sample, we can usually identify it with a corresponding feature of the distribution of X_i. Perhaps that feature is the value we would get as $n \to \infty$, or perhaps it is the average of the values we would get in a large number of independently generated samples of size n. This feature is necessarily a function of θ and so interest switches to learning θ from $X_1, \ldots, X_n$.

Any function of $X_1, \ldots, X_n$ that we compute is also a random variable with a distribution parameterized by θ. This allows us to quantify the uncertainty of our estimates within the parametric model. We may then seek a method of estimation that minimizes some measure of this uncertainty.

A good model can be a wonderfully effective tool for organizing statistical inference, but choosing a parametric model for applied work is a subtle task. Sometimes a parametric model is thought to be a faithful description of the mechanism generating the data. More often, a model is adopted because it is mathematically convenient and is thought to capture the important features of the problem.

One of the most vexing issues with parametric models is testing whether a given model is compatible with our data. When a goodness of fit test fails to reject our model, it may simply mean that the test was not powerful enough, perhaps because the sample size was too small. Conversely, when a test does reject our model, it may have identified a flaw with a negligible effect on the conclusions we would have drawn using the model. Finally, there are those cases in which a test shows that our model fits badly in a way that affects our answers. Then we may have to seek a correction term, or look for a new model.

One of the best blessings of nonparametric methods is that they reduce the need for goodness of fit testing. There can be a cost in using a nonparametric model for a problem that fits a parametric description. Some nonparametric methods make less efficient use of the data than do parametric methods. When using a nonparametric estimator or test, we should consider its efficiency or power, respectively.

A.3 Likelihood

Given a parametric model and some data, the likelihood function is the probability of getting the observed values from the assumed model, taken as a function of the parameter. The likelihood is thus

$$L(\theta; x_1, \ldots, x_n) = L(\theta) = \prod_{i=1}^{n} p(x_i; \theta)$$

for discrete data and

$$L(\theta; x_1, \ldots, x_n) = L(\theta) = \prod_{i=1}^{n} f(x_i; \theta) dx_i$$

for continuous data, where X_i has been observed to lie in a small set of volume dx_i near the value x_i. We also write $L(\theta; X_1, \ldots, X_n)$ for a random likelihood taking the value above when $X_i = x_i$.

In Bayesian analysis, our model takes θ to be a random variable with a prior distribution π_0, and then the posterior distribution of θ is

$$\pi_1(\theta \mid X_1, \ldots, X_n) \propto L(\theta)\pi_0(\theta).$$

Specialized numerical and sampling techniques are available to compute conclusions from $L(\theta)\pi_0(\theta)$. Just as with parametric probability models for the data, judgment is required to make a good choice of π_0.

A frequentist analysis postulates an unknown true value of θ. When we need to distinguish a generic value from the true value, we denote the latter by θ_0. The method of maximum likelihood estimates θ_0 by finding the value (or sometimes set of values) $\hat{\theta}$ that maximize L:

$$\hat{\theta} = \arg\max_{\theta \in \Theta} L(\theta; X_1, \ldots X_n).$$

A maximizer $\hat{\theta}$ is called a maximum likelihood estimate (MLE) of θ_0. Intuitively, it gives the best explanation of the data we got, by making that data most probable.

Usually it is easier to work with the log likelihood function

$$\ell(\theta) = c + \sum_{i=1}^{n} \log f(x_i; \theta),$$

where c depends on the dx_i but not, we assume, on θ. We will use the notation of the continuous data case; the discrete data case is very similar, starting with $\ell(\theta) = \sum_{i=1}^{n} \log p(x_i; \theta)$. For the most widely applied likelihoods, $\hat{\theta}$ is found by solving the score equation $\partial \ell(\theta)/\partial \theta = 0$, which may be written as solving estimating equations $\sum_{i=1}^{n} m(x_i, \theta) = 0$, where

$$m(x, \theta) = \frac{\frac{\partial}{\partial \theta} f(x; \theta)}{f(x; \theta)}.$$

For the Poisson distribution (A.5), $\hat{\theta}$ equals $\bar{X}$, so the sample mean is used to estimate the population mean. For the double exponential distribution, with density $f(x;\theta) = \exp(-|x-\theta|)/2$, the population mean and median coincide at θ, and we find that the median of X_i is the MLE of θ. Maximum likelihood estimation is also widely used to construct estimates of quantities that we might otherwise not know how to estimate.

The value $\hat{\theta}$ is unlikely to match θ_0 exactly, and furthermore, a value $\tilde{\theta}$ with $L(\tilde{\theta})$ very close to $L(\hat{\theta})$ would seem to be nearly as good as $\hat{\theta}$. We define the likelihood ratio function by $R(\theta) = L(\theta)/L(\hat{\theta})$. One way to separate reasonable from unreasonable values of θ is to order them by $R(\theta)$ and consider $C = \{\theta \mid R(\theta) > r\}$ to fit the data better than other values of θ.

The usual way to pick r is to aim for a given probability that $\theta_0 \in C$. Wilks's theorem has that $-2 \log R(\theta_0) \to \chi^2_{(p)}$ in distribution as $n \to \infty$, where p is the dimension of Θ. This result holds in considerable generality, though there are some exceptions. When it holds, we can use $C^{1-\alpha} = \{\theta \mid R(\theta) \geq r_0\}$ as an approximate $1 - \alpha$ confidence region for θ_0 with $r_0 = \exp(-\chi^{2,1-\alpha}_{(p)}/2)$. A Bartlett correction replaces r_0 by $\exp(-(1+a/n)\chi^{2,1-\alpha}_{(p)}/2)$, where a judiciously chosen scalar a can improve the rate at which $\Pr(\theta_0 \in C^{1-\alpha})$ tends to $1 - \alpha$ as $n \to \infty$.

Computational shortcuts to $C^{1-\alpha}$ are widely used. For large n the log likelihood ratio is very nearly quadratic around $\hat{\theta}$, under regularity conditions. A Taylor approximation gives $\ell(\theta) \doteq \ell(\hat{\theta}) - (1/2)(\theta - \hat{\theta})'\hat{I}(\theta - \hat{\theta})$ near $\hat{\theta}$, where $\hat{I}$ is the Hessian of ℓ at $\hat{\theta}$. This approximation motivates the use of the ellipsoid

$$\left\{ \theta \mid (\theta - \hat{\theta})'\hat{I}(\theta - \hat{\theta}) \leq \chi^{2,1-\alpha}_{(p)} \right\} \quad (A.6)$$

as a confidence region, and under standard assumptions, the set (A.6) has asymptotic probability $1 - \alpha$ of containing θ_0.

Equation (A.6) is one form of the Wald confidence region. The Fisher information in X_1 is defined as

$$I_1(\theta) = \int \left(\frac{\partial}{\partial \theta} \log f(x;\theta)\right)\left(\frac{\partial}{\partial \theta} \log f(x;\theta)\right)' f(x;\theta)dx.$$

As $n \to \infty$ we usually have $n^{-1}\hat{I} \to I_1(\theta_0)$, and another version of the Wald region uses $nI_1(\hat{\theta})$ instead of $\hat{I}$. Another important large sample confidence region due to Rao works with a Taylor approximation around a hypothesized value of θ, not the MLE.

Sometimes we are interested in some, but not all of the parameters. Let us retain θ for the parameters of interest, and suppose that there is an additional parameter ν, possibly a vector, used to complete the definition of the probability of the data. Then the likelihood is $L(\theta, \nu)$ involving θ and the nuisance parameter ν. The MLE of θ is found by maximizing L over θ and ν jointly yielding $\hat{\theta}$ and $\hat{\nu}$.

Likelihood ratios for θ alone are complicated by the presence of ν. If we knew the true value ν_0, then it would be natural to use $R_{\nu_0}(\theta) = L(\theta, \nu_0)/L(\hat{\theta}, \nu_0)$ as a likelihood ratio function. There is no practical difference between a likelihood with a known nuisance parameter and a likelihood without a nuisance parameter. But the idea of a known nuisance parameter suggests the following: we could plug in $\hat\nu$ for ν_0 and use $R_{\hat\nu}(\theta) = L(\theta,\hat\nu)/L(\hat\theta,\hat\nu)$ as a likelihood ratio function for θ. To see the problem with plugging in a value like $\hat\nu$, consider a parameter pair $(\tilde\theta,\tilde\nu)$ with $L(\tilde\theta,\tilde\nu)$ just slightly smaller than $L(\hat\theta,\hat\nu)$, and for which $L(\tilde\theta,\hat\nu) \ll L(\hat\theta,\hat\nu)$. A confidence region based on $R_{\hat\nu}$ might fail to include $\tilde\theta$ even though it belongs to a parameter pair that fits the data nearly as well as the MLE does. The profile likelihood function

$$R_{\mathrm{Pro}}(\theta) = \frac{\max_\eta L(\theta,\nu)}{L(\hat\theta,\hat\nu)},$$

remedies this flaw in $R_{\hat\nu}$, and is the most widely used way to construct a likelihood ratio function for an individual parameter in the presence of nuisance parameters. Letting $\hat\nu(\theta) = \arg\max_\nu L(\theta,\nu)$, we may write the profile likelihood as $R_{\mathrm{Pro}}(\theta) = L(\theta,\hat\nu(\theta))$. A version of Wilks's theorem holds for profile likelihoods.

For well-behaved parametric models, the MLE is a particularly good estimator. Typically there is an information inequality, one form of which shows that any unbiased estimator T of θ has a variance matrix at least as large as the Cramér-Rao bound $(nI_1(\theta))^{-1}$. Under the usual assumptions, $\sqrt{n}(\hat\theta - \theta_0)$ has asymptotic distribution $N(0, (nI_1(\theta))^{-1})$. The MLE $\hat\theta$ typically achieves the lower bound on variance, with an asymptotically negligible bias of order $1/n$.

A.4 The bootstrap idea

The bootstrap is a powerful method for estimating statistical uncertainties. Suppose that we have computed a statistic $\hat\theta = T(F_n)$ for which the corresponding true value is $T(F_0)$. The variance of $\hat\theta$ is yet another property of the unknown distribution F_0. Call it $\mathrm{VT}(F_0)$. If we knew F_0, then we might compute $\mathrm{VT}(F_0)$, by calculus, or by a numerical method such as Monte Carlo simulation.

Of course, if we knew F_0 we might instead use simulations to find $T(F_0)$, so the interesting case is what to do when we do not know F_0. The bootstrap is based on a plug-in principle: make your best guess $\hat F$ for F_0, and plug it in, using $\mathrm{VT}(\hat F)$ as the estimate of $\mathrm{VT}(F_0)$. Very often, the guess we use for F_0 is the ECDF F_n. Because this ECDF is a nonparametric maximum likelihood estimate (NPMLE), we find that the bootstrap estimate of VT is the NPMLE $\widehat{\mathrm{VT}}$.

To compute $\mathrm{VT}(F_n)$ by simulation, we sample with replacement, as follows. Let $J(i,b)$ be independent uniform random draws from the set $\{1, 2, \ldots, n\}$, for $i = 1, \ldots, n$ and $b = 1, \ldots, B$. Then put $X_i^{*b} = X_{J(i,b)}$, let F_n^{*b} be the empirical distribution of $X_1^{*b}, \ldots, X_n^{*b}$, and take $T^{*b} = T(F_n^{*b})$, the statistic T applied to resampled data $X_1^{*b}, \ldots, X_n^{*b}$. Now

$$\mathrm{VT}(F_n) \doteq \frac{1}{B} \sum_{b=1}^{B} (T^{*b} - \bar{T}^*)^2 \qquad (A.7)$$

where $\bar{T}^* = (1/B) \sum_{b=1}^{B} T^{*b}$. This computation may well be easier than that required to find $\mathrm{VT}(F)$ for an arbitrary known distribution F.

It is quite surprising that we can sample our original data over and over and obtain a useful error estimate. The reason that the bootstrap can work is that, for large enough n, F_n becomes close to F_0 and, if VT is continuous near F_0, then $\mathrm{VT}(F_n)$ is close to $\mathrm{VT}(F_0)$. The flatter VT is, as a function of F, the better we could expect the bootstrap to work. As an extreme case, if VT were constant in F we could compute the variance of $\hat{\theta}$ exactly.

The approximation in (A.7) converges to $\mathrm{VT}(F_n)$ as $B \to \infty$, by the law of large numbers, under the very mild assumption that $\mathrm{VT}(F_n)$ is finite. If, under sampling from F_n, the fourth moment of T is finite, then the error in (A.7) is $O_p(B^{-1/2})$. In practice, we might take B large enough that we feel safe that the error in (A.7) is negligible, either absolutely, or relative to $\mathrm{VT}(F_n) - \mathrm{VT}(F_0)$.

A.5 Bootstrap confidence intervals

Consider a scalar statistic $T(F_n) \in \mathbb{R}$ used to estimate $T(F_0)$. A $100(1-\alpha)\%$ confidence interval for T is a pair of random numbers $L = L(F_n)$ and $U = U(F_n)$ such that

$$\Pr\bigl(L(F_n) \le T(F_0) \le U(F_n)\bigr) = 1 - \alpha. \qquad (A.8)$$

Notice that equation (A.8) describes the probability that a random interval $[L, U]$ contains a nonrandom value $T(F_0)$. By contrast, in a prediction interval, a random quantity lies between two fixed endpoints with a given probability.

Equation (A.8) is supposed to hold for all F_0. Exact nonparametric confidence intervals do not exist outside of a few special cases. In particular, they do not exist for the mean. In practice, we use asymptotic confidence intervals with

$$\Pr(L(F_n) \le T(F_0) \le U(F_n)) = 1 - \alpha + o(1) \qquad (A.9)$$

as $n \to \infty$.

In Section A.4 an unknown variance under sampling from F_0 was estimated by the corresponding variance under sampling from F_n. Bootstrap confidence intervals can be constructed by estimating the distribution of an approximately pivotal quantity under F_0 by its distribution under F_n.

Let us write $\mathcal{L}(T(F_n) \mid F_0)$ for the distribution of $T(F_n)$ when $X_1, \ldots, X_n$ have distribution F_0. Similarly let $\mathcal{L}(T(F_n^*) \mid F_n)$ be the common distribution of each T^{*b} drawn on bootstrap samples from F_n.

The percentile method takes the resampled statistics at face value, using the

approximation
$$\mathcal{L}(T(F_n) \mid F_0) \doteq \mathcal{L}(T(F_n^*) \mid F_n).$$
Let us order the resampled T^{*b} values getting $T^{*(1)} \leq T^{*(2)} \leq \cdots \leq T^{*(B)}$. For large B essentially 95% of the resampled values are between $L_{\text{perc}} = T^{*(.025B)}$ and $U_{\text{perc}} = T^{*(.975B)}$, where for noninteger values $0.025B$ and $0.975B$, some rounding or interpolation is applied.

Having seen that
$$\Pr(L_{\text{perc}} \leq T(F_n^*) \leq U_{\text{perc}} \mid F_n) = 0.95 \tag{A.10}$$
we could estimate that
$$\Pr(L_{\text{perc}} \leq T(F_n) \leq U_{\text{perc}} \mid F_0) = 0.95. \tag{A.11}$$
In equation (A.10) we have found by simulation a prediction interval for $T(F_n^*)$ in sampling from $T(F_n)$. Equation (A.11) uses this prediction interval as an estimate of a prediction interval for $T(F_n)$ when sampling from F_0.

Because the endpoints L_{perc} and U_{perc} are computed from the data, they are indeed random. And it turns out that in reasonable generality equation (A.9) holds for L_{perc} and U_{perc}, so that the percentile interval can be used for confidence statements. We return to this point below.

The bias-corrected percentile method is based on the approximation
$$\mathcal{L}(T(F_n) - T(F_0) \mid F_0) \doteq \mathcal{L}(T(F_n^*) - T(F_n) \mid F_n). \tag{A.12}$$
Equation (A.12) describes an unknown quantity $T(F_n) - T(F_0)$ whose distribution we can approximate through simulation of known quantities $T(F_n^*) - T(F_n)$. Quantities containing the unknown, but having a known distribution, are called pivots. Where the distribution is approximately known, the quantities are approximate pivots.

The 0.025 and 0.975 quantiles of $T(F_n^*) - T(F_n)$ are, of course, $T^{*(.025B)} - T(F_n)$ and $T^{*(.975B)} - T(F_n)$. Therefore, the probability is approximately 0.95 that
$$T^{*(.025B)} - T(F_n) \leq T(F_n) - T(F_0) \leq T^{*(.975B)} - T(F_n), \tag{A.13}$$
and rearranging to get $T(F_0)$ in the middle, we get that
$$2T(F_n) - T^{*(.975B)} \leq T(F_0) \leq 2T(F_n) - T^{*(.025B)} \tag{A.14}$$
holds with approximate probability 0.95. The bias-corrected method has endpoints $L_{\text{bc}} = 2T(F_n) - U_{\text{perc}}$ and $U_{\text{bc}} = 2T(F_n) - L_{\text{perc}}$. It is the percentile interval flipped around $T(F_n)$.

We can use the bias-corrected interval to provide an explanation of the percentile interval. It commonly happens that for large n, the approximate pivot $T(F_n) - T(F_0)$ has a nearly symmetric distribution. Then the bias-corrected confidence interval is nearly unchanged by flipping it around $T(F_n)$. The percentile interval was derived using a more general symmetry assumption. Suppose that $\mathcal{L}(\phi(T(F_n)) - \phi(T(F_0)) \mid F_0)$ has a symmetric distribution, for some possibly unknown monotone transformation ϕ. Then the plug-in idea yields the percentile interval.

There are other choices for the pivot. In some settings it may be more natural to use

$$\mathcal{L}\left(\frac{T(F_n)}{T(F_0)} \mid F_0\right) \doteq \mathcal{L}\left(\frac{T(F_n^*)}{T(F_n)} \mid F_n\right) \tag{A.15}$$

or

$$\mathcal{L}\left(\frac{T(F_n) - T(F_0)}{S(F_n)} \mid F_0\right) \doteq \mathcal{L}\left(\frac{T(F_n^*) - T(F_n)}{S(F_n^*)} \mid F_n\right), \tag{A.16}$$

where $S(F_n)$ is an estimate, such as a standard error, of how large $T(F_n) - T(F)$ might be.

Inverting the pivot in equation (A.16) gives the bootstrap-t (or percentile-t) method. The bootstrap-t method provides especially accurate confidence intervals for the univariate mean. In that context $T(F_0) = E(X)$, $T(F_n) = \bar{X}$,

$$S(F_n)^2 = \frac{1}{n}\sum_{i=1}^{n}(X_i - \bar{X})^2, \quad \text{and}$$

$$S(F_n^*)^2 = \frac{1}{n}\sum_{i=1}^{n}(X_i^* - \bar{X}^*)^2.$$

The explanation is that Student's t statistic has a distribution that depends relatively weakly on F, making it a good approximate pivot.

The bootstrap-t interval can be applied in other settings, but the limiting factor is the availability of a good denominator statistic S. If one can afford to bootstrap the bootstrap, then $S(F_n)$ can be estimated as the square root of $\text{VT}(F_n)$ and each $S(F_n^{*b})$ can be estimated as a bootstrap variance $\text{VT}(F_n^{*b})$.

Bootstrap calibration of empirical log likelihood ratios works so well because the distribution of those ratios is only weakly dependent on the underlying data distribution.

A.6 Better bootstrap confidence intervals

A major focus of bootstrap research has been the construction of confidence intervals with one-sided coverage errors that are $O(n^{-1})$. The percentile and bias-corrected percentile methods are typically $O(n^{-1/2})$ accurate on one-sided inferences, though their two-sided inferences give $O(n^{-1})$ errors.

The bootstrap-t intervals are this accurate, but they do not respect transformations. The endpoints of the interval for $\exp(\theta)$ do not equal exponentiated endpoints for the bootstrap-t interval for θ.

The $100(1-\alpha)\%$ bias-corrected accelerated, or BC_a, interval is approximately transformation respecting and has one-sided coverage errors that are $O(n^{-1})$. The derivation is based on a complicated approximately normal pivot. See the references in Section A.7. The BC_a interval has endpoints $L_{\text{bca}} = T^{*(\alpha_1 B)}$ and

$U_{\text{bca}} = T^{*(\alpha_2 B)}$ where

$$\alpha_1 = \Phi\left(\hat{z}_0 + \frac{\hat{z}_0 + z^{\alpha/2}}{1 - \hat{a}(\hat{z}_0 + z^{\alpha/2})}\right) \quad (A.17)$$

$$\alpha_2 = \Phi\left(\hat{z}_0 + \frac{\hat{z}_0 + z^{1-\alpha/2}}{1 - \hat{a}(\hat{z}_0 + z^{1-\alpha/2})}\right), \quad (A.18)$$

wherein $\hat{z}_0$ and $\hat{a}$ are the bias correction and acceleration constant (defined below), Φ is the standard normal cumulative distribution function, and $z^p = \Phi^{-1}(p)$. For a 95% confidence interval $z^{\alpha/2} = -1.96$ and $z^{1-\alpha/2} = 1.96$.

The bias correction factor is

$$\hat{z}_0 = \Phi^{-1}\left(\frac{1}{B}\sum_{b=1}^{B} 1_{T^{*b} < T(F_n)}\right).$$

It vanishes if exactly half of the bootstrap samples were smaller than $T(F_n)$, otherwise it shifts the interval.

To define the acceleration constant, consider data sets $F_{n,-i}$ consisting of the $n-1$ original observations other than observation i. Let $T_{-i} = T(F_{n,-i})$ and $\bar{T}_{-\bullet} = (1/n)\sum_{i=1}^{n} T_{-i}$. Then the acceleration constant is

$$\hat{a} = \frac{\sum_{i=1}^{n}\left(T_{-i} - \bar{T}_{-\bullet}\right)^3}{6\left[\sum_{i=1}^{n}\left(T_{-i} - \bar{T}_{-\bullet}\right)^2\right]^{3/2}},$$

which makes a skewness adjustment.

The ABC method constructs approximate endpoints for the BC_a method. Instead of resampling, it makes a Taylor expansion based on the behavior of $T(F)$ for distributions F that are close to F_n and put all their probability on the sample. For $\epsilon \in (-1/n, 1)$, let $F_{i,\epsilon}$ be the distribution $(1-\epsilon)F_n + \epsilon\delta_{X_i}$. The α confidence limit is

$$T_{\text{ABC}}^{\alpha} = T\left(F_n + \frac{\tilde{z}_\alpha}{(1 - a\tilde{z}_\alpha)^2}\sum_{i=1}^{n} k_i \delta_{X_i}\right), \quad (A.19)$$

as described below. It is the value of T at a specially chosen reweighting of the data. For an asymptotic $100(1-\alpha)\%$ central confidence interval, take $L_{\text{ABC}} = T_{\text{ABC}}^{\alpha/2}$ and $U_{\text{ABC}} = T_{\text{ABC}}^{1-\alpha/2}$. The necessary quantities are defined through

$$l_i = \frac{d}{d\epsilon}T(F_{i,\epsilon})\Big|_{\epsilon=0}, \quad q_{ii} = \frac{d^2}{d\epsilon^2}T(F_{i,\epsilon})\Big|_{\epsilon=0}, \quad v_L = \frac{1}{n^2}\sum_{i=1}^{n} l_i^2,$$

$$a = \frac{1}{6}\frac{\sum_{i=1}^{n} l_i^3}{\left(\sum_{i=1}^{n} l_i^2\right)^{3/2}}, \quad b = \frac{1}{2n^2}\sum_{i=1}^{n} q_{ii}, \quad k_i = n^{-2}v_L^{-1/2} l_i,$$

and

$$\widetilde{z}_\alpha = z_\alpha + a + c - bv_L^{-1/2}, \quad c = \frac{1}{2v_L^{1/2}} \frac{d^2}{d\epsilon^2} T(F_n + \epsilon k)\big|_{\epsilon=0},$$

taking $F_n + \epsilon k$ as a shorthand for $F_n + \epsilon \sum_{i=1}^n k_i \delta_{X_i}$. The derivatives may be computed numerically as divided differences, and z_α is defined through $\Pr(Z \le z_\alpha) = \alpha$ for $Z \sim N(0,1)$. It is possible for $F_n + k\widetilde{z}_\alpha/(1 - a\widetilde{z}_\alpha)^2$ to put negative probability weight on some observations.

Both the BC_a and ABC methods achieve one-sided coverage errors that are $O(n^{-1})$. The formulas are less intuitive than those based on simple pivoting arguments. Similarly, adjustments to empirical or parametric likelihoods to attain $O(n^{-1})$ one-sided coverage errors give rise to more complicated expressions than for the corresponding unadjusted versions.

A.7 Bibliographic notes

The stochastic order notation, O_p and o_p, is due to Mann & Wald (1943).

Cox & Hinkley (1974) and Bickel & Doksum (2000) provide good coverage of the mathematics of parametric likelihood models. In higher order asymptotics, the profile likelihood ratio function does not have all the properties of an ordinary log likelihood function. There has been considerable work on modifying profile likelihoods to be more like likelihoods, starting with Barndorff-Nielsen (1983) and surveyed in Mukerjee & Reid (1999).

Standard references on the bootstrap are Efron & Tibshirani (1993), Hall (1992), and Davison & Hinkley (1997). The first two, as their titles indicate, are introductory and mathematical, respectively. The third has good coverage of computational issues. The discussion of the BC_a method is based on Efron & Tibshirani (1993), while that of the ABC method is based on Davison & Hinkley (1997).

References

Adimari, G. (1995), 'Empirical likelihood confidence intervals for the difference between means (Italian)', *Statistica* **55**, 87–94.

Adimari, G. (1997), 'Empirical likelihood type confidence intervals under random censorship', *Annals of the Institute of Statistical Mathematics* **49**, 447–466.

Akaike, H. (1973), Information theory and an extension of the maximum likelihood priciple, *in* 'Second international symposium on information theory', Academiai Kiado, Budapest, pp. 267–281.

Andersen, P. K., Borgan, O., Gill, R. D. & Keiding, N. (1993), *Statistical Models Based on Counting Processes*, Springer-Verlag, New York.

Anderson, T. W. (1994), *The Statistical Analysis of Time Series (Classics Edition)*, Wiley-Interscience, New York.

Anderson, T. W. & Darling, D. A. (1952), 'Asymptotic theory of certain goodness of fit criteria based on stochastic processes', *Annals of Mathematical Statistics* **23**, 193–212.

Azzalini, A. & Hall, P. (2000), 'Reducing variability using bootstrap methods with qualitative constraints', *Biometrika* **87**(4), 895–906.

Baggerly, K. A. (1998), 'Empirical likelihood as a goodness-of-fit measure', *Biometrika* **85**, 535–547.

Baggerly, K. A. (1999), Studentized empirical likelihood and maximum entropy, Technical report, Rice University, Department of Statistics.

Bahadur, R. R. & Savage, L. J. (1956), 'The nonexistence of certain statistical procedures in nonparametric problems', *Annals of Mathematical Statistics* **27**(4), 1115–1122.

Bailey, K. R. (1984), 'Asymptotic equivalence between the Cox estimator and the general Ml estimators of regression and survival parameters in the Cox model', *The Annals of Statistics* **12**, 730–736.

Banerjee, M. & Wellner, J. A. (2000), Likelihood ratio tests for monotone functions, Technical report, University of Washington, Department of Statistics.

Barndorff-Nielsen, O. E. (1983), 'On a formula for the distribution of the maximum likelihood estimator', *Biometrika* **70**, 343–365.

Barndorff-Nielsen, O. E. (1986), 'Inference on full or partial parameters based on the standardized signed log likelihood ratio', *Biometrika* **73**, 307–322.

Bates, D. M. & Watts, D. G. (1988), *Nonlinear Regression Analysis and Its Applications*, Wiley, New York.

Baum, L. (1972), 'An inequality and associated maximization technique in statistical estimation of probabilistic functions of a Markov process', *Inequalities* **3**, 1–8.

Beran, R. (1987), 'Prepivoting to reduce level error of confidence sets', *Biometrika* **74**, 457–468.

Beran, R. (1988), 'Prepivoting test statistics: A bootstrap view of asymptotic refinements', *Journal of the American Statistical Association* **83**, 687–697.

Berk, R. H. & Jones, D. H. (1978), 'Relatively optimal combinations of test statistics', *Scandinavian Journal of Statistics* **5**, 158–162.

Berk, R. H. & Jones, D. H. (1979), 'Goodness-of-fit test statistics that dominate the Kolmogorov statistics', *Zeitschrift für Wahrscheinlichkeitstheorie und verwandte Gebiete* **47**, 47–59.

Bhattacharya, R. N. & Ghosh, J. K. (1978), 'On the validity of the formal Edgeworth expansion', *The Annals of Statistics* **6**, 434–451.

Bickel, P. J. & Doksum, K. A. (2000), *Mathematical Statistics: Basic Ideas and Selected Topics, Vol I (Second Edition)*, Prentice-Hall, Englewood Cliffs, New Jersey.

Binney, J. & Merrifield, M. (1998), *Galactic Astronomy*, Princeton University Press, Princeton, NJ.

Bloch, D. A., Moses, L. E. & Michel, B. A. (1990), 'Statistical approaches to classification: methods for developing classification and other criteria rules', *Arthritis and Rheumatism* **33**(8), 1137–1144.

Box, G. E. P. (1949), 'A general distribution theory for a class of likelihood criteria', *Biometrika* **36**(3/4), 317–346.

Box, G. E. P., Jenkins, G. M. & Reinsel, G. C. (1994), *Time Series Analysis Forecasting and Control (Third Edition)*, Prentice-Hall, Englewood Cliffs, New Jersey.

Bratley, P., Fox, B. J. & Schrage, L. E. (1987), *A Guide to Simulation (Second Edition)*, Springer-Verlag, New York.

Breiman, L., Friedman, J. H., Olshen, R. A. & Stone, C. J. (1984), *Classification and Regression Trees*, Wadsworth, Belmont, CA.

Brown, B. M. & Chen, S. X. (1998), 'Combined and least squares empirical likelihood', *Annals of the Institute of Statistical Mathematics* **50**, 697–714.

Campbell, J. Y., Lo, A. W. & MacKinlay, A. C. (1996), *The Econometrics of Financial Markets*, Princeton University Press, Princeton, NJ.

Carlstein, E. (1986), 'The use of subseries values for estimating the variance of a general statistic from a stationary sequence', *The Annals of Statistics* **14**, 1171–1179.

Chatfield, C. (1989), *The Analysis of Time Series: An Introduction (Fourth Edition)*, Chapman & Hall, New York.

Chen, J. & Qin, J. (1993), 'Empirical likelihood estimation for finite populations and the effective usage of auxiliary information', *Biometrika* **80**, 107–116.

Chen, J. & Sitter, R. (1999), 'A pseudo-empirical likelihood approach to the effective usage of auxiliary information in complex surveys', *Statistica Sinica* **9**(2), 385–406.

Chen, S. X. (1993), 'On the accuracy of empirical likelihood confidence regions for linear regression model', *Annals of the Institute of Statistical Mathematics* **45**, 621–637.

Chen, S. X. (1994*a*), 'Comparing empirical likelihood and bootstrap hypothesis tests', *Journal of Multivariate Analysis* **51**, 277–293.

Chen, S. X. (1994*b*), 'Empirical likelihood confidence intervals for linear regression coefficients', *Journal of Multivariate Analysis* **49**, 24–40.

Chen, S. X. (1996), 'Empirical likelihood confidence intervals for nonparametric density estimation', *Biometrika* **83**, 329–341.

Chen, S. X. (1997), 'Empirical likelihood-based kernel density estimation', *The Australian Journal of Statistics* **39**, 47–56.

Chen, S. X. & Hall, P. (1993), 'Smoothed empirical likelihood confidence intervals for quantiles', *The Annals of Statistics* **21**, 1166–1181.

Chen, S. X. & Qin, Y. S. (2000), 'Empirical likelihood confidence intervals for local linear smoothers', *Biometrika* **87**(4), 946–953.

Chen, S. X. & Qin, Y. S. (2001), 'Confidence intervals based on a local linear smoother', *Scandinavian Journal of Statistics* **28**, To appear.

Chuang, C.-S. & Chan, N. H. (2001), 'Empirical likelihood for autoregressive models, with applications to unstable time series', *Statistica Sinica* **11**, To appear.

Chuang, C.-S. & Lai, T. L. (2000), 'Hybrid resampling methods for confidence intervals', *Statistica Sinica* **10**(1), 1–33.

Cochran, W. G. (1977), *Sampling Techniques (Third Edition)*, John Wiley & Sons, New York.

Coleman, T. F. (1984), *Large Sparse Numerical Optimization*, Springer-Verlag, Berlin.

Corcoran, S. A. (1998), 'Bartlett adjustment of empirical discrepancy statistics', *Biometrika* **85**, 967–972.

Corcoran, S. A., Davison, A. C. & Spady, R. H. (1995), Reliable inference from empirical likelihoods, Technical report, Oxford University, Department of Statistics.

Cosslett, S. R. (1981), 'Maximum likelihood estimator for choice-based samples', *Econometrica* **49**, 1289–1316.

Cover, T. M. & Thomas, J. A. (1991), *Elements of Information Theory*, Wiley, New York.

Cox, D. R. (1967), Some sampling problems arising in technology, *in* N. L. Johnson & H. Smith Jr., eds, 'New developments in survey sampling', Wiley, New York, pp. 506–527.

Cox, D. R. (1972), 'Regression models and life-tables (with discussion)', *Journal of the Royal Statistical Society, Series B, Methodological* **34**, 187–220.

Cox, D. R. (1975), 'Partial likelihood', *Biometrika* **62**, 269–276.

Cox, D. R. & Hinkley, D. V. (1974), *Theoretical Statistics*, Chapman & Hall, London.

Cox, D. R. & Oakes, D. O. (1984), *Analysis of Survival Data*, Chapman & Hall, London.

Cryer, J. D. (1986), *Time Series Analysis*, Duxbury Press, Boston.

Dahlhaus, R. & Janas, D. (1996), 'A frequency domain bootstrap for ratio statistics in time series analysis', *The Annals of Statistics* **24**, 1934–1963.

David, H. A. (1981), *Order Statistics (Second Edition)*, Wiley, New York.

Davidian, M. & Carroll, R. J. (1987), 'Variance function estimation', *Journal of the American Statistical Association* **82**, 1079–1091.

Davison, A. C. & Hinkley, D. V. (1997), *Bootstrap Methods and Their Application*, Cambridge University Press, Cambridge.

Davison, A. C., Hinkley, D. V. & Worton, B. J. (1992), 'Bootstrap likelihoods', *Biometrika* **79**(1), 113–130.

Davison, A. C., Hinkley, D. V. & Worton, B. J. (1995), 'Accurate and efficient construction of bootstrap likelihoods', *Statistics and Computing* **5**, 257–264.

Dembo, A. & Zeitouni, O. (1998), *Large Deviations Techniques and Applications*, Springer-Verlag, New York.

Dempster, A. P., Laird, N. M. & Rubin, D. B. (1977), 'Maximum likelihood from incomplete data via the EM algorithm', *Journal of the Royal Statistical Society, Series B, Methodological* **39**, 1–22.

DiCiccio, T. J. (1984), 'On parameter transformations and interval estimation', *Biometrika* **71**, 477–485.

DiCiccio, T. J., Hall, P. & Romano, J. (1991), 'Empirical likelihood is Bartlett-correctable', *The Annals of Statistics* **19**, 1053–1061.

DiCiccio, T. J. & Monti, A. C. (2001), Approximations to the profile empirical likelihood function for a scalar parameter in the context of M-estimation, Technical report, Cornell University, Department of Social Statistics.

DiCiccio, T. J. & Romano, J. P. (1989), 'On adjustments based on the signed root of the empirical likelihood ratio statistic', *Biometrika* **76**, 447–456.

DiCiccio, T. J. & Romano, J. P. (1990), 'Nonparametric confidence-limits by resampling methods and least favorable families', *International Statistical Review* **58**(1), 59–76.

Diggle, P. J., Liang, K.-Y. & Zeger, S. L. (1994), *Analysis of Longitudinal Data*, Clarendon Press, Oxford.

Donoho, D. (1982), Breakdown properties of multivariate location estimators, Harvard University, Qualifying paper.

Duffie, J. (1996), *Dynamic Asset Pricing Theory*, Princeton University Press, Princeton, NJ.

Efron, B. (1967), The two sample problem with censored data, *in* 'Proceedings of the Fifth Berkeley Symposium on Mathematical Statistics and Probability', Vol. 5, pp. 831–853.

Efron, B. (1981), 'Nonparametric standard errors and confidence intervals (with discussion)', *The Canadian Journal of Statistics* **9**, 139–172.

Efron, B. (1986), 'Double exponential families and their use in generalized linear regression', *Journal of the American Statistical Association* **81**, 709–721.

Efron, B. (1988), 'Bootstrap confidence intervals: Good or bad?', *Psychological Bulletin* **104**, 293–296.

Efron, B. & Petrosian, V. (1994), 'Survival analysis of the gamma-ray burst data', *Journal of the American Statistical Association* **89**, 452–462.

Efron, B. & Tibshirani, R. (1986), 'Bootstrap methods for standard errors, confidence intervals, and other measures of statistical accuracy', *Statistical Science* **1**, 54–75.

Efron, B. & Tibshirani, R. J. (1993), *An Introduction to the Bootstrap*, Chapman & Hall, New York.

Einmahl, J. H. J. & McKeague, I. W. (1999), 'Confidence tubes for multiple quantile plots via empirical likelihood', *Annals of Statistics* **27**(4), 1348–1367.

El Barmi, H. (1996), 'Empirical likelihood ratio test for or against a set of inequality constraints', *Journal of Statistical Planning and Inference* **55**, 191–204.

El Barmi, H. & Dykstra, R. L. (1994), 'Restricted multinomial maximum likelihood estimation based upon Fenchel duality', *Statistics & Probability Letters* **21**, 121–130.

Embury, S. H., Elias, L., Heller, P. H., Hood, C. E., Greenberg, P. L. & Schrier, S. L. (1977), 'Remission maintenance therapy in acute myelogenous leukemia', *Western Journal of Medicine* **126**, 267–272.

Fan, J. & Gijbels, I. (1996), *Local Polynomial Modelling and Its Applications*, Chapman & Hall, New York.

Fan, J. & Gijbels, I. (1999), Sieve empirical likelihood and extensions of the generalized least squares, Technical Report 9911, Universite Catholique de Louvain, Institute of Statistics.

Fan, J., Zhang, C. & Zhang, J. (2001), 'Generalized likelihood ratio statistics and Wilks phenomenon', *Annals of Statistics* **29**, To appear.

Fan, J. & Zhang, J. (2000), Sieve empirical likelihood ratio tests for nonparametric functions, Manuscript.

Fisher, N. I., Hall, P., Jing, B.-Y. & Wood, A. T. A. (1996), 'Improved pivotal methods for constructing confidence regions with directional data', *Journal of the American Statistical Association* **91**, 1062–1070.

Fleming, T. R. & Harrington, D. P. (1991), *Counting Processes and Survival Analysis*, Wiley,

New York.

Fletcher, R. (1987), *Practical Methods of Optimization (Second Edition)*, Wiley, New York.

Fokianos, K., Peng, A. & Qin, J. (1999), 'A generalized-moments specification test for the logistic link', *Canadian Journal of Statistics* **27**(4), 735–750.

Franke, J. & Härdle, W. (1992), 'On bootstrapping kernel spectral estimates', *The Annals of Statistics* **20**, 121–145.

Friedman, J. H. & Stuetzle, W. (1981), 'Projection pursuit regression', *Journal of the American Statistical Association* **76**, 817–823.

Fritts, H. C., Blasing, T. J., Hayden, B. P. & Kutzba, J. E. (1971), 'Multivariate techniques for specifying tree-growth and climate relationships and for reconstructing anomalies in paleoclimate', *Journal of Applied Meteorology* **10**(5), 845–864.

Gill, P. E., Murray, W., Saunders, M. A. & Wright, M. H. (1999), User's guide for NPSOL 5.0: a FORTRAN package for nonlinear programming, Technical report, Stanford University, Systems Optimization Laboratory.

Gill, P. E., Murray, W. & Wright, M. H. (1981), *Practical Optimization*, Academic Press, London.

Gill, R. D., Vardi, Y. & Wellner, J. A. (1988), 'Large sample theory of empirical distributions in biased sampling models', *The Annals of Statistics* **16**, 1069–1112.

Godambe, V. P. (1960), 'An optimum property of regular maximum likelihood estimation (ack: V32 p1343)', *The Annals of Mathematical Statistics* **31**, 1208–1212.

Godambe, V. P. & Thompson, M. E. (1974), 'Estimating equations in the presence of a nuisance parameter', *The Annals of Statistics* **2**, 568–571.

Greenwood, M. (1926), The errors of sampling of the survivorship tables, *in* 'Reports on public health and statistical subjects', number 33, HMSO, London.

Grenander, U. (1956), 'On the theory of mortality measurement, Part II', *Skand. Aktuar.* **39**, 125–153.

Grenander, U. (1981), *Abstract Inference*, Wiley, New York.

Groeneboom, P. & Wellner, J. A. (1992), *Information Bounds and Nonparametric Maximum Likelihood Estimation*, Birkhäuser, Basel.

Haberman, S. J. (1984), 'Adjustment by minimum discriminant information', *The Annals of Statistics* **12**, 971–988.

Hahn, G. J. & Meeker, W. Q. (1991), *Statistical Intervals. A Guide for Practitioners*, Wiley-Interscience, New York.

Hall, P. (1986), 'On the bootstrap and confidence intervals', *The Annals of Statistics* **14**, 1431–1452.

Hall, P. (1987), 'On the bootstrap and likelihood-based confidence regions', *Biometrika* **74**, 481–493.

Hall, P. (1990), 'Pseudo-likelihood theory for empirical likelihood', *The Annals of Statistics* **18**, 121–140.

Hall, P. (1992), *The Bootstrap and Edgeworth Expansion*, Springer-Verlag, New York.

Hall, P. & La Scala, B. (1990), 'Methodology and algorithms of empirical likelihood', *International Statistical Review* **58**, 109–127.

Hall, P. & LePage, R. (1996), 'On bootstrap estimation of the distribution of the studentized mean', *Annals of the Institute of Statistical Mathematics* **48**(3), 403–421.

Hall, P. & Martin, M. A. (1988), 'On bootstrap resampling and iteration', *Biometrika* **75**, 661–671.

Hall, P. & Owen, A. B. (1993), 'Empirical likelihood confidence bands in density estimation', *Journal of Computational and Graphical Statistics* **2**, 273–289.

Hall, P. & Presnell, B. (1999*a*), 'Biased bootstrap methods for reducing the effects of contamination', *Journal of the Royal Statistical Society Series B-Statistical Methodology* **61**(3), 661–680.

Hall, P. & Presnell, B. (1999*b*), 'Intentionally biased bootstrap methods', *Journal of the Royal Statistical Society, Series B, Methodological* **61**, 143–158.

Hansen, L. P. (1982), 'Large sample properties of generalized method of moments estimators', *Econometrica* **50**, 1029–1054.

Hansen, L. P., Heaton, J. & Yaron, A. (1996), 'Finite-sample properties of some alternative GMM estimators (pkg: P261-373)', *Journal of Business and Economic Statistics* **14**, 262–280.

Härdle, W. (1990), *Applied Nonparametric Regression*, Cambridge University Press, Cambridge.

Härdle, W., Hall, P. & Marron, J. S. (1988), 'How far are automatically chosen regression smoothing parameters from their optimum? (c/r: P96-101)', *Journal of the American Statistical Association* **83**, 86–95.

Härdle, W. & Marron, J. S. (1985), 'Optimal bandwidth selection in nonparametric regression function estimation', *The Annals of Statistics* **13**, 1465–1481.

Hartley, H. O. & Rao, J. N. K. (1968), 'A new estimation theory for sample surveys', *Biometrika* **55**, 547–557.

Hasminskii, R. Z. & Ibragimov, I. A. (1993), On asymptotic efficiency in the presence of an infinite dimensional nuisance parameter, *in* 'Lecture notes in mathematics, number 1021', Springer, New York, pp. 195–229.

Hastie, T. J. & Tibshirani, R. J. (1990), *Generalized Additive Models*, Chapman & Hall, London.

Heitjan, D. F. & Rubin, D. B. (1991), 'Ignorability and coarse data', *The Annals of Statistics* **19**, 2244–2253.

Hesterberg, T. (1995*a*), 'Tail-specific linear approximations for efficient bootstrap simulations', *Journal of Computational and Graphical Statistics* **4**, 113–133.

Hesterberg, T. (1995*b*), 'Weighted average importance sampling and defensive mixture distributions', *Technometrics* **37**(2), 185–194.

Hesterberg, T. (1999), Bootstrap tilting confidence intervals, Technical Report 84, MathSoft Inc., Seattle, WA.

Hipel, K. W. & McLeod, A. I. (1994), *Time Series Modelling of Water Resources and Environmental Systems*, Elsevier/North-Holland, Amsterdam.

Hoeffding, W. (1965), 'Asymptotically optimal tests for multinomial distributions', *The Annals of Mathematical Statistics* **36**, 369–400.

Hoff, P. D. (2000), 'Constrained nonparametric maximum likelihood via mixtures', *The Journal of Computational and Graphical Statistics*.

Hoffleit, D. & Warren, W. (1991), *The Bright Star Catalogue (Fifth Edition)*, Yale University Observatory, New Haven, CT.

Hollander, M., McKeague, I. W. & Yang, J. (1997), 'Likelihood ratio-based confidence bands for survival functions', *Journal of the American Statistical Association* **92**(437), 215–226.

Huber, P. J. (1967), The behavior of maximum likelihood estimates under nonstandard conditions, *in* 'Proceedings of the Fifth Berkeley Symposium on Mathematical Statistics and Probability', Vol. 1, pp. 221–233.

Huber, P. J. (1981), *Robust Statistics*, Wiley, New York.

Hull, J. C. (2000), *Options, Futures, and Other Derivatives*, Prentice-Hall, Englewood Cliffs, New Jersey.

Iles, T. C. (1993), Multiple regression, *in* J. C. Fry, ed., 'Biological Data Analysis. A Practical Approach', Oxford University Press, Oxford, pp. 127–172.

Imbens, G. W. & Lancaster, T. (1994), 'Combining micro and macro data in microeconometric models', *The Review of Economic Studies* **61**(4), 655–680.

Imbens, G. W., Spady, R. H. & Johnson, P. (1998), 'Information theoretic approaches to inference in moment condition models', *Econometrica* **66**(2), 333–357.

Jackson, R., Yee, R. L., Priest, P., Shaw, L. & Beaglehole, R. (1995), 'Trends in coronary heart-disease risk-factors in Auckland 1982–94', *New Zealand Medical Journal* **108**(1011), 451–454.

Jaeckel, L. (1972), The infinitesimal jackknife, Technical Report Memorandum MM72-1215-11, Bell Labs, Murray Hill, NJ.

James, G. (1951), 'The comparison of several groups of observations when the ratios of the population variances are unknown', *Biometrika* **38**, 324–329.

Janas, D. (1994), 'Edgeworth expansions for spectral mean estimates with applications to Whittle estimates', *Annals of the Institute of Statistical Mathematics* **46**, 667–682.

Jing, B.-Y. (1995*a*), 'Some resampling procedures under symmetry', *The Australian Journal of Statistics* **37**, 337–344.

Jing, B.-Y. (1995*b*), 'Two-sample empirical likelihood method', *Statistics & Probability Letters* **24**, 315–319.

Jing, B.-Y., Feuerverger, A. & Robinson, J. (1994), 'On the bootstrap saddlepoint approximations', *Biometrika* **81**, 211–215.

Jing, B.-Y. & Wood, A. T. A. (1996), 'Exponential empirical likelihood is not Bartlett correctable', *The Annals of Statistics* **24**, 365–369.

Johansen, S. (1978), 'The product limit estimator as maximum likelihood estimator', *Scandinavian Journal of Statistics* **5**, 195–199.

Jones, M. C. (1991), 'Kernel density estimation for length biased data', *Biometrika* **78**, 511–519.

Jorgensen, B. (1987), 'Exponential dispersion models', *Journal of the Royal Statistical Society, Series B, Methodological* **49**(2), 127–162.

Kalbfleisch, J. D. & Prentice, R. L. (1980), *The Statistical Analysis of Failure Time Data*, Wiley, New York.

Kaplan, E. L. & Meier, P. (1958), 'Nonparametric estimation from incomplete observations', *Journal of the American Statistical Association* **53**, 457–481.

Kauermann, G. & Carroll, R. (2000), The sandwich variance estimator: efficiency properties and coverage probability of confidence intervals, Technical Report 189, University of Munich, Department of Statistics.

Keiding, N. & Gill, R. D. (1990), 'Random truncation models and Markov processes', *The Annals of Statistics* **18**, 582–602.

Kiefer, J. & Wolfowitz, J. (1956), 'Consistency of the maximum likelihood estimator in the presence of infinitely many incidental parameters', *Annals of Mathematical Statistics* **27**, 887–906.

Kitamura, Y. (1997), 'Empirical likelihood methods with weakly dependent processes', *The Annals of Statistics* **25**, 2084–2102.

Kitamura, Y. (1999), Empirical likelihood and the bootstrap for time series regressions,

Technical report, University of Wisconsin, Department of Economics.

Kitamura, Y. (2001), 'Asymptotic optimality of empirical likelihood for testing moment restrictions', *Econometrica* **69**, To appear.

Kitamura, Y., Tripathi, G. & Ahn, H. (2000), Empirical likelihood-based inference in conditional moment restriction models, Technical report, University of Wisconsin, Department of Economics.

Kohavi, R., Brodley, C. E., Frasca, B., Mason, L. & Zheng, Z. (2001), 'Kdd-cup 2000 organizers' report: Peeling the onion', *SIGKDD Explorations* **2**(2), To appear.

Kolaczyk, E. D. (1994), 'Empirical likelihood for generalized linear models', *Statistica Sinica* **4**, 199–218.

Kolaczyk, E. D. (1995), An information criterion for empirical likelihood with general estimating equations, Technical Report 417, University of Chicago, Department of Statistics.

Künsch, H. R. (1989), 'The jackknife and the bootstrap for general stationary observations', *The Annals of Statistics* **17**, 1217–1241.

La Rocca, M. (1995*a*), Empirical likelihood and linear combinations of functions of order statistics, Technical Report Working paper 3.46, Universita Degli Studi di Salerno, Dipartimento di Scienze Economiche.

La Rocca, M. (1995*b*), L'uso del bootstrap nella verosimigliana empirica, Technical Report Working paper 3.47, Universita Degli Studi di Salerno, Dipartimento di Scienze Economiche.

La Rocca, M. (1996), L'uso della verosimiglianza empirica per il confronto di due parametri di posizione, Technical Report Working paper 3.49, Universita Degli Studi di Salerno, Dipartimento di Scienze Economiche.

La Rocca, M. (1998), Bootstrappling empirical likelihood for linear regression models, *in* 'NTTS'98 International Seminar on New Techniques & Technologies for Statistics', pp. 277–282.

Lazar, N. A. (2000), Bayesian empirical likelihood, Technical report, Carnegie Mellon University, Department of Statistics.

Lazar, N. A. & Mykland, P. A. (1999), 'Empirical likelihood in the presence of nuisance parameters', *Biometrika* **86**(1), 203–211.

Lazar, N. & Mykland, P. A. (1998), 'An evaluation of the power and conditionality properties of empirical likelihood', *Biometrika* **85**, 523–534.

LeBlanc, M. & Crowley, J. (1995), 'Semiparametric regression functionals', *Journal of the American Statistical Association* **90**(429), 95–105.

Lee, S. M. S. & Young, G. A. (1999), 'Nonparametric likelihood ratio confidence intervals', *Biometrika* **86**(1), 107–118.

Lesperance, M. L. & Kalbfleisch, J. D. (1992), 'An algorithm for computing the nonparametric MLE of a mixing distribution', *Journal of the American Statistical Association* **87**, 120–126.

Li, G. (1995*a*), 'Nonparametric likelihood ratio estimation of probabilities for truncated data', *Journal of the American Statistical Association* **90**, 997–1003.

Li, G. (1995*b*), 'On nonparametric likelihood ratio estimation of survival probabilities for censored-data', *Statistics & Probability Letters* **25**(2), 95–104.

Li, G., Hollander, M., McKeague, I. W. & Yang, J. (1996), 'Nonparametric likelihood ratio confidence bands for quantile functions from incomplete survival data', *Annals of Statistics* **24**(2), 628–640.

Li, G., Qin, J. & Tiwari, R. C. (1997), 'Semiparametric likelihood ratio-based inferences for truncated data', *Journal of the American Statistical Association* **92**, 236–245.

Liang, K.-Y. & Zeger, S. L. (1986), 'Longitudinal data analysis using generalized linear models', *Biometrika* **73**, 13–22.

Lindsay, B. G. (1980), 'Nuisance parameters, mixture-models, and the efficiency of partial likelihood estimators', *Philosophical Transactions of the Royal Society of London, Series A–Mathematical, Physical and Engineering Sciences* **296**(1427), 639–662.

Lindsay, B. G. (1995), *Mixture Models: Theory, Geometry and Applications*, Institute of Mathematical Statistics.

Liu, R. Y. & Singh, K. (1992), Moving blocks jackknife and bootstrap capture weak dependence, *in* 'Exploring the Limits of Bootstrap', pp. 225–248.

Loader, C. (1999), *Local Regression and Likelihood*, Springer-Verlag.

Loh, W. L. (1996), 'On Latin hypercube sampling', *Annals of Statistics* **24**(5), 2058–2080.

Loh, W.-Y. (1991), 'Bootstrap calibration for confidence interval construction and selection', *Statistica Sinica* **1**, 477–491.

Lohr, S. (1998), *Sampling: Design and Analysis*, Brooks/Cole Publishing Company, Pacific Grove, CA.

Lynden-Bell, D. (1971), 'A method for allowing known observational selection in small samples applied to 3CR quasars', *Monthly Notices of the Royal Astronomical Society* **155**, 95–118.

Mann, H. B. & Wald, A. (1943), 'On stochastic limit and order relationships', *Annals of Mathematical Statistics* **14**(3), 217–226.

McCullagh, P. (1984), 'Local sufficiency', *Biometrika* **71**, 233–244.

McCullagh, P. & Nelder, J. A. (1983), *Generalized Linear Models*, Chapman & Hall.

Miller, R. G., Gong, G. & Munoz, A. (1981), *Survival Analysis*, Wiley, New York.

Mittelhammer, R. C., Judge, G. G. & Miller, D. J. (2000), *Econometric Foundations*, Cambridge University Press, Cambridge.

Moeschberger, M. L. & Klein, J. P. (1985), 'A comparison of several methods of estimating the survival function when there is extreme right censoring', *Biometrics* **41**, 253–259.

Monahan, J. F. & Boos, D. D. (1992), 'Proper likelihoods for Bayesian analysis', *Biometrika* **79**, 271–278.

Monti, A. C. (1997), 'Empirical likelihood confidence regions in time series models', *Biometrika* **84**, 395–405.

Monti, A. C. & Ronchetti, E. (1993), 'On the relationship between empirical likelihood and empirical saddlepoint approximation for multivariate M-estimators', *Biometrika* **80**, 329–338.

Mukerjee, R. & Reid, N. (1999), 'On confidence intervals associated with the usual and adjusted likelihoods', *Journal of the Royal Statistical Society, Series B, Methodological* **61**(4), 945–953.

Murphy, S. A. (1995), 'Likelihood ratio-based confidence-intervals in survival analysis', *Journal of the American Statistical Association* **90**(432), 1399–1405.

Murphy, S. A. & van der Vaart, A. W. (1997), 'Semiparametric likelihood ratio inference', *The Annals of Statistics* **25**, 1471–1509.

Murphy, S. A. & van der Vaart, A. W. (2000), 'On profile likelihood', *Journal of the American Statistical Association* **95**(450), 449–465.

Muttlak, H. A. & McDonald, L. L. (1990), 'Ranked set sampling with size-biased probability of selection', *Biometrics* **46**, 435–445.

Mykland, P. A. (1995), 'Dual likelihood', *Annals of Statistics* **23**(2), 396–421.

Mykland, P. A. (1999), 'Bartlett identities and large deviations in likelihood theory', *Annals of Statistics* **27**(3), 1105–1117.

Nadaraya, E. A. (1965), 'On non-parametric estimates of density functions and regression curves', *Theory of Probability and its Applications* (Transl *of Teorija Verojatnostei i ee Primenenija)* **10**, 186–190.

Naik-Nimbalkar, U. V. & Rajarshi, M. B. (1997), 'Empirical likelihood ratio test for equality of k medians in censored data', *Statistics & Probability Letters* **34**, 267–273.

Nelder, J. A. & Pregibon, D. (1987), 'An extended quasi-likelihood function', *Biometrika* **74**, 221–232.

Nelder, J. A. & Wedderburn, R. W. M. (1972), 'Generalized linear models', *Journal of the Royal Statistical Society, Series A, General* **135**, 370–384.

Newton, M. A. & Raftery, A. E. (1994), 'Approximate Bayesian inference with the weighted likelihood bootstrap (disc: P26-48)', *Journal of the Royal Statistical Society, Series B, Methodological* **56**, 3–26.

Neyman, J. & Scott, E. (1948), 'Consistent estimates based on partially consistent observations', *Econometrica* **16**, 1–32.

Noé, M. (1972), 'The calculations of distributions of two-sided Kolmogorov-Smirnov type statistics', *The Annals of Mathematical Statistics* **43**, 58–64.

Owen, A. B. (1985), Nonparametric likelihood ratio intervals, Technical Report LCS-6, Stanford University, Department of Statistics.

Owen, A. B. (1987), Nonparametric conditional estimation, PhD thesis, Stanford University.

Owen, A. B. (1988*a*), Computing empirical likelihoods, *in* 'Computer Science and Statistics: Proceedings of the 20th Symposium on the Interface', pp. 442–447.

Owen, A. B. (1988*b*), 'Empirical likelihood ratio confidence intervals for a single functional', *Biometrika* **75**, 237–249.

Owen, A. B. (1990*a*), Empirical likelihood and small samples, *in* 'Computing Science and Statistics: Proceedings of the Symposium on the Interface', Springer-Verlag, Berlin, pp. 79–88.

Owen, A. B. (1990*b*), 'Empirical likelihood ratio confidence regions', *The Annals of Statistics* **18**, 90–120.

Owen, A. B. (1991), 'Empirical likelihood for linear models', *The Annals of Statistics* **19**, 1725–1747.

Owen, A. B. (1992), Empirical likelihood and generalized projection pursuit, Technical Report 393, Stanford University, Department of Statistics.

Owen, A. B. (1995), 'Nonparametric likelihood confidence bands for a distribution function', *Journal of the American Statistical Association* **90**, 516–521.

Pan, X.-R. & Zhou, M. (2000), Empirical likelihood ratio in terms of cumulative hazard function for censored data, Technical report, University of Kentucky, Department of Statistics.

Pawitan, Y. (2000), 'Computing empirical likelihood from the bootstrap', *Statistics & Probability Letters* **47**(4), 337–345.

Pearl, R. & Fuller, W. (1905), 'Variation and correlation in the earthworm', *Biometrika* **4**, 213–229.

Peto, R. (1973), 'Experimental survival curves for interval-censored data', *Applied Statistics* **22**, 86–91.

Politis, D. N. & Romano, J. P. (1992), 'A general resampling scheme for triangular arrays of

α-mixing random variables with application to the problem of spectral density estimation', *The Annals of Statistics* **20**, 1985–2007.

Politis, D. N. & Romano, J. P. (1994), 'The stationary bootstrap', *Journal of the American Statistical Association* **89**, 1303–1313.

Politis, D. N., Romano, J. P. & Wolf, M. (1999), *Subsampling*, Springer.

Press, W., Flannery, B., Teukolsky, S. & Vetterling, W. (1993), *Numerical Recipes in C: The Art of Scientific Computing*, Cambridge University Press.

Qin, J. (1992), Empirical likelihood and semiparametric models, PhD thesis, University of Waterloo.

Qin, J. (1993), 'Empirical likelihood in biased sample problems', *The Annals of Statistics* **21**, 1182–1196.

Qin, J. (1994), 'Semi-empirical likelihood ratio confidence intervals for the difference of two sample means', *Annals of the Institute of Statistical Mathematics* **46**, 117–126.

Qin, J. (1998), 'Semiparametric likelihood based method for goodness of fit tests and estimation in upgraded mixture models', *Scandinavian Journal of Statistics* **25**(4), 681–691.

Qin, J. (1999), 'Empirical likelihood ratio based confidence intervals for mixture proportions', *Annals of Statistics* **27**(4), 1368–1384.

Qin, J. (2000), 'Combining parametric and empirical likelihoods', *Biometrika* **87**, 484–490.

Qin, J. & Lawless, J. (1994), 'Empirical likelihood and general estimating equations', *The Annals of Statistics* **22**, 300–325.

Qin, J. & Wong, A. (1996), 'Empirical likelihood in a semi-parametric model', *Scandinavian Journal of Statistics* **23**(2), 209–219.

Qin, J. & Zhang, B. (1997), 'A goodness-of-fit test for logistic regression models based on case-control data', *Biometrika* **84**, 609–618.

Qin, Y. & Zhao, L. (1997), 'Empirical likelihood ratio statistics for the difference of two population quantiles (Chinese)', *Chinese Annals of Mathematics A (Chinese)* **18**, 687–694.

Qu, Z. (1995), Maximum empirical likelihood for symmetric linear models, Technical report, Memorial University of Newfoundland, Department of Mathematics and Statistics.

Quenouille, M. H. (1949), 'Approximate tests of correlation in time-series', *Journal of the Royal Statistical Society, Series B, Methodological* **11**, 68–84.

Ramos, E. (1989), Improved estimators for time series, *in* 'ASA Proceedings of the Statistical Computing Section', pp. 233–237.

Rawlings, J. O. (1988), *Applied Regression Analysis: A Research Tool*, Wadsworth.

Read, T. R. C. & Cressie, N. A. C. (1988), *Goodness-of-fit Statistics for Discrete Multivariate Data*, Springer-Verlag, New York.

Reid, N. (1988), 'Saddlepoint methods and statistical inference', *Statistical Science* **3**, 213–227.

Ren, J.-J. (2001), 'Weighted empirical likelihood ratio confidence intervals for the mean with censored data', *Annals of the Institute of Statistical Mathematics* **53**, To appear.

Rényi, A. (1961), On measures of entropy and information, *in* 'Proceedings of the Fourth Berkeley Symposium on Mathematical Statistics and Probability', Vol. 1, pp. 547–561.

Rice, J. A. (1988), *Mathematical Statistics and Data Analysis*, Wadsworth.

Romano, J. P. (1988), 'A bootstrap revival of some nonparametric distance tests', *Journal of the American Statistical Association* **83**, 698–708.

Rubin, D. B. (1981), 'The Bayesian bootstrap', *The Annals of Statistics* **9**, 130–134.

Schenker, N. (1985), 'Qualms about bootstrap confidence intervals', *Journal of the Ameri-*

can Statistical Association **80**, 360–361.

Seber, G. A. F. & Wild, C. J. (1989), *Nonlinear Regression*, Wiley, New York.

Shao, J. & Tu, D. (1995), *The Jackknife and Bootstrap*, Springer-Verlag.

Sheehy, A. (1987), Kullback-Leibler estimation of probability measures with an application to clustering, PhD thesis, University of Washington.

Shen, X. T., Shi, J. & Wong, W. H. (1999), 'Random sieve likelihood and general regression models', *Journal of the American Statistical Association* **94**(447), 835–846.

Shorack, G. R. & Wellner, J. A. (1986), *Empirical Processes With Applications to Statistics*, Wiley, New York.

Sitter, R. R. & Wu, C. (2000), Efficient estimation of quadratic finite population functions in the presence of auxiliary information, Technical Report 2000-09, University of Waterloo, Department of Statistics and Actuarial Science.

Stahel, W. A. (1981), Robuste Schätzungen: Infinitesimals Optimalität und Schätzungen von Kovarianzmatrizen, PhD thesis, Technical University, Graz, Austria.

Stapleton, J. H. (1995), *Linear Statistical Models*, Wiley, New York.

Stein, C. (1956), Efficient nonparametric testing and estimation, *in* 'Proceedings of the Third Berkeley Symposium on Mathematical Statistics and Probability', Vol. 1, pp. 187–195.

Stern, H. (1991), 'On the probability of winning a football game', *The American Statistician* **45**, 179–183.

Stigler, S. M. (1977), 'Do robust estimators work with real data? (c/r: P1078-1098)', *The Annals of Statistics* **5**, 1055–1077.

Stone, C. J. (1977), 'Consistent nonparametric regression (c/r: P620-645)', *The Annals of Statistics* **5**, 595–620.

Switzer, P. (1976), 'Geometrical measures of the smoothness of random functions', *Journal of Applied Probability* **13**, 86–95.

Therneau, T. M. & Grambsch, P. M. (2000), *Modeling Survival Data*, Springer-Verlag, New York.

Thomas, D. R. & Grunkemeier, G. L. (1975), 'Confidence interval estimation of survival probabilities for censored data', *Journal of the American Statistical Association* **70**, 865–871.

Tsai, W.-Y., Leurgans, S. & Crowley, J. (1986), 'Nonparametric estimation of a bivariate survival function in the presence of censoring', *The Annals of Statistics* **14**, 1351–1365.

Tsao, M. (2001), 'A small sample calibration method for the empirical likelihood ratio', *Statistics & Probability Letters*.

Tsao, M. & Zhou, J. (2001), 'On the robustness of empirical likelihood confidence intervals for location', *The Canadian Journal of Statistics* **28**, To appear.

Tsiatis, A. A. (1981), 'A large sample study of Cox's regression model', *The Annals of Statistics* **9**, 93–108.

Tukey, J. W. (1958), 'Bias and confidence in not quite large samples (abstract)', *Annals of Mathematical Statistics* **29**, 614.

Turnbull, B. W. (1976), 'The empirical distribution function with arbitrarily grouped, censored and truncated data', *Journal of the Royal Statistical Society, Series B, Methodological* **38**, 290–295.

Tusnády, G. (1977), 'On asymptotically optimal tests', *The Annals of Statistics* **5**, 385–393.

van der Laan, M. J. (1995), *Efficient and Inefficient Estimation in Semiparametric Models*, Centrum voor Wiskunde en Informatica, Math. Centrum.

van der Laan, M. J. (1996), 'Efficient estimation in the bivariate censoring model and repair-

REFERENCES

ing NPMLE', *The Annals of Statistics* **24**, 596–627.

van der Vaart, A. W. & Wellner, J. A. (1992), 'Existence and consistency of maximum likelihood in upgrade mixture models', *Journal of Multivariate Analysis* **43**, 133–146.

Vardi, Y. (1982), 'Nonparametric estimation in the presence of length bias', *The Annals of Statistics* **10**, 616–620.

Vardi, Y. (1985), 'Empirical distributions in selection bias models', *The Annals of Statistics* **13**, 178–203.

Vardi, Y. & Zhang, C.-H. (1992), 'Large sample study of empirical distributions in a random-multiplicative censoring model', *The Annals of Statistics* **20**, 1022–1039.

Wang, M.-C. (1987), 'Product limit estimates: A generalized maximum likelihood study', *Communications in Statistics, Part A – Theory and Methods* **16**, 3117–3132.

Watson, G. S. (1964), 'Smooth regression analysis', *Sankhya, Series A* **26**, 59–372.

Wedderburn, R. W. M. (1974), 'Quasi-likelihood functions, generalized linear models, and the Gauss-Newton method', *Biometrika* **61**, 439–447.

White, H. (1980), 'A heteroskedasticity-consistent covariance matrix estimator and a direct test for heteroskedasticity', *Econometrica* **48**, 817–838.

Whittle, P. (1953), 'Estimation and information in stationary time series', *Arkiv für Mathematik* **2**, 423–434.

Wood, A. T. A., Do, K.-A. & Broom, B. M. (1996), 'Sequential linearization of empirical likelihood constraints with application to U-statistics', *Journal of Computational and Graphical Statistics* **5**, 365–385.

Woodroofe, M. (1985), 'Estimating a distribution function with truncated data', *The Annals of Statistics* **13**, 163–177.

Wu, C. & Sitter, R. R. (2001), 'A model-calibration approach to using complete auxiliary information from survey data', *Journal of the American Statistical Association* **96**, To appear.

Yevjevich, V. M. (1963), Fluctuation of wet and dry years, 1, research data assembly and mathematical models, Technical report, Colorado State University, Hydrology Paper No. 1.

Yule, G. U. (1927), 'On a method of investigating periodicities in disturbed series, with special reference to Wolfer's sunspot numbers', *Philosophical Transactions of the Royal Society* **A226**, 267–298.

Zeger, S. L. & Liang, K.-Y. (1986), 'Longitudinal data analysis for discrete and continuous outcomes', *Biometrics* **42**, 121–130.

Zhang, B. (1996a), 'Confidence intervals for a distribution function in the presence of auxiliary information', *Computational Statistics and Data Analysis* **21**, 327–342.

Zhang, B. (1996b), 'On the accuracy of empirical likelihood confidence intervals for M-functionals', *Journal of Nonparametric Statistics* **6**, 311–321.

Zhang, B. (1997), 'Quantile processes in the presence of auxiliary information', *Annals of the Institute of Statistical Mathematics* **49**(1), 35–55.

Zhang, B. (1998), 'A note on kernel density estimation with auxiliary information', *Communications in Statistics, Part A – Theory and Methods* **27**, 1–11.

Zhang, B. (1999), 'Bootstrapping with auxiliary information', *Canadian Journal of Statistics* **27**(2), 237–249.

Zhong, B., Chen, J. & Rao, J. N. K. (2001), 'Empirical likelihood inference in the presence of measurement error', *The Canadian Journal of Statistics* **29**(1), To appear.

Zhong, B. & Rao, J. N. K. (2000), 'Empirical likelihood inference under stratified random sampling using auxiliary population information', *Biometrika* **87**(4), 929–938.

Zhou, M. (2000), Empirical envelope mle for a location problem, Technical Report 381, University of Kentucky, Department of Statistics.

Author index

Adimari, G., 107, 151
Ahn, H., 195, 198
Akaike, H., 108
Andersen, P.K., 153
Anderson, T.W., 160, 180
Azzalini, A, 214

Baggerly, K. A., 73, 74, 214, 257
Bahadur, R.R., 26, 28
Bailey, K.R., 152
Banerjee, M., 214
Barndorff-Nielsen, O.E., 258, 271
Bates, D.M., 108, 247
Baum, L., 151
Beaglehole, R., 111, 124
Beran, R., 70
Berk, R.H., 160
Bhattacharya, R.N., 257
Bickel, P.J., 271
Binney, J., 25
Blasing, T.J., 171, 181
Bloch, D.A., xv, 98, 108
Boos, D.D., 196
Borgan, O., 153
Box, G.E.P., 70, 180, 182
Bratley, P., 149
Breiman, L., 214
Brodley, C.E., 150
Broom, B.M., 70, 247
Brown, B.M., 213

Campbell, Y.J., 74
Carlstein, E., 181
Carroll, R., 27, 107
Chan, N.H., 181
Chatfield, C., 180

Chen, J., xv, 177, 179, 182, 183
Chen, S.X., 71, 72, 107, 124, 125, 213, 215, 257
Chuang, C.-S., 181, 197
Cochran, W.G., 149
Coleman, T.F., 247
Corcoran, S.A., 74, 198, 258
Cosslett, S.R., 150
Cover, T.M., 73
Cox, D.R., xv, 71, 147, 149, 151, 152, 271
Cressie, N.A.C., 73
Crowley, J., 151, 198
Cryer, J.D., 180

Dahlhaus, R., 182
Darling, D.A., 160
David, H.A., 72
Davidian, M., 107
Davison, A.C., 74, 108, 183, 197, 198, 271
Dembo, A., 258
Dempster, A., 151
DiCiccio, T.J., xiv, 70, 72, 197, 257, 258
Diggle, P.J., 183
Do, K.-A., 70, 247
Doksum, K.A., 271
Donoho, D.L., 6
Duffie, J.D., 74
Dykstra, R.L., 214

Eagleson, G.K., 72
Efron, B., 6, 27, 108, 151, 152, 181, 197, 198, 271
Einmahl, J.H.J., 161

El Barmi, H., 214
Elias, L., 144, 152
Embury, S.H., 144, 152

Fan, J., 124, 125, 195, 198
Feuerverger, A., 214
Fisher, N.I., 71
Flannery, B., 26
Fleming, T.R., 151, 153
Fletcher, R., 74, 247
Fokianos, K., 150
Fox, B.J., 149
Franke, J, 182
Frasca, B., 150
Friedman, J.H., 214
Fritts, H.C., 171, 181
Fuller, W., 2, 6

Ghosh, J.K., 257
Gijbels, I., 124, 125, 195, 198
Gill, P.E., xv, 74, 247
Gill, R.D., xv, 134, 139, 150, 152, 153
Glenn, N., xv
Godambe, V.P., 71
Gong, G., 152
Grambsch, P.M., 153
Greenberg, P.L., 144, 152
Greenwood, M., 151
Grenander, U., 6, 25, 214
Groeneboom, P.J., 214
Grunkemeier, G.L., xiv, 6, 26, 145, 151

Haberman, S.J., 72
Hahn, G.J., 72, 202
Hall, P., xiv, 6, 27, 70–73, 124, 197, 214, 215, 247, 257, 258, 271
Hansen, L.P., 72
Härdle, W., 124, 182
Harrington, D.F., 151, 153
Hartley, H.O., 6, 25, 182
Hasminskii, R.Z., 150

Hastie, T.J., 124
Hayden, B.P., 171, 181
Heaton, J., 72
Heitjen, D.F., 151
Heller, P.H., 144, 152
Hesterberg, T., 183, 197, 198
Hickernell, F., xv
Hinkley, D.V., xv, 71, 74, 108, 183, 197, 198, 271
Hipel, K.W., 181, 182
Hoeffding, W., 20, 27, 258
Hoff, P.D., 214
Hoffleit, D., 8, 25
Hollander, M., 160, 161
Hood, C.E., 144, 152
Huber, P.J., 71, 72, 74
Hull, J.C., 74

Ibragimov I.A., 150
Iles, T.C., 31, 74
Imbens, G.W., 27, 196

Jackson, R., 111, 124
Jaeckel, L., 198
James, G., 108
Janas, D., 182
Jenkins, G.M., 180, 182
Jennings, K., xv
Jing, B.-Y., 71, 107, 214, 257
Johansen, S., 25
Johnson, P., 27
Jones, D.H., 160
Jones, M.C., 149
Jorgensen, B., 108
Judge, G.G., xv, 73

Kalbfleisch, J.G., xiv, 151, 247
Kaplan, E.L., 25, 151
Kauermann, G., 27
Keiding, N, 139, 152, 153
Kiefer, J., 25
Kitamura, Y., xv, 27, 72, 170, 181, 182, 195, 198, 257, 258

Klein, J.P., 196
Kohavi, R., 150
Kolaczyk, E.D., 107, 108, 247
Künsch, H.R., 181
Kustra, R., 150
Kutzba, J.E., 171, 181

La Rocca, M., 70, 107, 213, 215
La Scala, B., 70, 247, 257, 258
Lai, T.-L., 197
Laird, N.M., 151
Lancaster, T., 196
Lawless, J.F., xiv, 55, 71, 72, 107, 160, 181
Lazar, N.A., 71, 196, 257
LeBlanc, M., 198
Lee, K.-A., xv
Lee, S.M.S., 198
LePage, R., 71
Lesperance, M., 247
Leurgans, S., 151
Li, G., xv, 145, 148, 151, 152, 161
Liang, K.-Y., 74, 183
Liao, L.Z., xv
Lindsay, B.G., 71, 214
Liu, R.Y., 181
Lo, A.W., 74
Loader, C., 124
Loh, W.-L., 183
Loh, W.-Y., 70
Lohr, S., 149
Lynden-Bell, D., 152

MacKinlay, A.C., 74
Mallows, C., 73
Mann, H.B., 271
Marron, S.J., 124
Martin, M.A., 71
Mason, L., 150
McCullagh, P., 108, 258
McDonald, L.L., 129, 149
McKeague, I.W., 160, 161
McLeod, A.I., 181, 182

Meeker, W.Q., 72, 202
Meier, P., 25, 151
Merrifield, M, 25
Michel, B.A., 98, 108
Miller, D.J., 73
Miller, R.G., xiv, 152
Mittelhammer, R.C., 73
Moeschberger, M.L., 196
Monahan, J.F., 196
Monti, A.C., 182, 198, 258
Moses, L.E., 98, 108
Mukerjee, R., 271
Munoz, A., 152
Murphy, S.A., xv, 145, 148, 149, 151–153
Murray, W., xv, 74, 247
Muttlak, H.A., 129, 149
Mykland, P.A., xiv, 71, 153, 181, 257

Nadaraya, E.A., 124
Naik-Nimbalkar, U.V., 212, 215
Narasimhan, B., xv
Nelder, J.A, 108
Newton, M.A., 197
Neyman, J., 26, 64
Noé, M., 157, 158, 160

Oakes, D., 151
Olkin, I., xv
Olshen, R.A., 214
Owen, A.B., 6, 26, 27, 70, 72, 73, 107, 124, 159, 160, 226, 247
Owen, G., 74

Pan, X.-R., 149, 151
Pawitan, Y, 197
Pearl, R., 2, 6
Peng, A., 150
Peto, R., 140, 150
Petrosian, V., 152
Picazo, J., 150
Politis, D.N., 180–182
Popescu, B., 150

Pregibon, D., 108
Prentice, R.L., xiv, 151
Presnell, B., 73, 197
Press, W., 26
Priest, P., 111, 124

Qin, J., xiv, 55, 71, 72, 107, 150, 152, 160, 177, 181, 182, 196, 226, 227
Qin, Y.S., 125, 214
Qu, Z., 72, 214
Quenouille, M.H., 198

Raftery, A.E., 197
Rajarshi, M.B., 212, 215
Ramos, E., 182
Rao, J.N.K., 6, 25, 182, 183
Rawlings, J.O., 95, 108
Read, T.R.C., 73
Reid, N., 198, 271
Reinsel, G.C., 180, 182
Ren, J.-J., 197
Renyi, A., 73
Rice, J.A., 80
Robinson, J., 214
Rokicki, T.G., xv
Romano, J.P., xiv, 70, 72, 180–182, 197, 214, 257, 258
Ronchetti, E., 198
Rubin, D.B., 6, 151, 197

Saunders, M.A., xv, 247
Savage, L.J., 26, 28
Schenker, N., 27
Schrage, L.E., 149
Schrier, S.L., 144, 152
Scott, E., 26
Seber, G.A.F., 108
Shao, J, 198
Shaw, L., 111, 124
Sheehy, A., 72
Shen, X.T., 198
Shi, J, 198
Shorack, G.R., 160

Singh, K., 181
Sitter, R.R., 179, 183
Spady, R.H., 27, 74, 198
Stahel, W.A., 6
Stapleton, J.H., 104, 108
Stein, C., 197
Stern, H., xv, 205
Stigler, S.M., 16, 17, 26, 58, 73
Stone, C.J., 124, 214
Stuetzle, W., 214
Switzer, P., 161

Teukolsky, S., 26
Therneau, T.M., xv, 153
Thomas, D.R., xiv, 6, 26, 145, 151
Thomas, J.A., 73
Thompson, M.E., 71
Tibshirani, R.J., xv, 124, 181, 271
Tiwari, R.C., 152
Tripathi, G., 195, 198
Tsai, W.-Y., 151
Tsao, M., 70, 73
Tsiatis, A.A., 152
Tu, D., 198
Tukey, J.W., 5, 198
Turnbull, B.W., 141, 151
Tusnády, G., 258

van der Laan, M.J., 141, 151
van der Vaart, A.W., xv, 148–150, 152, 153
Vardi, Y., 133, 134, 150
Vetterling, W., 26

Wald, A., 271
Wang, M.-C., 139, 152
Warren, W., 8, 25
Watson, G.S., 124
Watts, D.G., 108, 247
Wedderburn, R.W.M., 108
Wellner, J.A., 134, 150, 160, 214
White, H., 64, 74, 108
Whittle, P., 182

Wild, C.J., 108
Wolf, M., 180, 181
Wolfowitz, J, 25
Wong, A., 196
Wong, W.H., 198
Wood, A.T.A., 70, 71, 247, 257
Woodroofe, M., 152
Worton, B.J., 197
Wright, M.H., 74, 247
Wu, C., 183

Yang, J., 160, 161
Yaron, A., 72
Yee, R.L., 111, 124
Yee, T., xv
Yevjevich, V.M., 167, 181
Young, G.A., 198
Yule, G.U., 180

Zeger, S.L., 74, 183
Zeitouni, O., 258
Zhang, B., 70, 124, 150, 160, 161, 197, 257
Zhang, C., 198
Zhang, C.-H., 150
Zhang, J., 198
Zhao, L., 214
Zheng, Z., 150
Zhong, B., 183
Zhou, J., 73
Zhou, M., 149, 151, 214

Subject index

Aaron, Hank, *see* examples, home runs
acceptance sampling, 128, 133
additive model, 121–123, 198
Akaike's information criterion, 108
AML, *see* examples, leukemia remission
ancillarity, 85
Anderson-Darling statistic, 160
ANOVA, 87–90, 107
  decomposition, 225, 226
  Euclidean likelihood, 106–107
  unbalanced, 110
AR model, *see* time series, autoregressive
ARMA model, *see* time series, ARMA
arteritis, *see* examples, giant cell arteritis
astronomy, 136, 142, 152
  3CR quasars, 152
  bright star catalogue, 8, 25
  cosmological principle, 142, 152
  Lynden-Bell estimator, 143, 152, 153
autoregression, *see* time series, autoregressive
auxiliary information, *see* side information

Babe Ruth, *see* examples, home runs
back forecasting, 182
bandwidth, 111, 117, 118, 121
  selection, 124
  smoothed quantiles, 72
Bartlett correction, 5, 19, 31, 32, 38, 70, 73, 74, 249–251, 253, 257, 259
  ANOVA, 107

bootstrap, 19, 34
density estimate, 124
empirical likelihood-t, 210
estimated, 32, 72, 249
estimating equations, 257
jackknife, 19
linearized EL, 245
nuisance parameter, 257
other likelihoods, 214, 257
parametric likelihood, 253
pseudo-likelihood, 253
quantiles, 72, 257
regression, 107
smooth function of mean, 250
spectral method, 182
time series, 181
univariate mean, 250
vector mean, 250
Baseball-Reference.com, 104, 108
Bayes methods, 182, 188–189, 196, 197, 264
biased sampling, 5, 127–135, 139, 191
  length bias, 127, 128, 149
  multi-sample, 130–135, 150
  parameterized, 150
bisection, 22
bivariate survival, 150
blockwise bootstrap, *see* bootstrap, blockwise
blood pressure data, *see* examples, New Zealand blood pressure
bootstrap, 1, 4, 5, 27, 34, 108, 152, 163, 183, 214, 261, 266–271
  b-bootstrap, 197
  Bayesian, 188–189, 197

blocks of blocks, 182
blockwise, 169, 181
bootstrap-t, 210
calibration, *see* calibration, bootstrap
confidence interval, 18, 267–271
  ABC, 270
  BC_a, 269
  bias-corrected, 268
  bootstrap-t, 269
  percentile, 267
confidence region, 5
coverage errors, 18
from NPMLE, 6, 191–192, 197
iterated, 71, 269
leveraged, 197
likelihood, 191, 197
m of n, 71
nonparametric tilting, 189–190, 197–198
points illustrated, 193
time series, 181, 182
variance, 266, 269
bootstrap likelihood, *see* likelihood, bootstrap
Borel-Cantelli lemma, 218
boundedness
  estimating function, 15, 59, 60, 78, 207
  random variable, 13, 16, 26, 72
breast cancer, *see* examples, breast cancer mortality
Brent's method, 22, 26
bright star catalogue, 8, 25
bristlecone pine, *see* examples, bristlecone pine tree rings
butcher knife, 198

calcium uptake, *see* examples, calcium uptake
calibration
  bootstrap, 6, 31–34, 58, 70, 71, 103, 107, 114, 119, 120, 124, 254

  Bartlett, 34
  chi-bar-squared, 211, 214
  chisquared, *see* ELT
  F distribution, 30–34, 70
  finite population, 175, 179
  m of n bootstrap, 71
  noncentral chisquared, 20
  nonsmooth parameter, 55, 146, 201, 207
  parametric likelihood, 70
  permutation, 209
  simultaneous, 157
  time series, 167, 170
Campito Mountain, *see* examples, bristlecone pine tree rings
CART, 213
case-control study, 150
Cavendish data, *see* examples, Earth's density
censoring, 25, 26, 127, 135–138, 196
  bivariate, 140, 150
  counting process, 153
  double, 137
  interval, 137, 150, 197
  NPMLE for, 139
  right, 141–142
    EL, 143–145
    NPMLE, 151
  sieved likelihood, 141
  ties, 147
central limit theorem, 144, 175, 220, 223, 226
chi-bar squared, 211, 213, 214
choice-based sampling, 131, 150
Cinclus cinclus, *see* examples, Dipper survey data
classification, 98
clipping, 15, 59
cluster sampling, *see* finite population
coarsening at random, 139, 151, 199
  definition, 136
competing risks, 153
concomitants, 143

conditional likelihood, *see* likelihood, conditional
confidence band, 114, 119
  distribution function, 155–160, 197
  exact, 155–157
  kernel smooths, 124
  Kolmogorov-Smirnov, 156
  quantile function, 161, 182, 197
confidence ellipsoid
  infinitesimal jackknife, 244
  jackknife, 244
  normal theory, *see* Hotelling's T^2
confidence interval
  asymptotic, 18
  bootstrap, *see* bootstrap, confidence interval
  computation, 21–23
  degenerate, 13, 46, 48, 201
  efficiency, 20
  empirical likelihood, 7, 10, 14, 15, 24, 26
  exact, 17, 20, 26, 28, 184, 267
  finite population, 175
  inconsistency, 1
  invariance, 50
  kernel, 113
  length, 1, 14, 20, 24, 39, 184
    infinite, 13
    jacknife, 199
    outliers, 56
    simulation, 21, 27, 184
    simulations, 27
  mean, 16, 27
  non-existence, 26
  nonparametric, 1
  normal theory, 5, 39
  parametric, 18, 19, 24, 253
  quantile, 43–46, 72
  sample size, and, 83
  skewness, 184
  standard, 18, 19
  survival function, 26, 144
  tail probability, 45
  time series, 184

confidence region, 6
  Bartlett-corrected, 249
  bootstrap, 5
  bootstrap calibrated, 6
  center of symmetry, 204
  conservative, 204
  contour, 247
  convexity, 6, 30, 67, 95
  Cressie-Read, *see* Cressie-Read divergence
  degenerate, 41
  diameter, 15
  ellipsoid, *see* confidence ellipsoid
  empirical likelihood, 10, 11, 19
  estimating equations, 41
  improved, 253, 254
  linearized, 37, 93
  mean, 14, 30
  nonlinear least squares, 93, 94
  proportional hazards, 152
  regression, 81, 85, 107
  shape, 5, 19, 71, 251, 252
  smooth curve, 116
  smooth function of means, 253
  statistical function, 12
  time series, 173, 181, 182
  vs. convex hull, 209
  with ties, 11, 12
confidence sandwich, 119
confidence tube, 119, 120, 123
  QQ, 161
contamination, 60, 73
control variate, 51
convex dual, 60–63, 74
convex hull, 30, 60, 62, 63, 65, 73, 85, 86, 177
  condition, 85–87, 89, 115, 209–210, 219, 222
  Dipper survey, 34
  mean outside, 58
cosmological principle, *see* astronomy
counting process, 153
coverage, 17–19, 45, 209

asymptotic, 18, 27
Bartlett-corrected, *see* Bartlett correction
bootstrap, 27
central, 70
ellipsoids, 251
exact, 17, 26, 156, 158, 184
fixed regressors and lack of fit, 84
inconsistent, 5, 136, 163
jackknife, 18
multidimensional, 19
nonlinear least squares, 94
one-sided, *see* one-sided confidence interval
parametric, 19
quantiles, 44
   smoothed, 72
regression prediction, 106
simulations, 16, 27, 70, 107, 182, 214
simultaneous, 114, 118–120, 156, 159
smooth functions of means, 253
survival function, 144
time series, 173, 181, 182
vs. power, 20
cow data, *see* examples, milk production
Cramér's condition, 70, 249, 253, 257
Cramér-Rao bound, 266
Cressie-Read divergence, 66, 67, 73
   Bartlett correction, 257
   chisquared, 67
   contours, 68
   convex hull condition, 209
   ties, 78
   weights, 73, 78
curse of dimensionality, 113
curvature
   constrained empirical log likelihood, 52
   empirical log likelihood, 20, 239
   Euclidean log likelihood, 64
   intrinsic, 93, 94

log likelihood, 103
parameter effects, 93, 94
parametric log likelihood, 239
partial log likelihood, 152
curve estimation, 111

data fusion, 131
data mining, 150, 174
density estimation, *see* smoothing, kernel, density
Dippers, *see* examples, Dipper survey data
directional data, 71
Dirichlet prior, 182
distribution function, 72, 182
dominated convergence theorem, 217
double censoring, *see* censoring, double
double exponential family, 108
dual likelihood, *see* likelihood, dual

Earth's density, *see* examples, Earth's density
earthworms, *see* examples, earthworms
econometrics, 41, 64, 72, 73, 196
Edgeworth expansion, 250, 257
effective sample size, 83, 114
efficiency, 1, 20–21
   block size, 173
   least squares, 90
   moving average model, 173
   overlapping blocks, 181
   regression through origin, 91
   robust estimate, 56, 57
   sampling, 177
eigenvalue, 85, 87, 89, 115, 220
ELT, 11
   ANOVA, 88, 89
   blockwise, 181
   CDF, 157
   censoring and truncation, 151
   constrained mean, 51
   cumulative hazard, 151

SUBJECT INDEX

estimating equation, 41
hybrid likelihood, 186, 188, 196
infinite variance, 71
kernel estimates, 72, 115
lemmas, 217
linear model, 86
linearization, 37, 245
MANOVA, 88, 90
martingale, 153
multi-population, 208
multi-sample, 223–226
multiple biased samples, 135, 150
nuisance parameters, 42, 55
other likelihoods, 66, 67, 73
proofs, 219–226
smooth estimating equations, 54, 55, 71, 106
survival function, 145, 151
time series, 174
triangular array, 79, 83, 85–87, 89, 115
  proof, 222–223
univariate, 16, 26, 43
  bounded, 26
  proof, 219
vector, 30, 31, 52, 70, 219
  proof, 219
empirical distribution, 2, 3, 7, 29
empirical entropy, 67, 68, 73, 74, 198, 257
empirical information criterion, 108
empirical likelihood-t, 65, 209, 210, 214, 229
  simulations, 214
estimating equations, 39–42, 71–72
  generalized, 74
  Huber's, 56
  likelihood-based, 48–50
  overdetermined, 54
  robust, 59
  side information, 53
estimating function, 40
Euclidean likelihood, *see* likelihood, Euclidean

even part, 203
exact
  confidence band, 155–157
  nonexistence of confidence intervals, 26
region
  nonlinear least squares, 94
examples
  breast cancer mortality, 79, 81–83, 90–91, 109
  bristlecone pine tree rings, 170–172
  calcium uptake, 94–95, 108
  dipper survey, 30, 67
  Earth's density, 16–17
  earthworms, 2–5
  electronic device temperatures, 202
  football pointspreads, 204–207
  giant cell arteritis, 97–101, 108
  home runs, 101–103
  leukemia remission, 143–145
  milk production, 44–45, 74
  New Zealand blood pressure, 111, 116–124
  passage time of light, 57–59, 73, 215
  S and P 500 index, 38–39
  St. Lawrence river flow, 167–168
  star velocities, 8–10
  transect sampling, 129–130, 149
exponential empirical likelihood, *see* empirical entropy
exponential tilting, 190
extreme value theory, 119, 124

finite population, 128, 149, 163, 174–180, 182
  cluster sampling, 174, 178
  correction, 175
  jackknife, 182
  NPMLE, 25
  post-stratified, 182
  raking, 183

side information, 51
simple random sample, 25, 175, 176
    definition, 174
    inefficiency, 177
    likelihood, 179
    MELE, 183
    stratified sampling, 128, 174, 177, 178
    superpopulation, 174, 182
    survey sampling, 51, 174
finite population correction, 175
Fisher information, 20, 56, 197, 265
football, see examples, football point-spreads
Fréchet differentiable statistic, 26, 226
frailty models, 153
Freeman-Tukey statistic, 67

Galton-Watson process, 197
GARCH models, 183, 184
GCA, see examples, giant cell arteritis
generalized estimating equations, 74
generalized linear model, 79, 95–103, 108
generalized method of moments, 27, 64, 72, 73, 196, 258
generalized Neyman-Pearson criterion, 255, 258
generalized regression estimator, 178
Gibbs effect, 24
Glivenko-Cantelli theorem, 219
goodness of fit, 150, 263
graph condition, 134, 141
Greenwood's formula, 144, 151, 194

harmonic mean, 128, 129, 149
Hawsley random sphygmomanometer, 124
hazard function, 141–145, 147, 151
    cumulative, 141, 151, 152, 155
Heaviside function, 101

Hessian, 61, 63
heteroscedasticity, 64, 92, 106, 110
home runs, see examples, home runs
Horvitz-Thompson estimator, 128, 178
Hotelling's T^2, 32, 64, 65
Huber's estimator, see robust, estimator

IEEE floating point, 110
importance sampling, 127, 128, 149, 190
independence, 207–208
    approximate, 208
    constraint, 201
    testing, 201, 214
infinite variance ELT, 71
infinitesimal jackknife, 1, 192, 198, 244, 245
    bias estimate, 199
    points illustrated, 193
    variance estimate, 192
instrumental variables, 73, 195
interpolation, 18, 21, 23, 268
interquartile range, 55, 72, 212
interval censoring, see censoring, interval
intrinsic curvature, see curvature, intrinisic
invariance, 50–51, 94
isotonic, 210, 214
iterated bootstrap, see bootstrap, iterated
iterated least squares, 6, 60, 63, 66, 67, 93, 235

jackknife, 1, 192–193, 198
    Bartlett correctability, 19
    bias estimate, 192, 198
    butcher knife, 198
    confidence interval, 199
    coverage error, 18
    delete d, 198
    finite population, 182

SUBJECT INDEX

infinitesimal, *see* infinitesimal jackknife
  points illustrated, 193
  variance estimate, 192, 198
Jacobian, 50

Kaplan-Meier, 26, 142, 151
KDD CUP 2000, 150
kernel smoothing, *see* smoothing
knots, 122, 194
Kolmogorov-Smirnov
  band, 156
  conservativism, 156
  power, 158
  test, 160
  weighted, 156–158
Kullback-Liebler distance, 66, 67, 72, 190, 197, 214
kurtosis, 3, 5
  coverage error, and, 18

L-estimator, 212
lack of fit, 79, 81, 83, 109
Lagrange multiplier, 22, 61, 66, 134, 144, 153, 219, 222–224, 239
large deviations, 249, 255–257
Latin hypercube, 183
least favorable family, 24, 188–190
leukemia, *see* examples, leukemia remission
leveraged bootstrap, *see* bootstrap, leveraged
likelihood
  bootstrap, 191, 197
  conditional, 49, 71
    autoregressive, 165, 168
    censored and truncated, 139
    censored data, 137, 139
    coarsening at random, 151
    GARCH, 184
    hazards, 142
    logistic regression, as, 77
    NPMLE, 143, 152

  parametric, 186, 196
  regression, in, 83
  sieved empirical, 195
  truncated data, 138, 139, 142
  dual, 153, 166, 167, 181, 239
  EL and Bayes, 188, 196
  Euclidean, 30, 63–69, 72, 74, 106–108, 181, 189, 192, 209, 213, 229, 244, 255, 258
  local, 124
  marginal, 49, 71
    censored data, 137
    empirical, 186, 196
    truncated data, 138
  parametric, 1, 4–6
  partial, 71, 147, 152, 191
  profile, 10, 11, 14, 15, 23, 30, 36, 37, 39, 42, 249, 251
    Bartlett correctability, 257
    computation, 247
    linearized, 244
    modified, 271
    Wilks's, 266
  pseudo-, 20, 191, 251–253, 258
  quasi-, 53, 108
    extended, 108
  robust, 56, 59–60, 73, 78
  semi-empirical, 196
  working, 49, 97
linearization, 37, 93, 94, 244–247
local likelihood, 124
logistic regression, 77, 95–101
  accuracy, 98
  estimating equations, 97
longitudinal data, 180, 183
Lynden-Bell estimator, *see* astronomy

m of n bootstrap, *see* bootstrap, m of n
MA model, *see* time series, moving average
Mahalanobis distance, 210
MANOVA, 88, 90

marginal likelihood, *see* likelihood, marginal
median absolute deviation, 57
MELE, 52, 77, 176–177, 182
misclassification loss ratio, 98
missing data, 78, 109, 151, 153
misspecification, 1, 24, 50, 107, 136
   heteroscedasticity, 64, 92, 106, 110
   lack of fit, 79, 81, 83, 109
   overdispersion, 91, 97, 103, 108
   testing, 263
   underdispersion, 91
mixed effects, 195
mixing condition, 164, 168, 180
model selection, 107
Monte Carlo, 51, 128, 149, 183, 266
moving average, *see* time series, moving average
multinomial, 3, 7, 12–15, 207, 224
   Bayes, 188
   Cressie-Read, 66
   least favorable subfamily, 189, 197
   likelihood maximization, 214
   likelihood ratio, 20
   likelihood ratio test, 255
   Neyman's chisquared, 64

Nadaraya-Watson estimate, 111, 124
NaN, 110
nested algorithm, 23
New Zealand, *see* examples, New Zealand blood pressure
Newcomb's experiment, *see* examples, passage time of light
Newton's method, 22, 62, 232, 233, 247
   false convergence, 233
   quadratic convergence, 233
   step, 63
   step-halving, 233
Newton-Lagrange algorithm, 229, 237
Neyman's chisquared, 64, 67
Neyman-Scott paradox, 26

NFL, *see* examples, football point-spreads
Noé's recursion, 157, 158, 160
non-informative prior, 188, 189
noncentral chisquared, 20
nonlinear least squares, 79, 91–95, 194
nonparametric MLE, 2, 7–10, 25, 29
   biased sampling, 128, 131, 133, 150
   bootstrap as, 266
   censoring, 139–142, 150
   computation, 244
   curvature, 64
   degenerate, 133, 152, 187, 207
   empirical distribution, 29
   estimating equation, 129
   Euclidean, 192
   existence, 133
   finite population, 182
   inconsistent, 140, 150
   jittered, 47
   Kaplan-Meier, 25, 26, 142, 151
   Lynden-Bell, 143, 152, 153
   missing data, 110
   product-limit, 141–143, 151
   regression, 80
   resampling from, 6, 191, 197
   sieved, 141
   statistical function, 37
   sufficiency, 150
   truncation, 139–142, 151
   uniqueness, 128, 133–134, 139, 141
   variance, 36
nonsmooth parameter, 106
normal theory, 87
nuisance parameter, 6, 41, 42, 49, 50, 55, 62, 75, 106, 152, 188, 229, 244, 246, 257

odd part, 203
Ogdensburg, *see* examples, St. Lawrence river flow

SUBJECT INDEX

Old Faithful, 124
omnibus tests, 168
one-sided confidence interval, 19–20, 253
  Bartlett correction, 253
  bootstrap, 269, 271
  improved, 19, 249, 254, 269, 271
  nonparametric tilting, 197
  parametric, 271
one-tailed test, 19
optimization, 6, 66, 74, 101, 119, 182
  constrained, 247
  convex, 6
order restricted inference, 210, 214
outlier, 56–59, 73, 74
overdispersion, 91, 97, 103, 108
overflow, 28
overidentifying restrictions, 72
oversmoothing, 116, 124

parameter effects curvature, *see* curvature, parameter effects
partial likelihood, *see* likelihood, partial
passage time of light, *see* examples, passage time of light
Pearson's chisquared, 67
periodogram, 164, 173, 182
  smoothed, 165
permutation tests, 208–209, 214
  empirical likelihood in, 209
Pitman alternative, 255
pivot, 251, 268, 269, 271
plug-in principle, 266, 269
pointspreads, *see* examples, football pointspreads
Poisson regression, 101–103
power, 1, 5, 20–21
  bands, 155
  nonparametric methods, 263
  omnibus tests, 168
  permutation testing, 208
  Pitman alternative, 255

  simulations, 5, 21, 27, 258
  universal, 20, 158, 255, 257
  vs. bootstrap-t, 71
  vs. Kolmogorov-Smirnov, 158, 160
  vs. other nonparametric likelihoods, 255
  vs. parametric likelihood, 49, 71
power law, 92
pre-whitening, 181
probability simplex, 12, 66, 67, 188, 189, 197
  diagram, 63, 68, 193
profile
  bivariate, 232
  likelihood, *see* likelihood, profile
  trace, 231, 236, 247
projection pursuit regression, 213, 214
proportional hazard model, 147, 152
pseud-likelihood, *see* likelihood, pseudo-
pseudo-logarithm, 62
pure error variance, 109

QQ plot, 33, 155
  confidence tube, 161
  Dippers, 33
  discrepancies, 69
  log EL, 34, 36, 74
  pointspreads, 205
  S and P 500 returns, 38
  star velocities, 10
quantile
  conditional, 75
  function, 155, 161, 197
  median, 15, 43–46, 48, 53, 54, 56, 57, 72, 75, 76, 145, 146, 183, 204–206
  smoothed, 72
quasars, *see* astronomy
quasi-Newton, 233

radial velocity, *see* examples, star velocities

raking, *see* finite population
random effects, 153, 196, 198
random fields, 180
range restrictions, 65, 144, 176–179, 183, 229
range space, 229, 243, 247
Rao test, 198, 265
Rao-Blackwellization, 182
regression estimator, 51, 176, 182
regression through origin, 53, 54, 72, 79–83, 91, 92, 109
rejection sampling, *see* acceptance sampling
residual, 54, 80–82, 107, 181
response surface, 80, 95
retrospective sampling, 127
right censoring, *see* censoring, right
robust, 72
   confidence intervals, 73
   covariance, 64, 106, 108
   estimating equation, 60
   estimator, 15, 26, 56–59, 71
      conditional, 124
      Huber's, 42, 56, 237
      Huber's scale, 57, 237
   likelihood, 56, 59–60, 73, 78
rotational velocity, *see* examples, star velocities
rounding, 111, 121, 151, 268
Ruth, Babe, *see* examples, home runs

S-Plus, 108, 122
saddlepoint, 198
sample maximum, 215
sandwich estimator, 55–56, 64
score
   equation, 48, 264
      bivariate normal, 76
      optimality of, 71
   function, 48
self-consistency, 151
semi-empirical likelihood, *see* likelihood, semi-empirical

sensitivity, 100, 110, 212
sequential trials, 197
shrub widths, *see* examples, transect sampling
side information, 24, 41, 51–55, 72, 77, 160, 176–177, 183, 187, 191, 196, 202
sieves, 6, 125, 141, 185, 194–196, 208
   constraints, 194
   densities, 194
   empirical likelihood, and, 194
   NPMLE, 141
   random, 194, 195, 198
   sieved EL, 194
   sieved likelihood, 141
   smoothing, and, 198
signed moment, 204–206, 214
signed root correction, 27, 253–255, 258
   location and scale, 253
   multiparameter, 254
simple random sample, *see* finite population, simple random sample
simulations
   coverage, 16, 27, 70, 107, 214
      regression, 107
   EL and Bayes, 196
   empirical entropy, 198
   empirical likelihood-t, 214
   F calibration, 32
   interval length, 21, 27, 184
   power, 5, 21, 27, 258
   tilting bootstrap, 190
   time series, 182
   trimmed means, 215
skewness, 3, 5
   acceleration constant, 270
   coverage error, and, 18
   financial returns, 184
smooth function of means, 35–39, 42, 70, 75, 79–81, 249–254, 257
smoothing
   bandwidth, 118

SUBJECT INDEX

bias and variance, 113, 124
convergence rate, 114
Gaussian kernel, 118
kernel, 72, 85, 111–127
    density, 113, 124, 149, 155
    lobes, 112
    order, 112, 116
    local linear, 112, 124
    multivariate, 112
    spectrum, 181
    spline, 111, 121–123
somites, *see* examples, earthworms
space curve, 116, 118
specificity, 110, 212
sphygmomanometer, 124
St. Lawrence river, *see* examples, St. Lawrence river flow
standard intervals, 18, 19
starting value, 23, 92
Statlib, 171, 181
stochastic monotonicity, 214
stochastic order, 261–262
stratification, *see* finite population
sufficiency
    biasd sampling NPMLE, 150
    moments and entropy, 74
superpopulation, *see* finite population
survey sampling, *see* finite population
symmetry, 3, 53, 56, 72, 201–207, 214
    about line, 215, 216
    about point, 216
    approximate, 208, 216
    interval, 144, 268
    kernel, 72, 112, 114, 125
    testing, 201, 214

tail
    condition, 71
    heaviness, 3, 39, 184, 250
    probability, 10, 19, 40, 44, 47, 230, 231

    symmetry, 203
ties, 11–12, 147, 187
    breaking, 12, 27, 47
    censored data, 147
    Cressie-Read, 78
    ignoring, 12, 26
    multiple biased samples, 132
    quantiles, 46, 48, 76
    regression and pure error, 108
    sequential linearization, 245
    symmetry, 202
time series, 163–174, 180
    ARMA, 173, 182
    autoregressive, 165–168, 171, 173, 181, 230
    bootstrap, 181
    unstable, 181
    back forecasting, 182
    definitions, 163
    equilibrium, 165, 166
    explosive, 197
    GARCH, 183, 184
    moving average, 173, 182
    multiple, 180
    resampled periodogram, 182
    spectrum, 164, 173–174, 180–182
        Bartlett correction, 182
    tree rings, 181
    stationarity, 164, 168, 169, 172, 181
tolerance
    interval, 105, 212
    numerical, 22
triangular array ELT, *see* ELT, triangular array
trimmed mean, 212, 215
truncated power basis, 122
truncation, 127, 135–136
    astronomy, 152
    biased sampling, 135
    double, 152
    extrapolation, 135
    left, 136, 138, 139, 141, 142
    EL, 147–148
    NPMLE, 139

probability of, 152
sensitivity analysis, 136
vs. censoring, 135
tuna, 124

undata, 69
underdetermined estimating equation, 105
underdispersion, 91
underflow, 28
undersmoothed, 115, 124
unimodality, 152
universal optimality, 20, 158, 255, 257
upgraded mixture model, 150

variance modeling, 90–91, 110
vasculitis, *see* examples, giant cell arteritis

Wald test, 64, 194, 198, 265
weights, 11, 75, 140, 152, 153, 167, 177, 185, 188, 201, 202, 207, 209, 215, 223, 244–246, 271
 anova, 88
 censoring, 140
 convex hull, 62
 entropy, 78
 Euclidean, 74
 jittered, 47
 multiple biased samples, 132
Wilks's theorem, 10, 11, 265
 nonparametric, *see* ELT
 profile likelihood, 266
 smoothing, 198
working likelihood, *see* likelihood, working
worms, *see* examples, earthworms
Wye, river, *see* examples, Dipper survey data